Praise for *Power and Progress*

"America (and the world) is at a crossroads. Big business and the rich rewrote the rules of the US political economy since the 1970s, making it more grotesquely unfair than ever just as automation and offshoring jobs changed the game as well. Now with AI, renowned MIT economists Daron Acemoglu and Simon Johnson explain in their important and lucid book how the transformation of work could make life even worse for most people, or, possibly, *much* better—depending on the political and social and technological choices we make starting now. We must 'stop being mesmerized by tech billionaires,' they warn, because 'progress is never automatic.' With revealing, relevant stories from throughout economic history and sensible ideas for systemic reform, this is an essential guide for this crucial battle in the 'one-thousand-year struggle' between the powerful and everyone else."

—KURT ANDERSEN, author of *Evil Geniuses*

"One powerful thread runs through this breathtaking tour of the history and future of technology, from the Neolithic agricultural revolution to the ascent of artificial intelligence: Technology is not destiny, nothing is pre-ordained. Humans, despite their imperfect institutions and often-contradictory impulses, remain in the driver's seat. It is still our job to determine whether the vehicles we build are heading toward justice or down the cliff. In this age of relentless automation and seemingly unstoppable consolidation of power and wealth, *Power and Progress* is an essential reminder that we can, and must, take back control."

—ABHIJIT BANERJEE AND ESTHER DUFLO, 2019 Nobel laureates in economics and authors of *Poor Economics* and *Good Economics for Hard Times*

"Acemoglu and Johnson have written a sweeping history of more than a thousand years of technical change. They take aim at economists' mindless enthusiasm for technical change and their crippling neglect of power. An important book that is long overdue."

—SIR ANGUS DEATON, 2015 Nobel laureate in economics and coauthor of *Deaths of Despair*

"If you are not already an addict of Daron Acemoglu and Simon Johnson's previous books, *Power and Progress* is guaranteed to make you one. It offers their addictive hallmarks: sparkling writing and a big question that affects our lives. Are powerful new technologies guaranteed to benefit us? Did the industrial revolution bring happiness to our great-grandparents 150 years ago, and will artificial intelligence bring us more happiness now? Read, enjoy, and then choose your lifestyle!"

—JARED DIAMOND, Pulitzer Prize–winning author of *Guns, Germs, and Steel* and other international bestsellers

"Acemoglu and Johnson would like a word with the mighty tech lords before they turn over the entire world economy to artificial intelligence. The lesson of economic history is technological advances such as AI won't automatically lead to broad-based prosperity—they may end up benefiting only a wealthy elite. Just as the innovations of the Gilded Age of American industrialization had to be reined in by progressive politics, so too, in our Coded Age, we need not only trade unions, civil society, and trustbusters, but also legislative and regulatory reforms to prevent the advent of a new panopticon of AI-enabled surveillance. This book will not endear the authors to Microsoft executives, but it's a bracing wake-up call for the rest of us."

—NIALL FERGUSON, Milbank Family Senior Fellow, Hoover Institution, Stanford University, and author of *The Square and the Tower*

"A book you must read: compelling, beautifully written, and tightly argued, it addresses a crucially important problem with powerful solutions. Drawing on both historical examples and a deep dive into the ways in which artificial intelligence and social media depress wages and undermine democracy, Acemoglu and Johnson argue for a revolution in the way we manage and control technology. Throughout history, it has only been when elites have been forced to share power that technology has served the common good. Acemoglu and Johnson show us what this would look like today."

—REBECCA HENDERSON, John and Natty McArthur University Professor, Harvard University, and author of *Reimagining Capitalism in a World on Fire*

"The technology of artificial intelligence is moving fast and likely to accelerate. This powerful book shows we now need to make some careful choices to really share the benefits and reduce unintended, adverse consequences. Technology is too important to leave to the billionaires. Everyone everywhere should read Acemoglu and Johnson—and try to get a seat at the decision-making table."

—Ro Khanna, Silicon Valley member of Congress

"This singular book elevated my understanding of the present confluence of society, economics, and technology. Here we have a synthesis of history and analysis coupled with specific ideas about how the future can be improved. It pulls no punches but also inspires optimism."

—Jaron Lanier, author of *Ten Arguments for Deleting Your Social Media Accounts Right Now*

"Two of the best economists alive today are taking a closer look at the economics of technological progress in history. Their findings are as surprising as they are disturbing. This beautifully written and richly documented book marks a new beginning in our thinking about the political economy of innovation."

—Joel Mokyr, Robert H. Strotz Professor of Arts and Sciences, Northwestern University, and author of *The Lever of Riches: Technological Creativity and Economic Progress*

"Will the AI revolution increase the average worker's productivity while reducing their drudgery, or will it simply create more exploitative and heavily surveilled workplaces run by robotic overlords? That is the right question, and luckily Acemoglu and Johnson have set out to answer it, giving it profound historical context, combing through the economic incentives, and lighting a better path forward."

—Cathy O'Neil, author of *Weapons of Math Destruction* and *The Shame Machine*

"Technology is upending our world—automating jobs, deepening inequality, and creating tools of surveillance and misinformation that threaten democracy. But Acemoglu and Johnson show it doesn't have to be this way. The direction of technology is not, like the direction of the wind, a force of nature beyond human control. It's up to us. This humane and hopeful book shows how we can steer technology to promote the public good. Required reading for everyone who cares about the fate of democracy in a digital age."

—Michael J. Sandel, Robert M. Bass Professor of Government, Harvard University, and author of *The Tyranny of Merit: Can We Find the Common Good?*

"A remarkable analysis of the current drama of technology evolution versus human dignity, where the potent forces boosting inequality continue to destroy our belief in the nobility of work and the inevitability of egalitarian progress. Acemoglu and Johnson offer a fresh vision of how this drama unfolds by highlighting human capabilities and social skills. They are deeply informed, masters at synthesis, and passionate about shaping a better future where innovation supports equality."

—BEN SHNEIDERMAN, Distinguished University Professor, University of Maryland, and author of *Human-Centered AI*

"Our future is inevitable and determined by the acceleration of technologies like AI and Web3....Or so we are told. Here, from two of the greatest economists of our time, we have the definitive refutation of the techno-determinist story that has held us back from building a better future for the last four decades. With a bit of luck, we may look back at this as a turning point where we collectively once again took responsibility for defining the world we want technology to empower us to live in together."

—E. GLEN WEYL, research lead and founder, Decentralized Social Technology Collaboratory, Microsoft Research Special Projects

"In this brilliant, sweeping review of technological change past and present, Acemoglu and Johnson mean to grab us by the shoulders and shake us awake before today's winner-take-all technologies impose more violence on global society and the democratic prospect. This vital book is a necessary antidote to the poisonous rhetoric of tech inevitability. It reveals the realpolitik of technology as a persistent Trojan horse for economic powers that favor the profit-seeking aims of the few over the many. *Power and Progress* is the blueprint we need for the challenges ahead: technology *only* contributes to shared prosperity when it is tamed by democratic rights, values, principles, and the laws that sustain them in our daily lives."

—SHOSHANA ZUBOFF, Charles Edward Wilson Professor Emerita, Harvard Business School, and author of *The Age of Surveillance Capitalism*

POWER

AND

PROGRESS

POWER

AND

PROGRESS

Our Thousand-Year Struggle
over Technology and Prosperity

DARON ACEMOGLU
and SIMON JOHNSON

PUBLICAFFAIRS

New York

PublicAffairs
Hachette Book Group
1290 Avenue of the Americas, New York, NY 10104
www.publicaffairsbooks.com
@Public_Affairs

Printed in the United States of America

First Edition: May 2023

Published by PublicAffairs, an imprint of Perseus Books, LLC, a subsidiary of Hachette Book Group, Inc. The PublicAffairs name and logo is a trademark of the Hachette Book Group.

The Hachette Speakers Bureau provides a wide range of authors for speaking events. To find out more, go to www.hachettespeakersbureau.com or call (866) 376-6591.

PublicAffairs books may be purchased in bulk for business, educational, or promotional use. For information, please contact your local bookseller or Hachette Book Group Special Markets Department at special.markets@hbgusa.com.

The publisher is not responsible for websites (or their content) that are not owned by the publisher.

Print book interior design by Jeff Williams

Library of Congress Cataloging-in-Publication Data
Names: Acemoglu, Daron, author. | Johnson, Simon, author.
Title: Power and progress : our thousand-year struggle over technology and
    prosperity / Daron Acemoglu and Simon Johnson.
Description: First edition. | New York : PublicAffairs, 2023. | Includes
    bibliographical references and index.
Identifiers: LCCN 2022059230 | ISBN 9781541702530 (hardcover) | ISBN
    9781541702554 (ebook)
Subjects: LCSH: Technology—Social aspects. | Technology—Economic aspects.
    | Progress.
Classification: LCC T14.5 .A29 2023 | DDC 303.48/3--dc23/eng/20230111
LC record available at https://lccn.loc.gov/2022059230

ISBNs: 9781541702530 (hardcover), 9781541702554 (e-book)

LSC-C

Printing 1, 2023

Daron:
To Aras, Arda, and Asu, for a better future

Simon:
To Lucie, Celia, and Mary, always

If we combine our machine-potentials of a factory with the valuation of human beings on which our present factory system is based, we are in for an industrial revolution of unmitigated cruelty. We must be willing to deal in facts rather than in fashionable ideologies if we wish to get through this period unharmed.

—Norbert Wiener, 1949

Contents

Prologue

What Is Progress?

Every day, we hear from executives, journalists, politicians, and even some of our colleagues at MIT that we are heading relentlessly toward a better world, thanks to unprecedented advances in technology. Here is your new phone. There goes the latest electric car. Welcome to the next generation of social media. And soon, perhaps, scientific advances could solve cancer, global warming, and even poverty.

Of course, problems remain, including inequality, pollution, and extremism around the globe. But these are the birth pains of a better world. In any case, we are told, the forces of technology are inexorable. We couldn't stop them if we wanted to, and it would be highly inadvisable to try. It is better to change ourselves—for example, by investing in skills that will be valued in the future. If there are continuing problems, talented entrepreneurs and scientists will invent solutions—more-capable robots, human-level artificial intelligence, and whatever other breakthroughs are required.

People understand that not everything promised by Bill Gates, Elon Musk, or even Steve Jobs will likely come to pass. But, as a world, we have become infused by their techno-optimism. Everyone everywhere should innovate as much as they can, figure out what works, and iron out the rough edges later.

WE HAVE BEEN here before, many times. One vivid example began in 1791, when Jeremy Bentham proposed the *panopticon,*

1

a prison design. In a circular building and with the right lighting, Bentham argued, centrally positioned guards could create the impression of watching everyone all the time, without themselves being observed—supposedly a very efficient (low-cost) way of ensuring good behavior.

The idea at first found some traction with the British government, but sufficient funding was not forthcoming, and the original version was never built. Nevertheless, the panopticon captured the modern imagination. For the French philosopher Michel Foucault, it is a symbol of oppressive surveillance at the heart of industrial societies. In George Orwell's *1984*, it operates as the omnipresent means of social control. In the Marvel movie *Guardians of the Galaxy*, it proves to be a flawed design that facilitates an ingenious prison breakout.

Before the panopticon was proposed as a prison, it was a factory. The idea originated with Samuel Bentham, Jeremy's brother and an expert naval engineer then working for Prince Grigory Potemkin in Russia. Samuel's idea was to enable a few supervisors to watch over as many workers as possible. Jeremy's contribution was to extend that principle to many kinds of organizations. As he explained to a friend, "You will be surprised when you come to see the efficacy which this simple and seemingly obvious contrivance promises to be to the business of schools, manufactories, Prisons, and even Hospitals. . . ."

The panopticon's appeal is easy to understand—if you are in charge—and was not missed by contemporaries. Better surveillance would lead to more compliant behavior, and it was easy to imagine how this could be in the broader interest of society. Jeremy Bentham was a philanthropist, animated by schemes to improve social efficiency and help everyone to greater happiness, at least as he saw it. Bentham is credited today as the founder of the philosophy of utilitarianism, which means maximizing the combined welfare of all people in society. If some people could be squeezed a little in return for a few people gaining a great deal, that was an improvement worth considering.

The panopticon was not just about efficiency or the common good, however. Surveillance in factories implied inducing workers to labor harder, and without the need to pay them higher wages to motivate greater effort.

The factory system spread rapidly in the second half of the eighteenth century across Britain. Even though they did not rush to install panopticons, many employers organized work in line with Bentham's general approach. Textile manufacturers took over activities previously performed by skilled weavers and divided them up more finely, with key elements now done by new machines. Factory owners employed unskilled workers, including women and small children, to perform simple repetitive tasks, such as pulling a handle, for as many as fourteen hours per day. They also supervised this labor force closely, lest anyone slow down production. And they paid low wages.

Workers complained about conditions and the backbreaking effort. Most egregious to many were the rules they had to follow in factories. One weaver put it this way in 1834: "No man would like to work in a power-loom, they do not like it, there is such a clattering and noise it would almost make some men mad; and next, he would have to be subject to a discipline that a hand-loom weaver can never submit to."

New machinery turned workers into mere cogs. As another weaver testified before a parliamentary committee in April 1835, "I am determined for my part, that if they will invent machines to supersede manual labour, they must find iron boys to mind them."

To Jeremy Bentham, it was self-evident that technology improvements enabled better-functioning schools, factories, prisons, and hospitals, and this was beneficial for everyone. With his flowery language, formal dress, and funny hat, Bentham would cut an odd figure in modern Silicon Valley, but his thinking is remarkably fashionable. New technologies, according to this view of the world, expand human capabilities and, when applied throughout the economy, greatly increase efficiency and productivity. Then, the logic goes, society will sooner or later find a way of sharing these gains, generating benefits for pretty much everybody.

Adam Smith, the eighteenth-century founding father of modern economics, could also join the board of a venture capital fund or write for *Forbes*. In his view, better machines would lead to higher wages, almost automatically:

> In consequence of better machinery, of greater dexterity, and of a more proper division and distribution of work, all of which are the natural effects of improvement, a much smaller quantity of labour becomes requisite for executing any particular piece of work, and though, in consequence of the flourishing circumstances of the society, the real price of labour should rise very considerably. . . .

In any case, resistance is futile. Edmund Burke, contemporary of Bentham and Smith, referred to the laws of commerce as "the laws of nature, and consequently the laws of God."

How can you resist the laws of God? How can you resist the unstoppable march of technology? And anyway, why resist these advances?

ALL OF THIS optimism notwithstanding, the last thousand years of history are filled with instances of new inventions that brought nothing like shared prosperity:

- A whole series of technological improvements in medieval and early modern agriculture, including better plows, smarter crop rotation, more use of horses, and much improved mills, created almost no benefits for peasants, who constituted close to 90 percent of the population.

- Advances in European ship design from the late Middle Ages enabled transoceanic trade and created massive fortunes for some Europeans. But the same kinds of ships also transported millions of enslaved people from Africa to the New World and made it possible to build systems of oppression that lasted for generations and created awful legacies persisting today.

- Textile factories of the early British industrial revolution generated great wealth for a few but did not raise worker incomes for almost a hundred years. On the contrary, as the textile workers themselves keenly understood, work hours lengthened and conditions were horrible, both in the factory and in crowded cities.

- The cotton gin was a revolutionary innovation, greatly raising the productivity of cotton cultivation and turning the United States into the largest cotton exporter in the world. The same invention intensified the savagery of slavery as cotton plantations expanded across the American South.

- At the end of the nineteenth century, German chemist Fritz Haber developed artificial fertilizers that boosted agricultural yields. Subsequently, Haber and other scientists used the same ideas to design chemical weapons that killed and maimed hundreds of thousands on World War I battlefields.

- As we discuss in the second half of this book, spectacular advances in computers have enriched a small group of entrepreneurs and business tycoons over the last several decades, whereas most Americans without a college education have been left behind, and many have even seen their real incomes decline.

Some readers may object at this point: Did we not in the end hugely benefit from industrialization? Aren't we more prosperous than earlier generations, who toiled for a pittance and often died hungry, thanks to improvements in how we produce goods and services?

Yes, we are greatly better off than our ancestors. Even the poor in Western societies enjoy much higher living standards today than three centuries ago, and we live much healthier, longer lives, with comforts that those alive a few hundred years ago could not have even imagined. And, of course, scientific and technological progress is a vital part of that story and will have to be the bedrock of any future process of shared gains. But the broad-based prosperity of the past was not the result of any automatic, guaranteed gains of technological progress. Rather, shared prosperity emerged because, and only when, the direction of technological advances and society's approach to dividing the gains were pushed away from arrangements that primarily served a narrow elite. We are beneficiaries of progress, mainly because our predecessors made that progress work for more people. As the eighteenth-century writer and radical John Thelwall recognized, when workers congregated in factories and cities, it became easier for them to rally around common interests and make demands for more equitable participation in the gains from economic growth:

> The fact is, that monopoly, and the hideous accumulation of capital in a few hands, like all diseases not absolutely mortal, carry, in their own enormity, the seeds of cure. Man is, by his very nature, social and communicative—proud to display the little knowledge he possesses, and eager, as opportunity presents, to encrease his store. Whatever presses men together, therefore, though it may generate some vices, is favourable to the diffusion of knowledge, and ultimately promotive of human liberty. Hence every large workshop and manufactory is a sort of political society, which no act of parliament can silence, and no magistrate disperse.

Electoral competition, the rise of trade unions, and legislation to protect workers' rights changed how production was organized and wages were set in nineteenth-century Britain. Combined with the arrival of a new wave of innovation from the United States, they also forged a new direction of technology—focused on increasing worker productivity rather than just substituting machinery for the tasks they used to perform or inventing new ways of monitoring them. Over the next century, this technology spread throughout Western Europe and then the world.

Most people around the globe today are better off than our ancestors because citizens and workers in early industrial societies organized, challenged elite-dominated choices about technology and work conditions, and forced ways of sharing the gains from technical improvements more equitably.

Today we need to do the same again.

The good news is that incredible tools are available to us, including magnetic resonance imaging (MRI), mRNA vaccines, industrial robots, the internet, tremendous computational power, and massive amounts of data on things we could not measure before. We can use these innovations to solve real problems—but only if these awesome capabilities are focused on helping people. This is not the direction in which we are currently heading, however.

Despite what history teaches us, the predominant narrative today has shifted back toward something remarkably close to what was prevalent in Britain 250 years ago. We are living in an age that is even more blindly optimistic and more elitist about technology than the times of Jeremy Bentham, Adam Smith, and Edmund Burke. As we document in Chapter 1, people making the big decisions are once again deaf to the suffering created in the name of progress.

We wrote this book to show that progress is never automatic. Today's "progress" is again enriching a small group of entrepreneurs and investors, whereas most people are disempowered and benefit little.

A new, more inclusive vision of technology can emerge only if the basis of social power changes. This requires, as in the nineteenth century, the rise of counterarguments and organizations that can stand up to the conventional wisdom. Confronting the prevailing vision and wresting the direction of technology away from the control of a narrow elite may even be more difficult today than it was in nineteenth-century Britain and America. But it is no less essential.

1

Control over Technology

In the Fall as recorded in the book of Genesis, man underwent a loss of innocence and a weakening of his power over creation. Both of these losses can be to some extent made good, even in this life—the former by religion and faith, the latter by arts and sciences.

—FRANCIS BACON, *Novum Organum*, 1620

Instead, I saw a real aristocracy, armed with a perfected science and working to a logical conclusion the industrial system of to-day. Its triumph had not been simply a triumph over Nature, but a triumph over Nature and the fellow man.

—H. G. WELLS, *The Time Machine*, 1895

Since its first version in 1927, *Time* magazine's annual Man of the Year had almost always been a single person, typically a political leader of global significance or a US captain of industry. For 1960, the magazine chose instead a set of brilliant people: American scientists. Fifteen men (unfortunately, no women) were singled out for their remarkable achievements across a range of fields. According to *Time*, science and technology had finally triumphed.

The word *technology* comes from the Greek *tekhne* ("skilled craft") and *logia* ("speaking" or "telling"), implying systematic study of a technique. Technology is not simply the application of new methods to the production of material goods. Much more broadly, it concerns everything we do to shape our surroundings and organize production. Technology is the way that collective human knowledge is used to improve nutrition, comfort, and health, but often for other purposes, too, such as surveillance, war, or even genocide.

Time was honoring scientists in 1960 because unprecedented advances in knowledge had, through new practical applications, transformed everything about human existence. The potential for further progress appeared unbounded.

This was a victory lap for the English philosopher Francis Bacon. In *Novum Organum*, published in 1620, Bacon had argued that scientific knowledge would enable nothing less than human control over nature. For centuries, Bacon's writings seemed no more than aspirational as the world struggled with natural disasters, epidemics, and widespread poverty. By 1960, however, his vision was no longer fantastical because, as *Time's* editors wrote, "The 340 years that have passed since *Novum Organum* have seen far more scientific change than all the previous 5,000 years."

As President Kennedy put it to the National Academy of Sciences in 1963, "I can imagine no period in the long history of the world where it would be more exciting and rewarding than in the field today of scientific exploration. I recognize with each door that we unlock we see perhaps 10 doors that we never dreamed existed and, therefore, we have to keep working forward." Abundance was now woven into the fabric of life for many people in the United States and Western Europe, with great expectations for what would come next both for those countries and the rest of the world.

This upbeat assessment was based on real achievement. Productivity in industrial countries had surged during the preceding decades so that American, German, or Japanese workers were now producing on average a lot more than just twenty years before.

New consumer goods, including automobiles, refrigerators, televisions, and telephones, were increasingly affordable. Antibiotics had tamed deadly diseases, such as tuberculosis, pneumonia, and typhus. Americans had built nuclear-powered submarines and were getting ready to go to the moon. All thanks to breakthroughs in technology.

Many recognized that such advances could bring ills as well as comforts. Machines turning against humans has been a staple of science fiction at least since Mary Shelley's *Frankenstein*. More practically but no less ominously, pollution and habitat destruction wrought by industrial production were increasingly prominent, and so was the threat of nuclear war—itself a result of astonishing developments in applied physics. Nevertheless, the burdens of knowledge were not seen as insurmountable by a generation becoming confident that technology could solve all problems. Humanity was wise enough to control the use of its knowledge, and if there were social costs of being so innovative, the solution was to invent even more useful things.

There were lingering concerns about "technological unemployment," a term coined by the economist John Maynard Keynes in 1930 to capture the possibility that new production methods could reduce the need for human labor and contribute to mass unemployment. Keynes understood that industrial techniques would continue to improve rapidly but also argued, "This means unemployment due to our discovery of means of economising the use of labour outrunning the pace at which we can find new uses for labour."

Keynes was not the first to voice such fears. David Ricardo, another founder of modern economics, was initially optimistic about technology, maintaining that it would steadily increase workers' living standards, and in 1819 he told the House of Commons that "machinery did not lessen the demand for labour." But for the third edition of his seminal *Principles of Political Economy and Taxation* in 1821, Ricardo added a new chapter, "On Machinery," in which he wrote, "It is more incumbent on me to declare my opinion on this question, because they have, on further reflection, undergone a considerable change." As he explained in a private letter that year, "If machinery could do all the work that labour now does, there would be no demand for labour."

But Ricardo's and Keynes's concerns did not have much impact on mainstream opinion. If anything, optimism intensified after personal computers and digital tools started spreading rapidly in the 1980s. By the late 1990s, the possibilities for economic and social advances seemed boundless. Bill Gates was speaking for many in the tech industry at the time when he said, "The [digital] technologies involved here are really a superset of all communications technology that has come along in the past, e.g., radio, newspaper. All of those things will be replaced by something that is far more attractive."

Not everything might go right all the time, but Steve Jobs, cofounder of Apple, captured the zeitgeist perfectly at a conference in 2007 with what became a famous line: "Let's go and invent tomorrow rather than worrying about yesterday."

In fact, both *Time* magazine's upbeat assessment and subsequent techno-optimism were not just exaggerated; they missed entirely what happened to most people in the United States after 1980.

In the 1960s, only about 6 percent of American men between the ages of 25 and 54 were out of the labor market, meaning they were long-term unemployed or not seeking a job. Today that number is around 12 percent, primarily because men without a college degree are finding it increasingly difficult to get well-paid jobs.

American workers, both with and without college education, used to have access to "good jobs," which, in addition to paying decent wages, provided job security and career-building opportunities. Such jobs have largely disappeared for workers without a college degree. These changes have disrupted and damaged the economic prospects for millions of Americans.

An even bigger change in the US labor market over the past half century is in the structure of wages. During the decades following World War II, economic growth was rapid and widely shared, with workers from all backgrounds and skills experiencing rapid growth in real incomes (adjusted for inflation). No longer. New digital technologies are everywhere and have made vast fortunes for entrepreneurs, executives, and some investors, yet real wages for most workers have scarcely increased. People without college education have seen their real earnings decline, on average, since 1980, and even workers with a college degree but no postgraduate education have seen only limited gains.

The inequality implications of new technologies reach far beyond these numbers. With the demise of good jobs available to most workers and the rapid growth in the incomes of a small fraction of the population trained as computer scientists, engineers, and financiers, we are on our way to a truly two-tiered society, in which workers and those commanding the economic means and social recognition live separately, and that separation grows daily. This is what the English writer H. G. Wells anticipated in *The Time Machine*, with a future dystopia where technology had so segregated people that they evolved into two separate species.

This is not just a problem in the United States. Because of better protection for low-paid workers, collective bargaining, and decent minimum wages, workers with relatively low education levels in Scandinavia, France, or Canada have not suffered wage declines like their American counterparts. All the same, inequality has risen, and good jobs for people without college degrees have become scarce in these countries as well.

It is now evident that the concerns raised by Ricardo and Keynes cannot be ignored. True, there has been no catastrophic technological unemployment, and throughout the 1950s and 1960s workers benefited from productivity growth as much as entrepreneurs and business owners did. But today we are seeing a very different picture, with skyrocketing inequality and wage earners largely left behind as new advances pile up.

In fact, a thousand years of history and contemporary evidence make one thing abundantly clear: there is nothing automatic about new technologies bringing widespread prosperity. Whether they do or not is an economic, social, and political choice.

This book explores the nature of this choice, the historical and contemporary evidence on the relationship among technology, wages, and inequality, and what we can do in order to direct innovations to work in service of shared prosperity. To lay the groundwork, this chapter addresses three foundational questions:

- What determines when new machines and production techniques increase wages?

- What would it take to redirect technology toward building a better future?

- Why is current thinking among tech entrepreneurs and visionaries pushing in a different, more worrying direction, especially with the new enthusiasm around artificial intelligence?

The Bandwagon of Progress

Optimism regarding shared benefits from technological progress is founded on a simple and powerful idea: the "productivity bandwagon." This idea maintains that new machines and production methods that increase productivity will also produce higher wages. As technology progresses, the bandwagon will pull along everybody, not just entrepreneurs and owners of capital.

Economists have long recognized that demand for all tasks, and thus for different types of workers, does not necessarily grow at the same rate, so inequality may increase because of innovation. Nevertheless, improving technology is generally viewed as the tide lifting all boats because everyone is expected to derive some benefits. Nobody is supposed to be completely left behind by technology, let alone be impoverished by it. According to the conventional wisdom, to rectify the rise in inequality and build even more solid foundations for shared prosperity, workers must find a way to acquire more of the skills they need to work alongside new technologies. As succinctly summarized by Erik Brynjolfsson, one of the foremost experts on technology, "What can we do to create shared prosperity? The answer is not to slow down technology. Instead of racing against the machine, we need to race with the machine. That is our grand challenge."

The theory behind the productivity bandwagon is straightforward: when businesses become more productive, they want to expand their output. For this, they need more workers, so they get busy with hiring. And when many firms attempt to do so at the same time, they collectively bid up wages.

This is what happens, but only sometimes. For example, in the first half of the twentieth century, one of the most dynamic sectors of the US economy was car manufacturing. As Ford Motor Company and then General Motors (GM) introduced new electrical machinery, built more-efficient factories, and launched better models, their productivity soared, as did their employment. From a few thousand workers in 1899, producing just 2,500 automobiles, the industry's employment rose to more than 400,000 by the 1920s. By 1929, Ford and GM were each selling around 1.5 million cars every year. This unprecedented expansion of automobile production pulled up wages throughout the economy, including for workers without much formal education.

For most of the twentieth century, productivity rose rapidly in other sectors as well, as did real wages. Remarkably, from the end of World War II to the mid-1970s, the wages of college graduates in the US grew at roughly the same rate as the wages of those workers with only a high school education.

Unfortunately, what subsequently occurred is not consistent with the notion that there is any kind of unstoppable bandwagon. How productivity benefits are shared depends on how exactly technology changes and on the rules, norms, and expectations that govern how management treats workers. To understand this, let us unpack the two steps that link productivity growth to higher wages. First, productivity growth increases the demand for workers as businesses attempt to boost profits by expanding output and hiring more people. Second, the demand for more workers increases the wages that need to be offered to attract and retain employees. Unfortunately, neither step is assured, as we explain in the next two sections.

Automation Blues

Contrary to popular belief, productivity growth need not translate into higher demand for workers. The standard definition of productivity is average output per worker—total output divided by total employment. Obviously, the hope is that as output per worker grows, so will the willingness of businesses to hire people.

But employers do not have an incentive to increase hiring based on average output per worker. Rather, what matters to companies is *marginal productivity*—the additional contribution that one more worker brings by increasing production or by serving more customers. The notion of marginal productivity is distinct from output or revenue per worker: output per worker may increase while marginal productivity remains constant or even declines.

To clarify the distinction between output per worker and marginal productivity, consider this often-repeated prediction: "The factory of the future will have only two employees, a man and a dog. The man will be there to feed the dog. The dog will be there to keep the man from touching the equipment." This imagined factory could churn out a lot of output, so average productivity—its output divided by the one (human) employee—is very high. Yet worker marginal productivity is minuscule; the sole employee is there to feed the dog, and the implication is that both the dog and the employee could be let go without much reduction in output. Better machinery might further increase output per worker, but it is reasonable to expect that this factory would not rush to hire more workers and their dogs, or increase the pay of its lonely employee.

This example is extreme, but it represents an important element of reality. When a car company introduces a better vehicle model, as Ford and GM did in the first half of the twentieth century, this tends to increase the demand for the company's cars, and both revenues per worker and worker marginal productivity rise. After all, the company needs more workers, such as welders and painters, to meet the additional demand, and it will pay them more, if necessary. In contrast, consider what happens when the same automaker installs industrial robots. Robots can perform most welding and painting tasks, and can do so more cheaply than production methods employing a larger number of workers. As a result, the company's average productivity increases significantly, but it has less need for human welders and painters.

This is a general problem. Many new technologies, like industrial robots, expand the set of tasks performed by machines and algorithms, displacing workers who used to be employed in these tasks. Automation raises average productivity but does not increase, and in fact may reduce, worker marginal productivity.

Automation is what Keynes worried about, and it was not a new phenomenon when he was writing early in the twentieth century. Many of the iconic innovations of the British industrial revolution in textiles were all about substituting new spinning and weaving machines for the labor of skilled artisans.

What is true of automation is true of many aspects of globalization as well. Major breakthroughs in communication tools and shipping logistics have enabled a massive wave of offshoring over the last several decades, with production tasks such as assembly or customer service being transferred to countries where labor is cheaper. Offshoring has reduced costs and boosted profits for companies such as Apple, whose products are made of parts produced in many countries and are almost entirely assembled in Asia. But in industrialized nations it has also displaced workers who used to perform these tasks domestically and has not activated a powerful bandwagon.

Automation and offshoring have raised productivity and multiplied corporate profits, but have brought nothing resembling shared prosperity to the United States and other developed countries. Replacing workers with machines and moving work to lower-wage countries are not the only options for improving economic efficiency. There are multiple ways of increasing output per worker—and this has been true throughout history, as we explain in chapters 5 through 9. Some innovations boost how much individuals contribute to production, rather than automating or offshoring work. For example, new software tools that aid the tasks of car mechanics and enable greater precision work increase worker marginal productivity. This is completely different from installing industrial robots with the goal of replacing people.

Even more important for raising worker marginal productivity is the creation of new tasks. There was plenty of automation in car manufacturing during the momentous reorganization of the industry led by Henry Ford starting in the 1910s. But mass-production methods and assembly lines simultaneously introduced a range of new design, technical, machine-operation, and clerical tasks, boosting the industry's demand for workers (as we will detail in Chapter 7). When new machines create new uses for human labor, this expands the ways in which workers can contribute to production and increases their marginal productivity.

New tasks were vital not just in early US car manufacturing but also in the growth of employment and wages over the last two centuries. Many of the fastest-growing occupations in the last few decades—MRI radiologists, network engineers, computer-assisted machine operators, software programmers, IT security personnel, and data analysts—did not exist eighty years ago. Even people in occupations that have been around for quite a while, such as bank tellers, professors, or accountants, now work on a variety of tasks that did not exist before World War II, including all of those that involve the use of computers and modern communication devices. In almost all these cases, new tasks were introduced as a result of technological advances and have been a major driver of employment growth. These new tasks have also been an integral part of productivity growth, for they have helped launch new products and more efficient reorganization of the production process.

The reason that Ricardo's and Keynes's worst fears about technological unemployment did not come to pass is intimately linked to new tasks. Automation was rapid throughout the twentieth century but did not reduce the demand for workers because it was accompanied by other improvements and reorganizations that produced new activities and tasks for workers.

Automation in an industry can also push up employment—in that sector or in the economy as a whole—if it reduces costs or increases productivity by enough. New jobs in this case may come either from nonautomated tasks in the same industry or from the expansion of activities in related industries. In the first half of the twentieth century, the rapid increase in car manufacturing raised the demand for a range of nonautomated technical and clerical functions. Just as important, productivity growth in car factories during these decades was a major driver for the expansion of the oil, steel, and chemical industries (think gasoline, car bodies, and tires). Car manufacturing at mass scale also revolutionized the possibilities for transportation, enabling the rise of new retail, entertainment, and service activities, especially as the geography of cities transformed.

There will be few new jobs created, however, when the productivity gains from automation are small—what we call "so-so automation" in Chapter 9. For example, self-checkout kiosks in grocery stores bring limited productivity benefits because they shift the work of scanning items from employees to customers. When self-checkout kiosks are introduced, fewer cashiers are employed, but there is no major productivity boost to stimulate the creation of new jobs elsewhere. Groceries do not become much cheaper, there is no expansion in food production, and shoppers do not live differently.

The situation is similarly dire for workers when new technologies focus on surveillance, as Jeremy Bentham's panopticon intended. Better monitoring of workers may lead to some small improvements in productivity, but its main function is to extract more effort from workers and sometimes also reduce their pay, as we will see in chapters 9 and 10.

There is no productivity bandwagon from so-so automation and worker surveillance. The bandwagon is also weak, even from new technologies that generate nontrivial productivity gains, when these tasks predominantly focus on automation and cast workers aside. Industrial robots, which have already revolutionized modern manufacturing, generate little or no gains for workers when they are not accompanied by other technologies that create new tasks and opportunities for human labor. In some cases, such as the industrial heartland of the American economy in the Midwest, the rapid adoption of robots has instead contributed to mass layoffs and prolonged regional decline.

All of this brings home perhaps the most important thing about technology: *choice*. There are often myriad ways of using our collective knowledge for improving production and even more ways of directing innovations. Will we use digital tools for surveillance? For automation? Or for empowering workers by creating new productive tasks for them? And where will we put our efforts toward future advances?

When the productivity bandwagon is weak and there are no self-acting correction mechanisms ensuring shared benefits, these choices become more consequential—and those who make them become more powerful, both economically and politically.

In sum, the first step in the productivity bandwagon causal chain depends on specific choices: using existing technologies and developing new ones for increasing worker marginal productivity— not just automating work, making workers redundant, or intensifying surveillance.

Why Worker Power Matters

Unfortunately, even an increase in worker marginal productivity is not enough for the productivity bandwagon to boost wages and living standards for everyone. Recall that the second step in the causal chain is that an increase in the demand for workers induces firms to pay higher wages. There are three main reasons why this may not happen.

The first is a coercive relationship between employer and employed. Throughout much of history, most agricultural workers were unfree, either working as slaves or in other forms of forced labor. When a master wants to obtain more labor hours from his slaves, he does not have to pay them more money. Rather, he can intensify coercion to extract greater effort and more output. Under such conditions, even revolutionary innovations such as the cotton gin in the American South do not necessarily lead to shared benefits. Even beyond slavery, under sufficiently oppressive conditions, the introduction of new technology can increase coercion, further impoverishing slaves and peasants alike, as we will see in Chapter 4.

Second, even without explicit coercion, the employer may not pay higher wages when productivity increases if she does not face competition from rivals. In many early agricultural societies, peasants were legally tied to the land, which meant that they could not seek or accept employment elsewhere. Even in eighteenth-century Britain, employees were prohibited from seeking alternative employment and were often jailed if they tried to take better jobs. When your outside option is prison, employers do not typically offer you generous compensation.

History provides plenty of confirmation. In medieval Europe, windmills, better crop rotation, and increased use of horses boosted agricultural productivity. However, there was little or no improvement in the living standards of most peasants. Instead, most of the additional output went to a small elite, and especially to a massive construction boom during which monumental cathedrals were built throughout Europe. When industrial machinery and factories started spreading in Britain in the 1700s, this did not initially increase wages, and there are many instances in which it worsened living standards and conditions for workers. At the same time, factory owners became fabulously wealthy.

Third and most important for today's world, wages are often negotiated rather than being simply determined by impersonal market forces. A modern corporation is often able to make sizable profits thanks to its market position, scale, or technological expertise. For example, when Ford Motor Company pioneered new mass-production techniques and started producing good-quality, cheap cars in the early twentieth century, it also became massively profitable. This made its founder, Henry Ford, into one of the richest businessmen of the early twentieth century. Economists call such megaprofits "economic rents" (or just "rents") to signify that they are above and beyond the prevailing normal return on capital expected by shareholders given the risks involved in such an investment. Once there are economic rents in the mix, wages for workers are not simply determined by outside market forces but also by potential "rent sharing"—their ability to negotiate some part of these profits.

One source of economic rents is market power. In most countries, there is a limited number of professional sports teams, and entry into the sector is typically constrained by the amount of capital required. In the 1950s and 1960s, baseball was a profitable business in the US, but players were not highly paid, even as revenues from television broadcasts poured in. This changed starting in the late 1960s because the players found ways to increase their bargaining power. Today, the owners of baseball teams still do well, but they are forced to share much more of their rents with the athletes.

Employers may also share rents to cultivate goodwill and motivate employees to work harder, or because prevailing social norms convince them to do so. On January 5, 1914, Henry Ford famously introduced a minimum pay of five dollars per day to reduce absenteeism, to improve retention of workers, and presumably to reduce the risk of strikes. Many employers have since tried something similar, particularly when it is hard to hire and retain people or when motivating employees turns out to be critical for corporate success.

Overall, Ricardo and Keynes may not have been right on every detail, but they correctly understood that productivity growth does not necessarily, automatically deliver broad-based prosperity. It will do so only when new technologies increase worker marginal productivity and the resulting gains are shared between firms and workers.

Even more fundamentally, these outcomes depend on economic, social, and political choices. New techniques and machines are not gifts descending unimpeded from the skies. They can focus on automation and surveillance to reduce labor costs. Or they can create new tasks and empower workers. More broadly, they can generate shared prosperity or relentless inequality, depending on how they are used and where new innovative effort is directed.

In principle, these are decisions a society should make, collectively. In practice, they are made by entrepreneurs, managers, visionaries, and sometimes political leaders, with defining effects on who wins and who loses from technological advances.

Optimism, with Caveats

Even though inequality has skyrocketed, many workers have been left behind, and the productivity bandwagon has not come to the rescue in recent decades, we have reasons to be hopeful. There have been tremendous advances in human knowledge, and there is ample room to build shared prosperity based on these scientific foundations—if we start making different choices about the direction of progress.

Techno-optimists have one thing right: digital technologies have already revolutionized the process of science. The accumulated knowledge of humanity is now at our fingertips. Scientists have access to incredible measurement tools, ranging from atomic force microscopes to magnetic resonance imagery and brain scans. They also have the computing power to crunch vast amounts of data in a way that even thirty years ago would have seemed like fantasy.

Scientific inquiry is cumulative, with inventors building on each other's work. Unlike today, knowledge used to diffuse slowly. In the 1600s, scholars such as Galileo Galilei, Johannes Kepler, Isaac Newton, Gottfried Wilhelm Leibniz, and Robert Hooke shared their scientific discoveries in letters that took weeks or even months to reach their destination. Nicolaus Copernicus's heliocentric system, which correctly placed Earth in the orbit of the sun, was developed during the first decade of the sixteenth century. Copernicus had written out his theory by 1514, even if his most widely read book, *On the Revolutions of the Celestial Spheres*, was published only in 1543. It took almost a century from 1514 for Kepler and Galileo to build on Copernicus's work and more than two centuries for the ideas to become widely accepted.

Today, scientific discoveries travel at lightning speed, especially when there is a pressing need. Vaccine development usually takes years, but in early 2020 Moderna, Inc., invented a vaccine just forty-two days after receiving the recently identified sequence of the SARS-CoV-2 virus. The entire development, testing, and authorization process took less than one year, resulting in remarkably safe and effective protection against severe illness caused by COVID. The barriers to sharing ideas and spreading technical know-how have never been lower, and the cumulative power of science has never been stronger.

However, to build on these advances and turn them to work for the betterment of billions of people around the world, we need to redirect technology. This must start by confronting the blind techno-optimism of our age and then developing new ways to use science and innovation.

The good and the bad news is that how we use knowledge and science depends on vision—the way that humans understand how they can turn knowledge into techniques and methods targeted at solving specific problems. Vision shapes our choices because it specifies what our aspirations are, what means we will pursue to achieve them, what alternative options we will consider and which ones we will ignore, and how we perceive the costs and benefits of our actions. In short, it is how we imagine technologies and their gifts, as well as the potential damage.

The bad news is that even at the best of times, the visions of powerful people have a disproportionate effect on what we do with our existing tools and the direction of innovation. The consequences of technology are then aligned with their interests and beliefs, and often prove costly to the rest. The good news is that choices and visions can change.

A shared vision among innovators is critical for the accumulation of knowledge and is also central to how we use technology. Take the steam engine, which transformed Europe and then the world economy. Rapid innovations from the beginning of the eighteenth century built on a common understanding of the problem to be solved: to perform mechanical work using heat. Thomas Newcomen built the first widely used steam engine, sometime around 1712. Half a century later, James Watt and his business partner Matthew Boulton improved Newcomen's design by separating the condenser and producing a more effective and commercially much more successful engine.

The shared perspective is visible in what these innovators were trying to achieve and how: using steam to push a piston back and forth inside a cylinder to generate work and then increasing the efficiency of these engines so that they could be used in a variety of different applications. A shared vision not only enabled them to learn from each other but meant that they approached the problem in similar ways. They predominantly focused on what is called the atmospheric engine, in which condensed steam creates a vacuum inside the cylinder, allowing atmospheric pressure to push the piston. They also collectively ignored other possibilities, such as high-pressure steam engines, first described by Jacob Leupold in 1720. Contrary to the eighteenth-century scientific consensus, high-pressure engines became the standard in the nineteenth century.

The early steam engine innovators' vision also meant that they were highly motivated and did not pause to reflect on the costs that the innovations might impose—for example, on very young children sent to work under draconian conditions in coal mines made possible by improved steam-powered drainage.

What is true of steam engines is true of all technologies. Technologies do not exist independent of an underlying vision. We look for ways of solving problems facing us (this is vision). We imagine what kind of tools might help us (also vision). Of the multiple paths open to us, we focus on a handful (yet another aspect of vision). We then attempt alternative approaches, experimenting and innovating based on that understanding. In this process, there will be setbacks, costs, and almost surely unintended consequences, including potential suffering for some people. Whether we are discouraged or even decide that the responsible thing is to abandon our dreams is another aspect of vision.

But what determines which technology vision prevails? Even though the choices are about how best to use our collective knowledge, the decisive factors are not just technical or what makes sense in a pure engineering sense. Choice in this context is fundamentally about power—the power to persuade others, as we will see in Chapter 3—because different choices benefit different people. Whoever has greater power is more likely to persuade others of their perspective, which is most often aligned with their interests. And whoever succeeds in turning their ideas into a shared vision gains additional power and social standing.

Do not be fooled by the monumental technological achievements of humankind. Shared visions can just as easily trap us. Companies make the investments that management considers best for their bottom line. If a company is installing, say, new computers, this must mean that the higher revenues they generate more than make up for the costs. But in a world in which shared visions guide our actions, there is no guarantee that this is indeed the case. If everybody becomes convinced that artificial-intelligence technologies are needed, then businesses will invest in artificial intelligence, even when there are alternative ways of organizing production that could be more beneficial. Similarly, if most researchers are working on a particular way of advancing machine intelligence, others may follow faithfully, or even blindly, in their footsteps.

These issues become even more consequential when we are dealing with "general-purpose" technologies, such as electricity or computers. General-purpose technologies provide a platform on which myriad applications can be built and potentially generate benefits—but sometimes also costs—for many sectors and groups of people. These platforms also allow widely different trajectories of development.

Electricity, for instance, was not just a cheaper source of energy; it also paved the way to new products, such as radios, household appliances, movies, and TVs. It introduced new electrical machinery. It enabled a fundamental reorganization of factories, with better lighting, dedicated sources of power for individual machinery, and the introduction of new precision and technical tasks in the production process. Advances in manufacturing based on electricity increased demand for raw materials and other industrial inputs, such as chemicals and fossil fuels, as well as retail and transport services. They also launched novel products, including new plastics, dyes, metals, and vehicles, that were then used in other industries. Electricity has also paved the way for much greater levels of pollution from manufacturing production.

Although general-purpose technologies can be developed in many different ways, once a shared vision locks in a specific direction, it becomes difficult for people to break out of its hold and explore different trajectories that might be socially more beneficial. Most people affected by those decisions are not consulted. This creates a natural tendency for the direction of progress to be socially biased—in favor of powerful decision makers with dominant visions and against those without a voice.

Take the decision of the Chinese Communist Party to introduce a social credit system that collects data on individuals, businesses, and government agencies to keep track of their trustworthiness and whether they abide by the rules. Initiated at the local level in 2009, it aspires to blacklist people and companies nationally because of their speech or social media posts that go against the party's preferences. This decision, which affects the lives of 1.4 billion people, was taken by a few party leaders. There was no consultation with those whose freedom of speech and association, education, government jobs, ability to travel, and even likelihood of getting government services and housing are now being shaped by the system.

This is not something that happens only in dictatorships. In 2018 Facebook founder and CEO Mark Zuckerberg announced that the company's algorithm would be modified to give users "meaningful social interactions." What this meant in practice was that the platform's algorithm would prioritize posts from other users, especially family and friends, rather than news organizations and established brands. The purpose of the change was to increase user engagement because people were found to be more likely to be drawn to and click on posts by their acquaintances. The main consequence of the change was to amplify misinformation and political polarization, as lies and misleading posts spread rapidly from user to user. The change did not just affect the company's then almost 2.5 billion users; billions more people who were not on the platform were also indirectly affected by the political fallout from the resulting misinformation. The decision was made by Zuckerberg; the company's chief operating officer, Sheryl Sandberg; and a few other top engineers and executives. Facebook users and citizens of affected democracies were not consulted.

What propelled the Chinese Communist Party's and Facebook's decisions? In neither case were they dictated by the nature of science and technology. Nor were they the obvious next step in some inexorable march of progress. In both cases you can see the ruinous role of interests—to quash opposition or to increase advertising revenues. Equally central was their leadership's vision for how communities should be organized and what should be prioritized. But even more important was how technology was used for control: over the political views of the population in the Chinese case, and people's data and social activities for Facebook.

This is the point that, with the advantage of an additional 275 years of human history to draw on, H. G. Wells grasped and Francis Bacon missed: technology is about control, not just over nature but often over other humans. It is not simply that technological change benefits some more than others. More fundamentally, different ways of organizing production enrich and empower some people and disempower others.

The same considerations are equally important for the direction of innovation in other contexts. Business owners and managers may often wish to automate or increase surveillance because this enables them to strengthen their control over the production process, save on wage costs, and weaken the power of labor. This demand then translates into incentives to focus innovation more on automation and surveillance, even when developing other, more worker-friendly technologies could increase output more and pave the way to shared prosperity.

In these instances, society may even become gripped by visions that favor powerful individuals. Such visions then help business and technology leaders pursue plans that increase their wealth, political power, or status. These elites may convince themselves that whatever is good for them is also best for the common good. They may even come to believe that any suffering that their virtuous path generates is a price well worth paying for progress—especially when those bearing the brunt of the costs are voiceless. When thus inspired by a selfish vision, leaders deny that there are many different paths with widely different implications. They may even become incensed when alternatives are pointed out to them.

Is there no remedy against ruinous visions imposed on people without their consent? Is there no barrier against the social bias of technology? Are we locked in a constant cycle of one overconfident vision after another shaping our future while ignoring the damage?

No. There is reason to be hopeful because history also teaches us that a more inclusive vision that listens to a broader set of voices and recognizes the effects on everyone is possible. Shared prosperity is more likely when countervailing powers hold entrepreneurs and technology leaders accountable—and push production methods and innovation in a more worker-friendly direction.

Inclusive visions do not avoid some of the thorniest questions, such as whether the benefits that some reap justify the costs that others suffer. But they ensure that social decisions recognize their full consequences and without silencing those who do not gain.

Whether we end up with selfish, narrow visions or something more inclusive is also a choice. The outcome depends on whether there are countervailing forces and whether those who are not in the corridors of power can organize and have their voices heard. If we want to avoid being trapped in the visions of powerful elites, we must find ways of countering power with alternative sources of power and resisting selfishness with a more inclusive vision. Unfortunately, this is becoming harder in the age of artificial intelligence.

Fire, This Time

Early human life was transformed by fire. In Swartkrans, a South African cave, the earliest excavated layers show ancient hominid bones that were eaten by predators—big cats or bears. To the apex predators of the day, humans must have seemed like easy prey. Dark places in caves were particularly dangerous places, to be avoided by our ancestors. Then the first evidence of fire appears inside that cave, with a layer of charcoal about a million years old. Subsequently, the archaeological record shows a complete reversal: from that time forward, the bones are mostly those of nonhuman animals. Control of fire gave hominins the ability to take and hold caves, turning the tables on other predators.

No other technology in the last ten thousand years can claim to approach this type of fundamental impact on everything else we do and who we are. Now there is another candidate, at least according to its boosters: artificial intelligence (AI). Google's CEO Sundar Pichai is explicit when he says that "AI is probably the most important thing humanity has ever worked on. I think of it as something more profound than electricity or fire."

AI is the name given to the branch of computer science that develops "intelligent" machines, meaning machines and algorithms (instructions for solving problems) capable of exhibiting high-level capabilities. Modern intelligent machines perform tasks that many would have thought impossible a couple of decades ago. Examples include face-recognition software, search engines that guess what you want to find, and recommendation systems that match you to the products that you are most likely to enjoy or, at the very least, purchase. Many systems now use some form of natural-language processing to interface between human speech or written enquiries and computers. Apple's Siri and Google's search engine are examples of AI-based systems that are used widely around the world every day.

AI enthusiasts also point to some impressive achievements. AI programs can recognize thousands of different objects and images and provide some basic translation among more than a hundred languages. They help identify cancers. They can sometimes invest better than seasoned financial analysts. They can help lawyers and paralegals sift through thousands of documents to find the relevant precedents for a court case. They can turn natural-language instructions into computer code. They can even compose new music that sounds eerily like Johann Sebastian Bach and write (dull) newspaper articles.

In 2016 the AI company DeepMind released AlphaGo, which went on to beat one of the two best Go players in the world. The chess program AlphaZero, capable of defeating any chess master, followed one year later. Remarkably, this was a self-taught program and reached a superhuman level after only nine hours of playing against itself.

Buoyed by these victories, it has become commonplace to assume that AI will affect every aspect of our lives—and for the better. It will make humankind much more prosperous, healthier, and able to achieve other laudable goals. As the subtitle of a recent book on the subject claims, "artificial intelligence will transform everything." Or as Kai-Fu Lee, the former president of Google China, puts it, "Artificial Intelligence (AI) could be the most transformative technology in the history of mankind."

But what if there is a fly in the ointment? What if AI fundamentally disrupts the labor market where most of us earn our livelihoods, expanding inequalities of pay and work? What if its main impact will not be to increase productivity but to redistribute power and prosperity away from ordinary people toward those controlling data and making key corporate decisions? What if along this path, AI also impoverishes billions in the developing world? What if it reinforces existing biases—for example, based on skin color? What if it destroys democratic institutions?

The evidence is mounting that all these concerns are valid. AI appears set on a trajectory that will multiply inequalities, not just in industrialized countries but everywhere around the world. Fueled by massive data collection by tech companies and authoritarian governments, it is stifling democracy and strengthening autocracy. As we will see in chapters 9 and 10, it is profoundly affecting the economy even as, on its current path, it is doing little to improve our productive capabilities. When all is said and done, the newfound enthusiasm about AI seems an intensification of the same optimism about technology, regardless of whether it focuses on the automation, surveillance, and disempowerment of ordinary people that had already engulfed the digital world.

Yet these concerns are not taken seriously by most tech leaders. We are continuously told that AI will bring good. If it creates disruptions, those problems are short-term, inevitable, and easily rectified. If it is creating losers, the solution is more AI. For example, DeepMind's cofounder, Demis Hassabis, not only thinks that AI "is going to be the most important technology ever invented," but he is also confident that "by deepening our capacity to ask how and why, AI will advance the frontiers of knowledge and unlock whole new avenues of scientific discovery, improving the lives of billions of people."

He is not alone. Scores of experts are making similar claims. As Robin Li, cofounder of the Chinese internet search firm Baidu and an investor in several other leading AI ventures, states, "The intelligent revolution is a benign revolution in production and lifestyle and also a revolution in our way of thinking."

Many go even further. Ray Kurzweil, a prominent executive, inventor, and author, has confidently argued that the technologies associated with AI are on their way to achieving "superintelligence" or "singularity"—meaning that we will reach boundless prosperity and accomplish our material objectives, and perhaps a few of the nonmaterial ones as well. He believes that AI programs will surpass human capabilities by so much that they will themselves produce further superhuman capabilities or, more fancifully, that they will merge with humans to create superhumans.

To be fair, not all tech leaders are as sanguine. Billionaires Bill Gates and Elon Musk have expressed concern about misaligned, or perhaps even evil, superintelligence and the consequences of uncontrolled AI development for the future of humanity. Yet both of these sometime holders of the title "richest person in the world" agree with Hassabis, Li, Kurzweil, and many others on one thing: most technology is for good, and we can and must rely on technology, especially digital technology, to solve humanity's problems. According to Hassabis, "Either we need an exponential improvement in human behavior—less selfishness, less short-termism, more collaboration, more generosity—or we need an exponential improvement in technology."

These visionaries do not question whether technological change is always progress. They take it for granted that more technology is the answer to our social problems. We do not need to fret too much about the billions of people who are initially left behind; they will soon benefit as well. We must continue to march onward, in the name of progress. As LinkedIn cofounder Reid Hoffman puts it, "Could we have a bad twenty years? Absolutely. But if you're working toward progress, your future will be better than your present."

Such faith in the beneficent powers of technology is not new, as we already saw in the Prologue. Like Francis Bacon and the foundational story of fire, we tend to see technology as enabling us to turn the tables on nature. Rather than being the weakling prey, thanks to fire we became the planet's most devastating predator. We view many other technologies through the same lens—we conquer distance with the wheel, darkness with electricity, and illness with medicine.

Contrary to all these claims, we should not assume that the chosen path will benefit everybody, for the productivity bandwagon is often weak and never automatic. What we are witnessing today is not inexorable progress toward the common good but an influential shared vision among the most powerful technology leaders. This vision is focused on automation, surveillance, and mass-scale data collection, undermining shared prosperity and weakening democracies. Not coincidentally, it also amplifies the wealth and power of this narrow elite, at the expense of most ordinary people.

This dynamic has already produced a new vision oligarchy—a coterie of tech leaders with similar backgrounds, similar world-views, similar passions, and unfortunately similar blind spots. This is an oligarchy because it is a small group with a shared mind-set, monopolizing social power and disregarding its ruinous effects on the voiceless and the powerless. This group's sway comes not from tanks and rockets but because it has access to the corridors of power and can influence public opinion.

The vision oligarchy is so persuasive because it has had brilliant commercial success. It is also supported by a compelling narrative about all the abundance and control over nature that new technologies, especially the exponentially increasing capabilities of artificial intelligence, will create. The oligarchy has charisma, in its nerdy way. Most importantly, these modern oligarchs mesmerize influential custodians of opinion: journalists, other business leaders, politicians, academicians, and all sorts of intellectuals. The vision oligarchy is always at the table and always at the microphone when important arguments are being made.

It is critical to rein in this modern oligarchy, and not just because we are at a precipice. This is the time to act because these leaders have one thing right: we have amazing tools at our disposal, and digital technologies could amplify what humanity can do. But only if we put these tools to work for people. And this is not going to happen until we challenge the worldview that prevails among our current global tech bosses. This worldview is based on a particular—and inaccurate—reading of history and what that implies about how innovation affects humanity. Let us start by reassessing this history.

Plan for the Rest of the Book

In the rest of this book we develop the ideas introduced in this chapter and reinterpret the economic and social developments of the last thousand years as the outcome of the struggle over the direction of technology and the type of progress—and who won, who lost, and why. Because our focus is on technologies, most of this discussion centers on the parts of the world where the most important and consequential technological changes were taking place. This means first Western Europe and China for agriculture, then Britain and the US for the Industrial Revolution, and then the US and China for digital technologies. Throughout we also emphasize how at times different choices were made in different countries, as well as the implications of technologies in the leading economies on the rest of the world, as they spread, sometimes voluntarily, sometimes forcefully, across the globe.

Chapter 2 ("Canal Vision") provides a historical example of how successful visions can lead us astray. The success of French engineers in building the Suez Canal stands in remarkable contrast to their spectacular failure when the same ideas were brought to Panama. Ferdinand de Lesseps persuaded thousands of investors and engineers into the unworkable plan of building a sea-level canal at Panama, resulting in the deaths of more than twenty thousand people and financial ruin for many more. This is a cautionary tale for any history of technology: great disaster often has its roots in powerful visions, which in turn are based on past success.

Chapter 3 ("Power to Persuade") highlights the central role of persuasion in how we make key technology and social decisions. We explain how the power to persuade is rooted in political institutions and the ability to set the agenda, and emphasize how countervailing powers and a wider range of voices can potentially rein in overconfidence and selfish visions.

Chapter 4 ("Cultivating Misery") applies the main ideas of our framework to the evolution of agricultural technologies, from the beginning of settled agriculture during the Neolithic Age to the major changes in the organization of land and techniques of production during the medieval and early modern eras. In these momentous episodes, we find no evidence of an automatic productivity bandwagon. These major agricultural transitions have tended to enrich and empower small elites while generating few benefits for agricultural workers: peasants lacked political and social power, and the path of technology followed the visions of a narrow elite.

Chapter 5 ("A Middling Sort of Revolution") reinterprets the Industrial Revolution, one of the most important economic transitions in world history. Although much has been written about the Industrial Revolution, what is often underemphasized is the emergent vision of newly emboldened middle classes, entrepreneurs, and businesspeople. Their views and aspirations were rooted in institutional changes that started empowering the middling sort of English people from the sixteenth and seventeenth centuries onward. The Industrial Revolution may have been propelled by the ambitions of new people attempting to improve their wealth and social standing, but theirs was far from an inclusive vision. We discuss how changes in political and economic arrangements came about, and why these were so important in producing a new concept of how nature could be controlled and by whom.

Chapter 6 ("Casualties of Progress") turns to the consequences of this new vision. It explains how the first phase of the Industrial Revolution was impoverishing and disempowering for most people, and why this was the outcome of a strong automation bias in technology and a lack of worker voice in technology and wage-setting decisions. It was not just economic livelihoods that were adversely affected by industrialization but also the health and autonomy of much of the population. This awful picture started changing in the second half of the nineteenth century as regular people organized and forced economic and political reforms. The social changes altered the direction of technology and pushed up wages. This was only a small victory for shared prosperity, and Western nations would have to travel along a much longer, contested technological and institutional path to achieve shared prosperity.

Chapter 7 ("The Contested Path") reviews how arduous struggles over the direction of technology, wage setting, and more generally politics built the foundations of the most spectacular period of economic growth in the West. During the three decades following World War II, the United States and other industrial nations experienced rapid economic growth that was broadly shared across most demographic groups. These economic trends went together with other social improvements, including expansions in education, health care, and life expectancy. We explain how and why technological change did not just automate work but also created new opportunities for workers, and how this was embedded in an institutional setting that bolstered countervailing powers.

Chapter 8 ("Digital Damage") turns to our modern era, starting with how we lost our way and abandoned the shared-prosperity model of the early postwar decades. Central to this volte-face was a change in the direction of technology away from new tasks and opportunities for workers and toward a preoccupation with automating work and cutting labor costs. This redirection was not inevitable but rather resulted from a lack of input and pressure from workers, labor organizations, and government regulation. These social trends contributed to the undermining of shared prosperity.

Chapter 9 ("Artificial Struggle") explains that the post-1980 vision that led us astray has also come to define how we conceive of the next phase of digital technologies, artificial intelligence, and how AI is exacerbating the trends toward economic inequality. In contrast to claims made by many tech leaders, we will also see that in most human tasks existing AI technologies bring only limited benefits. Additionally, the use of AI for workplace monitoring is not just boosting inequality but also disempowering workers. Worse, the current path of AI risks reversing decades of economic gains in the developing world by exporting automation globally. None of this is inevitable. In fact, this chapter argues that AI, and even the emphasis on machine intelligence, reflects a very specific path for the development of digital technologies, one with profound distributional effects—benefiting a few people and leaving the rest behind. Rather than focusing on machine intelligence, it is more fruitful to strive for "machine usefulness," meaning how machines can be most useful to humans—for example, by complementing worker capabilities. We will also see that when it was pursued in the past, machine usefulness led to some of the most important and productive applications of digital technologies but has become increasingly sidelined in the quest for machine intelligence and automation.

Chapter 10 ("Democracy Breaks") argues that the problems facing us may be even more severe because massive data collection and harvesting using AI methods are intensifying surveillance of citizens by governments and companies. At the same time, AI-powered advertisement-based business models are propagating misinformation and amplifying extremism. The current path of AI is neither good for the economy nor for democracy, and these two problems, unfortunately, reinforce each other.

Chapter 11 ("Redirecting Technology") concludes by outlining how we can reverse these pernicious trends. It provides a template for redirecting technological change based on altering the narrative, building countervailing powers, and developing technical, regulatory, and policy solutions to tackle specific aspects of technology's social bias.

2

Canal Vision

Walk carefully, do not wake the envy of the happy gods,
Shun Hubris.

—C. S. LEWIS, "A Cliché Came Out of Its Cage," 1964

If the committee had decided to build a lock canal,
I would have put on my hat and gone home.

—FERDINAND DE LESSEPS, 1880,
speaking of plans to build the Panama Canal

On Friday, May 23, 1879, Ferdinand de Lesseps rose to address the Congrès International d'Études du Canal Interocéanique. Delegates from around the world had converged on Paris to discuss how best to proceed with one of the most ambitious construction projects of the age—linking the Atlantic and Pacific Oceans with a canal across Central America.

On the first day of the conference, several days earlier, Lesseps had addressed the delegates certain that his preferred scheme, a sea-level canal through Panama, would prevail. He reportedly concluded the first session with a quip: "Gentlemen, we are going to rush this thing *à l'Américaine*: we shall get through by next Tuesday."

The US representatives were not amused. They preferred a canal through Nicaragua that would, in their assessment, have major engineering and economic advantages. They and many of the other experts in attendance were also far from convinced that a sea-level canal was practical for any part of Central America. There were multiple calls for more substantive discussion of alternatives. Lesseps dug in his heels. The canal must be built in Panama and at sea level, entirely without locks.

The vision guiding Lesseps was rooted in three strongly held tenets. The first was a nineteenth-century version of techno-optimism. Progress would benefit everybody, and transoceanic canals, one of the most important applications of the technological advances of the age, would drive progress by reducing the time needed to ship goods around the world. If there were obstacles to building such infrastructure, technology and science would come to the rescue. The second was a belief in markets: even the largest projects could be financed with private capital, and the returns from the projects would benefit investors and constitute another way of serving the common good. Third was a set of blinders. Lesseps's focus was on European priorities, and the fate of non-Europeans mattered little.

Lesseps's story is as relevant in our age of digital technologies as it was a century and a half ago because it illustrates how a compelling vision takes hold and pushes the frontiers of technology, for good and bad.

Lesseps was backed by French institutions and at times the power of the Egyptian state. He was persuasive because of his previous magnificent success at Suez, where he was able to cajole French investors and Egyptian leaders to accept his plan for a canal and demonstrate how new technologies could rise to the challenge of solving thorny problems along the way.

Even at the height of its success, however, Lesseps's version of progress was not for everybody. Egyptian workers who were coerced to toil on the Suez Canal were likely not among the main beneficiaries of this technological feat, and Lesseps's vision appeared unbothered about their plight.

The Panama project also illustrates how powerful visions can fail spectacularly, even on their own terms. Gripped by confidence and optimism, Lesseps refused to admit the difficulties in Panama even when they became all too obvious to everybody else. French engineering suffered a humbling failure, investors lost their fortunes, and more than twenty thousand people died to no avail.

We Must Go to the Orient

In early 1798, Napoleon Bonaparte, a twenty-eight-year-old general, had just defeated the Austrians in Italy. Now he was looking for his next big adventure, preferably one that would strike a blow against France's public enemy number one, the British Empire.

Realizing that French naval forces were too weak to support an invasion of Britain itself, Napoleon proposed instead to undermine British interests in the Middle East and open new trade routes to Asia. Besides, as he put it to a colleague, "We must go to the Orient; all great glory has always been acquired there."

The "Orient" was a stage upon which the European ambitions could be played out. Invading Egypt would, in Napoleon's condescending view, help Egyptians modernize (or at the very least, this provided a good excuse).

In July 1798, not far from the pyramids, Napoleon's force of twenty-five thousand confronted about six thousand highly trained Mamluk cavalry supported by fifteen thousand infantry. The Mamluks, descendants of slave soldiers, had ruled Egypt as a warrior aristocracy since the Middle Ages. They were renowned for their fierce fighting skills, and each horseman was impeccably dressed and equipped with a carbine (a short gun), two or three pairs of pistols, several lances, and a scimitar (a short curved sword).

The Mamluks' charge, when it arrived, was impressive and terrifying. But Napoleon's experienced infantry, organized in squares and backed by mobile cannon, easily withstood the attack and prevailed. The Mamluks lost several thousand men, while French casualties were only 29 killed and 260 wounded. The capital, Cairo, quickly fell.

Napoleon was bringing new ideas to Egypt, whether the Egyptians wanted them or not. The expedition included 167 scientists and scholars, with the mission of understanding one of the most ancient civilizations. Their cumulative work, *Description de l'Égypte*, ran to 23 volumes, published from 1809 to 1829, and founded modern Egyptology, deepening European fascination with the region.

Napoleon's remit from the French government included the charge of exploring the potential for a canal connecting the Red Sea with the Mediterranean:

> The general in chief of the Army of the Orient will seize Egypt; he will chase the English from all their possessions in the Orient; and he will destroy all of their settlements on the Red Sea. He will then cut the Isthmus of Suez and take all necessary measures in order to assure the free and exclusive possession of the Red Sea for the French Republic.

After some wandering in the desert, Napoleon supposedly stumbled on a long disused route linked to ancient canal banks. French experts took on the task of surveying the remains of canals that had apparently operated, on and off, for thousands of years, though not over the previous six hundred years. Soon they established the basic geographic facts: the Red Sea and the Mediterranean were separated by an isthmus not more than a hundred miles long.

The historical route had been indirect, via the Nile, and used small canals: north from Suez on the Red Sea to the Bitter Lakes, situated about halfway up the isthmus, and then west to the Nile. A direct north-south route had never been attempted. Still, European war and the pursuit of glory intervened, and the canal project was shelved for a generation.

Capital Utopia

To understand Lesseps's vision, we must first turn to the ideas of the French social reformer Henri de Saint-Simon and his colorful followers. Saint-Simon was an aristocratic writer who maintained that human progress is driven by scientific invention and the application of new ideas to industry. But he also thought that the right leadership was critical for this progress: "All enlightened peoples will adopt the view that men of genius should be given the highest social standing."

Power should be in the hands of those who worked for a living and particularly the "men of genius," not those whom he referred to as "idlers," which included his own aristocratic family. This meritocracy would naturally facilitate industrial and technological development, broadly sharing the resulting prosperity, not just in France but also around the world. Some regard him as an early socialist, but Saint-Simon was a firm believer in private property and the importance of free enterprise.

Saint-Simon was largely ignored during his lifetime, but soon after his death in 1825 his ideas started to gain traction, in part because of effective proselytizing by Barthélemy Prosper Enfantin. Enfantin was a graduate of an elite engineering school, École Polytechnique, and he pulled many smart young engineers into his orbit. This group elevated Saint-Simon's belief in industry and technology to an almost religious creed.

Canals and, later, railways were the main places they applied these ideas. In Enfantin's view, investments of this kind should be organized by entrepreneurs, backed by privately owned capital. The government role should be limited to providing the necessary "concession," which would grant the rights needed to build and operate a particular piece of infrastructure for long enough to generate an attractive return to investors.

Canals were on the European mind long before Saint-Simon and Enfantin. Among the most famous engineering achievements of the ancien régime in France was the Canal du Midi. This 240-kilometer (150-mile) canal, opened in 1681, crossed a summit approximately 190 meters (620 feet) above sea level and connected the city of Toulouse to the Mediterranean. It provided the first direct waterway connection between the Atlantic and the Mediterranean, and significantly reduced travel time for boats.

By the second half of the eighteenth century, early British industrialization was fueled by a "transport revolution," with scores of new canals linking English rivers to the sea. Waterborne transportation was important in North America as well, epitomized by the high-profile success of the Erie Canal, which opened in 1825.

By the 1830s, Enfantin believed that a canal at Suez would provide the type of infrastructure that would bring shared global prosperity. He argued that not just France and Britain would benefit from the canal but also Egypt and India. Underscoring both the religious mysticism of his group's philosophy and their Orientalism, Enfantin also maintained that the West (Europe) was male and the East (India and elsewhere) was female, so the canal could actually join the world in a form of mutually beneficial global matrimony!

Following the French withdrawal from Egypt in 1801, the Ottoman Empire sent one of its generals, Mohammed Ali, to reassert control. He became the official viceroy in 1805, and for the next half dozen years there was a tense standoff between Mohammed Ali's forces and the Mamluk aristocracy.

On March 1, 1811, Mohammed Ali invited the Mamluk elite to a reception in the Cairo Citadel. The atmosphere was cordial and the food outstanding, but as the aristocracy filed down a narrow medieval pathway, they were shot.

Ali went on to establish himself as an autocratic modernizer, strengthening his grip on power by importing modern technology and ideas from Western Europe. Throughout Ali's forty-three-year reign, he made extensive use of European engineers for public works, including in irrigation projects and health campaigns. Arriving in 1833, Enfantin's group fit right in and had no difficulty making itself useful by working on several projects, including a barrage (a type of diversion dam) that would use a system of gates to control flooding on the Nile.

However, Enfantin could not convince Ali to grant the right to build a canal across Egypt. The Egyptian strongman grasped that his position required a delicate balance between the declining regional power of his Ottoman overlord and the rising global force represented by Britain and France. A canal at Suez could upset the geopolitical dance that kept the Europeans and the sultan at bay. Worse, directly linking the Mediterranean and the Red Sea would bypass Egyptian population centers and potentially undermine Egypt's prosperity.

Enfantin and his friends eventually achieved impressive success in business back home, most notably in the 1840s with the formation of French railroad companies and joint-stock banks able to support sizable stock issues. Whereas the French government attempts to build long-distance railroads floundered, the private sector had much greater success. Another big new idea took hold: small investors could combine resources to finance even the largest industrial projects.

As for a potential Suez Canal, the keys to the isthmus were firmly in the hands of the ruler of Egypt, and Ali's answer was an adamant no, right up to his death in 1848. Near the end of his life in 1864, Enfantin admitted: "In my hands, the canal affair was a failure. I did not have the necessary flexibility to deal with all of the adversities, to fight simultaneously in Cairo, London, and Constantinople. . . . In order to succeed, one must have, like Lesseps, a devil's determination and ardor that doesn't know fatigue or obstacles."

Lesseps Finds Vision

In 1832, so the story goes, Lesseps read the Napoleonic survey team's account of the canal that existed between the Red Sea and the Mediterranean, running across ancient Egypt. He met Enfantin shortly afterward and was smitten with the idea that the Suez Canal would be a glorious and profitable way of connecting the world.

Lesseps was infused with the ideas of his time. His diplomatic background and social circle made him a natural Orientalist, seeing the world from an unflinching European viewpoint. He spent the first twenty years of his career representing French interests around the Mediterranean, and an implicit belief in the superiority of European thinking is evident throughout his memoir,

Recollections of Forty Years. The French had, in his view, a civilizing mission that justified taking over Algeria in the 1820s and other colonial expansions.

Lesseps also internalized Saint-Simon's ideas on the importance of large public infrastructure projects to unite the world and make long-distance trade easier and cheaper. If anything, Lesseps went even further, stressing that public-private partnership was essential for such projects: "Governments can encourage such enterprises; they cannot execute them. It is the public then on whom we must call. . . ."

Lesseps further reckoned that technological ingenuity would always come to the rescue. By the 1850s, technology had advanced far beyond what was available in Saint-Simon's time. Steam engines had been improved to make ever-more-powerful machines, and advances in metallurgy had brought many new and sturdier materials, especially steel, which revolutionized construction.

Lesseps found most engineers lacking in imagination; they were too keen to tell him what could not happen. He sought out instead experts who could think big—new equipment for dredging waterways, new ways to shift hard rock out of the way, and new measures to protect against infectious disease. He saw his role as imagining the solution and arranging enough financing. One of his favorite aphorisms was very Saint-Simonian: "Men of genius always arise." To Lesseps, this meant some bright person would find a technological solution to any problem—once he, Lesseps, had driven everyone to the point where the problem to be solved had become fully apparent.

Since the first investigation by Napoleon's team, there had been an active technical discussion around what form the canal at Suez should take.

Most inland canals need locks. A rectangular chamber with gates at both ends, a lock allows boats to climb steep hills. When the water in a lock between two bodies of water is at the lower level, the gates at that level open, and a boat enters. Once the gate on the lower side is closed, water from the higher level fills the chamber, raising the boat to the level of its destination. The procedure repeats in reverse when traveling from the higher to the lower level.

The Chinese pioneered the development of effective locks more than a thousand years ago. Later improvements included the fifteenth-century invention of the miter gate, often attributed to Leonardo da Vinci, with two leaves at each end, which swing out from the side and meet at an angle pointing toward the upper level, making for easier opening and closing. Further advances came with French-designed valves that could regulate the flow of water into and out of the lock. The marvelous Erie Canal, linking Albany on the Hudson River and Buffalo on the Great Lakes, originally had 83 locks that enabled barges to climb a total of 566 feet in elevation.

Enfantin's team had figured out that the Mediterranean Sea and the Red Sea had the same level on average, even if the Red Sea had a larger tide. This implied that a sea-level canal was theoretically possible, although locks could be helpful for reducing the impact of tides on any canal at Suez.

Lesseps would have none of it. In his view, locks would significantly slow down traffic. He viewed this as an unacceptable impediment to the flow of ships promised by opening the Suez route, consistently holding fast to a principle that he would later articulate as "a ship must not now be delayed."

However, he did like the idea of using the dried-up lakes. This became the plan: connect dried-up lakes to the Mediterranean in the north and the Red Sea in the south, and then let water flood in to help with the rest of the work.

Little People Buy Small Shares

In 1849 Lesseps's promising diplomatic career ended suddenly after a major falling-out with the French government. At the age of forty-three, he retired to a family estate, apparently finished with public service. For several years he enjoyed the life of a French country gentleman, working on agricultural improvements and corresponding with leading Saint-Simonians about their fanciful projects. In 1853 personal tragedy struck. His wife and one of his sons died, likely from scarlet fever. Lesseps was desolated and desperate for a distraction. Little did he know that events in Egypt would soon provide much more than just a distraction.

In 1848 a seriously ill Mohammed Ali had been pushed from power. His successor, his eldest son, Ibrahim Pasha, passed away the same year. The next viceroy died unexpectedly in July 1854, and Mohammed Said, the fourth son of Mohammed Ali, became the ruler of Egypt.

When Lesseps had been a senior French representative in Egypt in the 1830s, Mohammed Ali had asked him to help the teenage Mohammed Said lose weight. Not only did Lesseps impress Mohammed Ali by accepting this unusual assignment; he also managed to stay on the good side of Said by combining a program of vigorous horse riding (a passion for both of them) with generous plates of pasta.

In late 1854, pausing only to consult with some leading Saint-Simonians and borrow their maps, Lesseps rushed to Egypt. He was warmly welcomed and invited to camp in the desert with the new viceroy, which was a great honor and an augur of things to come. According to Lesseps, he exited his tent one morning to see the sun rising over the eastern horizon. Suddenly a rainbow arose from the west and spanned the sky—an omen, he said later, that he would be personally able to unite East and West.

That evening he painted a persuasive spoken picture for Mohammed Said of how modern technology could be used to build a canal that would excel all ancient achievements. In Lesseps's account, his pitch included these lines: "The names of those Egyptian sovereigns who built the Pyramids, those monuments to human pride, are forgotten. The name of the Prince who opens the great maritime canal will be blessed from century to century until the end of time."

Mohammed Said granted Lesseps a concession very much along the lines of the one that the Saint-Simonians had received to build long-distance French railways. The viceroy provided land to the project for ninety-nine years and in return would receive 15 percent of the profits. Lesseps would promote, raise funding for, and run the canal. At least on paper, all the financial risk would fall on private shareholders to be named later.

By 1856, the legal framework and a rough design was in place, based on detailed work by two French engineers in Egyptian service who knew local conditions well. Lesseps consulted a bevy of international engineering experts, all of whom agreed that a north-south canal was technically feasible. Now Lesseps had to convince people to put up the cash for the canal and the British to stay out of the way.

In the mid-1850s, most cargo between England and India moved by sea, taking up to six months around the hazardous coast of Africa. In 1835 the East India Company had launched a mail route through the Red Sea, which transferred passengers by donkey- or horse-drawn wagon for eighty-four miles across the desert from Suez to Cairo, then down the Nile and along a small canal to Alexandria. This overland route cut the travel time to less than two months but was suitable only for higher-value and less-bulky cargo. In 1858, to aid this kind of transshipment and make it more appealing to travelers, a railway line opened between Suez and Alexandria.

The winds and currents of the Red Sea were not well suited to long-distance European sailing ships, and towing large ships along a 120-mile-long canal would not have been a winning proposition. But Lesseps correctly presaged the next stage of long-distance transport technology—large steamships, for which a Suez Canal would be perfect.

By early 1857, Lesseps had a well-honed pitch about how the canal at Suez would reduce travel time and transform global commerce. But a vision is nothing if it is not shared. This is where Lesseps excelled, partly because of his determination and charisma, and more importantly because he could talk to the right people and confer with his network of influential connections.

Lesseps toured Britain in the spring and summer of 1857, speaking at twenty meetings across sixteen cities and meeting as many prominent industrialists as he could. He was a big hit in places such as Manchester and Bristol, where the business community grasped the value of faster transportation for raw Indian cotton heading to British mills, and for manufactured goods and (when needed) soldiers moving in the other direction.

Armed with their statements of support, Lesseps paid one of his regular visits to the prime minister, Lord Palmerston. Disappointingly, however, Palmerston was consistently not well disposed to the canal, which he saw as continuing the Napoleonic tradition of trying to cut Britain out of lucrative global trade routes. The British government remained deeply skeptical and worked hard to throw up obstacles in Cairo, Constantinople, and anywhere else it had influence.

Undeterred, in October 1858, after two years of intense publicity, Lesseps was finally ready to sell stock. Lesseps resolved to get as many investors as possible directly involved, bypassing all intermediaries. He offered 400,000 shares at 500 francs each.

The price per share was slightly more than the average annual income in France at that time, making the shares expensive but plausibly affordable to members of the fast-growing French middle class. Shares were also offered in all Western European countries, the United States, and the Ottoman Empire. On the final road show, Lesseps himself visited Odessa, Trieste, Vienna, Barcelona, and Turin, as well as Bordeaux and Marseilles in France.

By the end of November 1858, twenty-three thousand people had bought shares, and twenty-one thousand of these investors were French. Demand elsewhere was tepid, and investors based in Britain, Russia, Austria, and the United States bought a grand total of zero shares.

The British newspapers sneered that the shares had been bought by hotel waiters, priests, and grocery-store employees. As Palmerston quipped, "Little men have been induced to buy small shares."

But he had been outsmarted by Lesseps, who got the backing of the French urban professional class—engineers, judges, bankers, teachers, priests, civil servants, merchants, and the like bought shares—as well as the ruler of Egypt, who stepped up to buy up all the shares unwanted by others. Said's stake ended up at 177,000 shares, costing more than his total annual revenue. The Egyptian state was all in.

One Cannot Say That They Are
Exactly Forced Labor

Visionaries derive their power partly from the blinders that they have on—including the suffering that they ignore. It was no different for Lesseps, who cared foremost about European commerce, European industry, and of course his overall Eurocentric vision of trade expansion. The viceroy of Egypt and the sultan of the Ottoman Empire needed to be managed and cajoled, but outcomes for ordinary Egyptians were not really part of his calculus. Egyptians could be left behind or even coerced as necessary, and this was still consistent with the notion of "progress" that Lesseps and many of his contemporaries shared.

When digging began in 1861, most of the workforce was supplied by the Egyptian government under a system of corvée labor, where peasants were forced to work on public projects.

Over the next three years, roughly sixty thousand men were engaged on the canal at any given time, of whom thousands might be on their way from the Nile Valley to the construction area, thousands were digging, and the rest were on their way home. Officials had to fill recruiting quotas by assigning peasants who would otherwise have been working on their own land or on local projects, and the Egyptian military was charged with bringing the workers to the canal site and supervising their manual labor.

Conditions were harsh and uncompromising. Huge amounts of rock were moved by pickax and basket year-round, even during Ramadan, the month of fasting for Muslims. Workers slept in the open desert, were provided with minimal rations, and lived in unsanitary conditions. Wages were less than half of the market rate and paid only at the end of a month's service, to discourage desertion. Corporal punishment was routine, although the company was careful not to release details. Once the compulsory labor period was over, workers had to find their own way home.

British critics argued that Lesseps was running an operation based on essentially slave labor. As one member of Parliament put it, "A great evil was being perpetrated by that [Suez] company in an unblushing manner." A senior British official went further:

"This forced labour system degrades and demoralizes the population and strikes at the root of the productive resources of the country."

Lesseps's response illustrates his general approach. He countered that this simply was how things were done in Egypt:

> It is true that without the intervention of the Government no public works can be undertaken in an oriental country, but while remembering that the workers on the isthmus are regularly paid and well fed, one cannot say that they are exactly forced labor. On the isthmus they live much better than they do when they are engaged in their usual occupation.

In 1863 Lesseps's good luck ran out. Mohammed Said, still only in his early forties, died suddenly, and Ismail, his successor, listened much more closely to London. British critics had long argued that the sultan had banned forced labor throughout the Ottoman Empire, so Lesseps's corvée labor arrangement with the viceroy of Egypt was illegal. The British government now redoubled its diplomatic efforts to frustrate the canal project and seemed to win over Ismail. After much diplomatic back-and-forth, in 1864 the French emperor, Louis Napoleon, was called to arbitrate the dispute between the canal company and the ruler of Egypt.

Louis Napoleon, a nephew of Napoleon Bonaparte, known to his supporters as "Saint-Simon on horseback" but mocked by Victor Hugo as "Napoleon the Small," was inclined to side with Lesseps. He was married to the daughter of Lesseps's cousin, but even without this personal connection the emperor loved grand projects that boosted French prestige. The medieval streets of central Paris were in the process of being transformed into the tree-lined, wide grand boulevards for which the city is now famous, and thousands of miles of new rail tracks were being laid.

As the British government was vying to shut down Lesseps's pesky project, Lesseps could count on the support of his small shareholders. On top of his personal connection with Lesseps, Louis Napoleon also had no desire to antagonize French investors.

He decided to strike a compromise, and he ruled that the corvée could be withdrawn, but only if the viceroy paid generous compensation.

Lesseps now had a substantial amount of cash, yet he had lost most of his indigenous workforce. He could not persuade European workers, or any others for that matter, to engage in the kind of backbreaking labor that the Egyptians had been coerced into doing, certainly not for what he could afford to pay.

Frenchmen of Genius

Visions are powered by optimism. For Lesseps, this optimism centered on technology and (French) men of genius who would save the day. Luckily, in his hour of need, two such men stepped up. In December 1863 Paul Borel and Alexandre Lavalley, both graduates of École Polytechnique, had formed a dredging company. Borel had experience building the French railways and had started manufacturing train engines. Lavalley had worked in Britain on the design of specialized machinery, becoming an expert on metallurgy, and had worked in Russia on deepening harbors. Together they formed a dream team, capable of greatly increasing the productivity of labor at the canal site.

Lesseps's original dredgers were designed to work on the Nile, where the task was primarily removing silt. In contrast, the canal project needed to move large amounts of heavy sand and rock. Each excavator had to be carefully calibrated for local conditions, which varied significantly along the canal route. Borel and Lavalley's company built new and more capable machines for dredging and excavation. They quickly came to supply and maintain the bulk of the expanded dredging fleet, which reached three hundred machines by 1869.

Of the 74 million cubic meters excavated for the main canal, it is estimated that the Borel-Lavalley dredgers were responsible for 75 percent, with most of this achieved between 1867 and 1869. By the time the canal opened in November 1869, French industry led the world in its ability to move earth even in the most difficult conditions.

Lesseps had been proved right on every issue that mattered. A sea-level canal was better than feasible; it was ideal. On-site technological progress had conquered all obstacles. Strategically, the canal was transformational, strengthening the grip of European commerce on the world.

For some years it seemed that the investors' capital remained at risk: canal traffic initially grew more slowly than predicted. But soon Lesseps proved just as prescient on financial matters. Steam displaced sail, steamships became larger, and the volume of global trade rose rapidly. The advantages of a sea-level canal at Suez became obvious to all Europeans. By the end of the 1870s, passenger ships carrying up to two thousand people were steaming through the canal, day and night. With no locks to slow them down, the trip could be made in less than one day. From a European perspective, Lesseps's vision had been brought to fruition in its entirety.

Even more miraculously, Lesseps's hopes that Britain would come around to supporting his canal turned out to be right as well. By the mid-1870s, around two-thirds of the traffic in the canal was British, and continuing to keep the ships moving was viewed as a strategic priority by London. In 1875, taking advantage of the Egyptian government's financial distress, Prime Minister Benjamin Disraeli acquired a significant stake in the canal company. The Suez Canal was now effectively under the protection of the world's most powerful navy.

Lesseps's shareholders were ecstatic. It did not matter that the work had been expected to take six years but took ten, or that the initial forecast of five million tons of shipping per year through the canal was not realized until well into the 1870s. The future belonged to ever-larger steamships, for which the canal was well suited.

By 1880, the value of shares in the Suez Canal company had more than quadrupled, and the company was paying an annual dividend of around 15 percent. Lesseps was not just a great diplomat and an audacious innovator but also a financial genius, now known to contemporaries as *Le Grand Français*.

Panama Dreaming

The idea of a canal across Central America had long been a European dream, dating back at least to 1513, when explorers wanted to move cargo quickly between the two oceans. There was an arduous route around South America, past Cape Horn. But by the mid-nineteenth century, most passengers preferred to take a ship to Panama and then a roughly fifty-mile train ride across the isthmus.

The Spanish government took nominal steps toward building a canal in 1819, but nothing came of this, and for half a century various other European schemes went nowhere. By 1879, with expanding trade through the Pacific, a canal across Central America was on the agenda again. There were two main contenders for a location, each backed by its own set of explorers and their alleged facts.

An American group strongly preferred a route through Nicaragua. A set of locks would lift boats up from the Caribbean to a large lake and back down the other side. The obvious drawback was that, with so many locks, travel time would be slowed. There was also some concern regarding volcanic activity, and Lesseps was quick to point out that a volcanic eruption would not be good for canal locks.

The alternative route was through Panama, and for this location the supposed parallels with Suez appealed to Lesseps. From the beginning of his involvement, Lesseps distinguished himself by his emphasis on the need to build the canal at sea level, entirely without locks, just as in Suez.

In 1878 Lesseps's agents received a concession from the government of Colombia, which controlled the relevant territory at that time. Lesseps received terms and conditions that resembled the arrangements in Suez—a long lease of land and participation by the government in the revenues for the project. He would also organize the work and bring the necessary capital, as he had done in Egypt.

One significant difference was that there could be no corvée workers in Panama, for there was an insufficient local labor supply. Lesseps was not deterred; workers could be brought in from Jamaica and other island colonies in the Caribbean. Relative to Europeans, West Indian workers were willing to work at lower wages and in more difficult conditions. Lesseps was also confident that, just like in Suez, machines would boost productivity and that whenever needed, technological advances would come to the rescue.

As had been the case with the Suez project, Lesseps sought the opinion of international experts, although this time around he was mostly interested in public expressions of support that would help him raise money. Still, having convened the May 1879 Congress in Paris, Lesseps had to ensure that the assembled experts recommended what he already wanted to do.

All day and long into the evenings, the Americans and French argued engineering facts and economic implications. The Panama route would require more excavation, costing 50 percent more and exposing a larger number of workers to the risk of disease for longer. The rainfall in Panama was higher, posing serious problems of watershed management. The locks needed on the Nicaragua route would be prone to damage in earthquakes. And so on.

The congress was in no way intended to be a free and fair competition between ideas; Lesseps had carefully handpicked many of the delegates to stack the deck in his favor. All the same, by May 23, it was clear that he and his allies were losing their grip on the debate. With a perfect sense of timing, Lesseps rose to address the core issues head-on. He spoke without notes, demonstrating remarkable command of the relevant details, and he quickly had the audience eating out of his hand. Suez had taught him, he said, that great achievements required great efforts. Of course there would be difficulties—surely there was little point in any undertaking that would be easy. Nevertheless, technology and men of genius would again rise to solve such problems. In his telling of events, "I do not hesitate to declare that the Panama Canal will be easier to begin, to finish, and to maintain, than the Canal of Suez."

When the capital to fund Suez had run short, new sources of financial support had appeared. When labor for digging became scarce, new excavating equipment was invented. When the fatal grip of cholera closed around the necks of its people, the Suez company responded with an effective public health response. From these successes, Lesseps learned the lesson that audacity paid. Vision demanded ambition. Or, as he put it,

> To create a harbor in the Gulf of Pelusium; to cross the morasses of the Lake of Menzaleh, and to mount the threshold of El-Guisr; to dig through the sands of the desert; to establish workshops at a distance of twenty-five leagues from any village; to fill the basin of the Bitter Lakes; to prevent the sands from encroaching on the canal—what a dream of madness it all was!

As an American delegate observed, Lesseps "is the great canal digger; his influence with his countrymen is legitimate and universal; he is kindhearted and obliging, but he is ambitious also. . . ."

At the final vote of the congress, the seventy-three-year-old Lesseps stated categorically that he would manage the endeavor personally. The delegates were impressed, and a majority voted as he wished. Panama was on.

Waking the Envy of the Happy Gods

Following the Paris Congress, Lesseps traveled to Panama, finally inspecting the terrain for himself. Arriving at the end of 1879, he and his family were received as visiting royalty. People turned out to cheer at every opportunity and to attend a string of celebratory balls.

Lesseps arrived during the healthy, dry season, and he left before it started to rain. He therefore failed to see for himself what he had been warned about at the Paris Congress and what his engineers would soon grapple with: the rapidly rising river level and the calamitous mudslides. Lesseps was also dismissive of concerns about potentially rampant infectious disease. He quipped to reporters that the only health issue during the trip had been his wife's mild sunburn.

The careless lack of attention to detail on this first trip contributed to the foundational error of the project: a massive underestimate of the amount of soil and rock that needed to be moved. The original Paris Congress estimate was that 45 million cubic meters of earth (mostly rock) needed to be excavated in Panama. This was increased to 75 million cubic meters by a technical commission made up of nine men who accompanied Lesseps to Panama.

In fact, the French dug out at least 50 million cubic meters over the next eight years. The Americans, who took up the mantle twenty-five years after the French abandoned the Panama project, ended up moving another 259 million cubic meters between 1904 and 1914—and this was without trying to dig down to anywhere near sea level.

Until too late, Lesseps refused to acknowledge the geographic reality: a serious mountain range, everywhere at least three hundred feet above sea level, blocked the way, and a dangerous, flood-prone river intersected the presumed canal route. Digging down to sea level, one expert later estimated, would take about two centuries.

The Suez Canal took ten years to complete; Lesseps was consistently optimistic that the one in Panama could be built in six or eight years at the outmost. His role was to imagine what was possible, not to worry about what could go wrong. As he wrote to one of his sons after the Panama trip, "Now that I have gone over the various localities in the Isthmus with our engineers, I cannot understand why they hesitated so long in declaring that it would be practicable to build a maritime canal between the two oceans at sea level, for the distance is as short as between Paris and Fontainebleau."

Another major miscalculation followed. At the Paris Congress, the consensus was that the Panama Canal would cost about 1.2 billion francs, about three times the ultimate cost of the Suez Canal. The technical commission accompanying Lesseps to Panama lowered this cost estimate to 847 million francs, on rather dubious grounds. But in early 1880, on the boat journey from

Panama to the United States, Lesseps further cut total projected costs down to just over 650 million francs.

After returning to Paris, he determined, once again confident that his project was going well, to raise much less equity capital than even he had himself thought necessary previously: just 300 million francs. Once more, there was no one to tell him to do it differently. Lesseps liked to quote what Viceroy Mohammed Ali had supposedly said to him early his career: "Remember, when you have anything important to accomplish, that if there are two of you, there is one too many."

In December 1880, Lesseps's company issued 600,000 shares with a face value of 500 francs each. This time around, Lesseps agreed to pay some big banks a 4 percent commission to help stimulate interest in the public subscription. More than 1.5 million francs were spent to ensure positive press coverage.

It played well that Lesseps had recently toured Panama in person and returned in good health. More than 100,000 people applied for shares, requesting double the number available. Eighty thousand investors bought between one and five shares each.

Unfortunately, building the Panama Canal needed at least four or five times as much capital as was raised in this first round, and the company was perpetually short of funds and scrambled to raise more almost every year. As costs began to exceed initial estimates, Lesseps's credibility started crumbling.

In Suez there were financial backstops: Mohammed Said, who was willing to buy up extra shares when the initial subscription faltered, and then Louis Napoleon, who provided a generous arbitration settlement. Eventually, Louis Napoleon also lent his political support to a large lottery loan—attractive to the public because of the cash prizes some bondholders would win. This injected an extra 100 million francs at a critical moment, when a conventional bond offering had failed. But Louis Napoleon was out of office in 1870, defeated by Prussia on the battlefield. The elected politicians who ran the Third French Republic proved much less inclined to bail out Lesseps and the shareholders in his Panama company.

Death on the Chagres

Work on the ground started in February 1881, and initially there was reasonable progress dredging harbors and rivers. But as the work began to shift to the high ground, the excavation became more difficult. Once the rains came, everything started falling apart.

In the summer, yellow fever arrived. The first canal worker died in June. According to one estimate, about sixty people died later that year, including some senior managers, from either malaria or yellow fever—it was hard to keep track.

In October of that year, Lesseps was still denying there were epidemics in Panama; he insisted that the only yellow fever cases were among people who arrived already infected. This became a familiar pattern: deny the existence of any difficulty. After a sizable earthquake in September 1882, Lesseps even publicly asserted that there would be no future earthquakes.

More warning signs started to appear. In 1882 the general contractor overseeing construction decided to pull out. Still undeterred, Lesseps had his company take over the excavation and, in March 1883, sent in a new general director.

Despite Lesseps's promises, the problems from disease continued to intensify. The new general director's family soon perished, most likely from yellow fever. Lesseps pressed on, increasing the workforce to nineteen thousand in 1884. Malaria and yellow fever continued to cut down the French and the native workforce in heartbreaking numbers.

None of this was inevitable. Measures that the French, the British, and other Europeans had developed over more than a century for military operations in tropical countries could have been adopted in Panama and would have reduced death rates by an order of magnitude. But this would have meant significantly less digging per year. Lesseps was warned in no uncertain terms about these risks, including during the Paris Congress. Yet he chose to regard all reports of ill health in Central America as disinformation spread by his enemies.

From 1881 to 1889 the cumulative death toll was estimated at twenty-two thousand, of which around five thousand were

French. In some years, more than half the people who came out from France died. One-third of the workforce may have been sick at any one time.

People employed directly by Lesseps's company were eligible to receive free medical attention, although this was a mixed blessing; conditions in the hospital included standing water that allowed mosquitoes to breed, and epidemics spread mercilessly in the wards. Men who worked for contractors had it even worse; if they could not pay the daily hospital fees, they were essentially abandoned on the streets when they fell ill.

Even this human suffering, much more dramatic and visible than the coercion that Egyptian workers suffered at Suez, did not dent Lesseps's determination. He remained committed to what he imagined was reality, and aloof from the day-to-day problems. In the critical years of 1882–1885, he consistently refused to listen to well-informed feedback from his own people, even as conditions became dire.

By the mid-1880s, Lesseps had already tapped the bond market multiple times, and he was having to pay a hefty risk premium in terms of promised interest payments. In May 1885 he raised the possibility of issuing lottery bonds, which had proved an effective technique in the last year of the Suez Canal project. But issuing lottery bonds required permission of the legislative assembly. To shore up political support, in February 1886 Lesseps arrived for his second visit to Panama. His stay lasted two weeks. Again, it was pageantry and all about Lesseps. As one of his top engineers observed, "Any homage paid to any other personality but himself seemed to steal a ray from his crown of glory."

Lesseps himself came away just as confident that a sea-level canal could be built on time and under the expanded budget. Yet this time around, three experts, one sent by the French legislature and two working for the company itself, independently determined that a sea-level canal was infeasible. Despite Lesseps's extraordinary persuasive powers, the legislative assembly started paying attention to the facts, and enough deputies dug their heels in.

In October 1887 Lesseps finally conceded and began to shift toward an interim plan that included locks, which would be designed by Alexandre-Gustave Eiffel, then at work on his eponymous tower. Eventually, after many twists and turns, he received permission to borrow another 720 million francs through a lottery bond issue. By December 1888, however, the bond issue had failed to raise enough money to meet minimum requirements. The Panama Canal company was placed into receivership.

Lesseps died in disgrace a few years later. His son and other associates were sentenced to prison for fraud. The canal was abandoned. But it was not Lesseps who paid the real price. Investors had contributed around one billion francs, and the lives of five thousand Frenchmen had been lost; another seventeen thousand workers, mostly from the West Indies, had also perished. All this to build essentially nothing.

Panama à l'Américaine

When the Americans took up the project in earnest in 1904, the railway and dredging equipment they started using were almost the same as had been available to the French. And the Americans made many of the same mistakes early on, including triggering a yellow fever epidemic.

Ultimately, the French failed because they were trapped in a deluded vision that did not allow them to see the alternative paths for using the available know-how and technology—and accept the difficulties. They did not change course when the evidence, and the bodies, mounted, showing the folly of their ways. It was Lesseps's vision through and through, with its techno-optimism and false sense of confidence. In this instance, it did not just impose costs on the disempowered, in the name of progress. It was deep in the throes of hubristic indifference to contrary evidence; unencumbered by facts, it marched toward disaster.

The Americans naturally had their own preconceptions. Like Lesseps, they did not pay much attention to the locals, and conditions for the immigrant labor force were hard. But one big difference was that without the overconfident vision that Lesseps imposed, setbacks meant something, especially to the politicians back home. When early efforts faltered, the senior leadership of the canal was replaced, and new people, ideas, and techniques were brought in. When the excavation lagged and disease threatened, President Theodore Roosevelt transferred control over the project to American executives who were based locally and much more responsive to local conditions, including the crucial issue of keeping workers healthy.

The Americans had learned a good deal about tropical health from their occupation of Cuba, and they brought newly understood mosquito-elimination techniques to Panama. Vegetation was stripped out of the way, and having standing water on a property was forbidden. Roads and drains were improved to remove breeding grounds.

Scientific knowledge about canals and excavation had not advanced from the time of the French efforts, but when freed from Lesseps's vision, Americans used that knowledge differently and more effectively. New engineers brought in the best thinking on how to best organize drilling, excavation, and logistics from America's extensive experience with railroad construction. The French had struggled with how to remove enough dirt and rock fast enough. The American chief executive saw this as a railway scheduling problem, laying and relaying track at phenomenal speed to keep the trains running.

There was also a big new idea that can be traced back, ironically, to what had been implemented at Suez and previously proposed for the Panama Canal. A sea-level canal required far too much digging, so why not divert the troublesome Chagres River to flood the highlands, creating a large artificial lake? Then large locks could allow boats to rise up to the level of the lake and sail across to locks that would enable descent to the other side.

The Suez Canal has no locks even today, but look closely at a map and you will see a structure with striking similarities to Panama. Lesseps's engineers dug a canal from the Mediterranean to the Great Bitter Lake and then filled it with ocean water to turn a dry salt bed into a (small) inland sea. Lesseps had drawn the wrong lesson from Suez. Instead of resisting locks, he could have emulated how the natural terrain was used to reduce the amount of digging required. Unfortunately, by the time the Suez Canal was complete, Lesseps was locked into a way of thinking that ignored all other options.

What you do with technology depends on the direction of progress you are trying to chart and what you regard as an acceptable cost. It also depends on how you learn from setbacks and the evidence on the ground. This is where the Americans' vision, even if faulty and equally callous in some respects, proved superior.

Vision Trap

Lesseps was charismatic, entrepreneurial, ambitious. He had connections, with the power of the French state and sometimes the Egyptian state behind him. His past success was mesmerizing to many of his contemporaries. Most importantly, Lesseps was peddling a nineteenth-century version of techno-optimism: big public infrastructure investments and technological advances would benefit everybody, in Europe and globally. This vision brought the French public and the French and Egyptian decision makers on board. Without it, Lesseps would not have had the sheer force of will that made him build a canal across 120 miles of Egyptian desert, even when things started going against his initial plans. Technology is nothing without vision.

But vision also implies distorted lenses, limiting what people can see. Although we may celebrate Lesseps's farsightedness at Suez and his commitment to technological advances, his use of thousands of coerced Egyptian laborers was as central to his approach as his insistence on a sea-level canal—and his brand of progress never intended to include these laborers. Even in its own terms, Lesseps's vision was a colossal failure, precisely because his greatest strength, rooted in confidence and a clear sense of purpose, was also his fateful weakness. His vision made it difficult for him to recognize failure and adapt to changing circumstances.

The tale of two canals illustrates the most pernicious aspect of this dynamic. To Panama, Lesseps brought the same beliefs, the same French expertise and capital, and essentially the same institutional support from Europe. But this time he failed to understand what was needed, and he resolutely refused to update his plans in the face of facts on the ground that contradicted his original view.

Lesseps's sensibilities were remarkably modern in some regards. His penchant for big projects, his techno-optimism, his belief in the power of private investors, and his indifference to the fate of all of those who were voiceless would put him in good company with many contemporary corporate boardrooms.

The lessons from the Panama Canal debacle resonate today, on an even grander scale. As one American delegate to the 1879 Paris Congress put it, "The failure of this Congress will teach the people the salutary lesson that under the republic they must think for themselves, and not follow the lead of any man." Alas, it is difficult to argue that this lesson has so far been learned.

Before we come to discuss our current afflictions and failure to learn from past catastrophes imposed on people in the name of progress, important questions still await answers: Why was it Lesseps's vision that prevailed? How did he convince others? Why were other voices, and those who were suffering as a result, effectively unheard? The answers are rooted in social power and whether we indeed still live, in any meaningful sense, "under the republic."

3

Power to Persuade

Power in this narrow sense is the priority of output over intake, the ability to talk instead of listen. In a sense, it is the ability to afford not to learn.

—KARL DEUTSCH, *The Nerves of Government*, 1963

We are governed, our minds molded, our tastes formed, our ideas suggested, largely by men we have never heard of.

—EDWARD BERNAYS, *Propaganda*, 1928

The direction of progress, and consequently who wins and who loses, depends on which visions society follows. For example, it was Ferdinand de Lesseps's vision, combined with a good dose of hubris, that caused the Panama Canal debacle. What explains, then, how his vision became so dominant? Why did Lesseps's views convince others to risk their money and lives against the odds? The answer is social power, and particularly his power to persuade thousands of small investors.

Lesseps acquired enormous credibility because of his social status, his political connections, and his spectacular success in leading the effort to build the Suez Canal. He had charisma, backed by a compelling story. He persuaded the French public and investors, as well as people in positions of political power, that building a canal in Panama would generate both wealth and broader benefits for the nation. His vision was credible in part because it appeared to draw on the best possible engineering expertise. Lesseps was also quite clear, and fully in alignment with his financial backers, concerning whose interests really mattered: his focus was French priorities and prestige, along with financial returns for European investors.

In sum, Lesseps had the power to persuade. He was famous for his success, he was listened to, he had the confidence to push his views, and he had the ability to set the agenda.

Power is about the ability of an individual or group to achieve explicit or implicit objectives. If two people want the same loaf of bread, power determines who will get it. The objective in question need not be a material one. It will sometimes be about whose vision of the future of technology will prevail.

You may think power is ultimately all about coercion. That is not exactly right. True, constant friction between and within societies, punctuated by invasions and subjugations, has made violence endemic throughout human history. Even during periods of peace, threats of war and violence hang over people's heads. You do not have much chance to claim a loaf of bread, or express your opinion, when being run over by hordes.

But modern society turns on persuasion power. Not many presidents, generals, or chieftains are strong enough to coerce their soldiers into battle. Few political leaders can just decree a change in laws. These leaders are obeyed because institutions, norms, and beliefs confer great standing and prestige on them. They are followed because people are persuaded to follow.

You Can Shoot Your Emperor If You Dare

A set of French republican political institutions emerged from the first ten years of the 1789 revolution. But there was also a great deal of chaos and disorder, including repeated coups and executions.

Napoleon Bonaparte came to power in 1799, seen as someone who would preserve key principles of the revolution, such as equality before the law, a commitment to science, and the abolition of aristocratic privilege, while also bringing greater stability.

In 1804, following a string of military triumphs, Napoleon crowned himself emperor. From then on, he was both a faithful son of the revolution (arguably) and supreme ruler (definitely), with complete political control backed by a huge amount of prestige within French society. Hundreds of thousands of French conscripts and volunteers followed Napoleon to Italy, across Europe, and deep into Russia. This was not because he had any special economic power. And it was not simply because he was the emperor or because the French army, under his command, had an impressive array of artillery.

Napoleon's persuasion power is clearly visible in his final return to France. After a series of defeats, he was deposed and exiled to the island of Elba, in the Mediterranean. In early 1815 he escaped from the island and landed on the south coast of France with a small number of trusted soldiers. Heading north, he was intercepted near Grenoble by the 5th Regiment of the Line. At this point, Napoleon had no formal political power, no money, and no coercion power to speak of.

But he still had his personal appeal. He dismounted from his horse and advanced toward the soldiers who were there to arrest him. When he was within gunshot range, he spoke firmly: "Soldiers of the 5th, you can shoot your Emperor if you dare! Do you not recognize me as your Emperor? Am I not your old general?" The troops rushed forward, shouting "Vive l'Empereur!" In Napoleon's subsequent assessment, "Before Grenoble, I was an adventurer; at Grenoble, I was a ruling prince." Within eight weeks the reinstated emperor had 280,000 soldiers in the field and was once more maneuvering against his European enemies.

Napoleon wielded great coercion and political power because of his ability to persuade. Over the next two hundred years the power and importance of persuasion only increased, as the power of the US financial sector vividly illustrates.

Wall Street on Top

Like coercion and political power, economic power also relies on being able to persuade others. Today, it is everywhere around us, especially in the United States. A small group of people are fabulously wealthy, and this wealth grants them great status and considerable say in political and social affairs. One of the most visible nexuses of economic power is Wall Street—the largest banks and the bankers who control them.

Where does Wall Street's power come from? The events preceding and during the global financial crisis of 2007–2008 provide a clear answer.

Historically, the banking industry in the US was fragmented, with many small financial firms and few powerful national players. After a wave of deregulation in the 1970s, a few of the larger banks, such as Citigroup, began to expand and joined up with others to form conglomerates that spanned almost all kinds of financial transactions. Bigger was more efficient, according to the private and official thinking of the time, so very large banks could provide better services at lower cost.

There was also a dimension of international competition. As the European economy became more unified, financial companies based there grew larger and more able to operate across international borders. The captains of large US banks argued that they too should be allowed to operate freely around the world to reap the same benefits from larger size and global reach. Journalists, ministers of finance, and the people running international financial regulatory bodies bought into this narrative.

On the eve of the global financial crisis in 2008, some of these banks had taken on a great deal of risk, betting that housing prices could only go up. Their profits and the bonuses of their executives and traders became inflated because of these excessive risks and because of their heavy borrowing, which generated high profits compared to the capital invested in these institutions—but only as long as things went well. Complex financial transactions known as derivatives became a potent source of profits for the industry as well. Trading options, swaps, and other instruments boosted measured profits during the boom years. In the first half of the 2000s the finance sector alone accounted for over 40 percent of total US corporate profits. But it soon became painfully clear that the same financial structure had greatly magnified the losses that some firms would face as housing prices and other asset prices fell.

On both sides of the Atlantic, finance ministry and central bank officials recommended protecting banks and bankers against financial loss, even when executives were deeply involved in questionable and potentially illegal activities, such as misleading borrowers or misrepresenting risks to the market and the regulators. According to top officials at the US Department of Justice, it was hard to bring criminal cases against the responsible parties, in effect making these banks "too big to jail." This effective immunity from prosecution and eventual access to unprecedented levels of public financial support had nothing to do with bank executives' ability to use force.

Not just too big to jail, these banks were also "too big to fail." Generous bailouts were provided because, amid the crisis, the banks and other large financial corporations convinced policy makers that what was good for these firms and their executives was good for the economy. After the collapse of Lehman Brothers in September 2008, the prevailing argument became that further failures among leading financial firms would translate into system-wide problems, harming the entire economy.

Hence, it was critical to protect the big banks and other large financial firms—their shareholders, creditors, executives, and traders—as much as possible and with few conditions. This narrative was powerful because it was persuasive. And it was persuasive because it came to be viewed by policy makers as sensible economics rather than a sweet insider deal for banks. Almost everybody who mattered, including financial journalists and academics, believed in and started espousing this view of what needed to be done. For long after these decisions, leading decision makers boasted how they saved the American economy, as well as the global economy, by helping the big banks.

At first, persuasion power may appear elusive. Political power comes from political institutions (the rules of the game for legislation and determining who has executive authority) and the ability of different individuals and groups to form effective political coalitions. Economic power comes from controlling economic resources and what you are allowed to do with them. Coercion capabilities are rooted in command over the means of violent action. But where does persuasion power come from?

The rescue of big banks, their executives, and creditors clarifies the two sources of persuasion: the power of ideas and agenda setting.

The Power of Ideas

Some ideas, especially when expressed in the right context and with conviction, have ample ability to convince. Ideas spread and become influential if they self-replicate, meaning if they convince and persuade many people, who then repeat and further propagate these concepts: a repeated idea is a strong idea.

Whether an idea is accepted, gets repeated, and spreads depends on many factors—some of them institutional, others related to social status and the networks that propagate it, and yet others about the qualities of the individuals promoting it, such as their charisma. All else equal, an idea is more likely to spread if it is simple, is backed by a nice story, and has a ring of truth to it. It also helps if it is advocated by individuals with the right type of social status—for example, those who have a demonstrated ability to lead and who are supported by respected cheerleaders, such as the *Institut de France* for Napoleon and finance and law school professors for Wall Street.

Ideas played a role in Wall Street's ability to influence policy and regulation. The executives who built these financial conglomerates advanced the notion that the entire modern economy depends on the smooth functioning of a few large financial firms, with little regulation by the government. The big-finance-is-good idea was made more plausible because the financial industry was growing as a share of the economy and gaining status, with lavish salaries and lifestyles that movies and newspapers depicted with relish.

The envy and prestige that this engendered can be seen from the way that Michael Lewis's best-selling 1989 book about bond traders, *Liar's Poker: Rising Through the Wreckage on Wall Street*, was received. Lewis wrote the book based on his own experience in bond trading, in part as a critique of big finance's practices, values, and arrogant attitudes. Lewis says that he hoped that the book would discourage people from joining such financial firms. Yet by the time it appeared, the allure of Wall Street had increased so much that when ambitious university students read the book, they were apparently not bothered by the ruthless characters and the soulless culture of finance. Some wrote to Lewis asking if he had any more career tips. In Lewis's own assessment, the book became a recruitment tool for Wall Street.

Where do compelling ideas come from? What determines whether an individual or group has the charisma or the resources to push such ideas? It is safe to say that quite a bit of this process is random. Creativity and talent matter, of course, and societies and their rules deeply influence who has social status and charisma and who can develop their talents and creativity.

In many societies, minorities, women, and those who are economically or politically disempowered are discouraged not just from voicing their ideas but even from having original thoughts. As an extreme but telling example, in parts of the British West Indies, at the height of the plantation economy, teaching enslaved people to read was forbidden. For much of history, women have been discouraged and deliberately excluded from leadership positions in science and business.

Even charisma depends on institutions and conditions. It is not just something you are born with; it depends on self-confidence and on your social networks. For example, when it came to the power of big banks, it was not only ideas and stories. Bank executives and board members belonged to social networks that had enormous economic power and propagated these ideas.

The big-finance-is-good idea was being repeated by economists and lawmakers, who were eager to provide theories and supportive evidence.

A huge amount of creativity, charisma, and hard work is no guarantee that an academic or entrepreneur will come up with an impactful idea. Prevailing beliefs and the attitudes of powerful individuals and organizations determine which ideas will appear compelling, rather than wacky or so ahead of their time as to be safely ignored. You are enormously lucky if you get the right idea, with just the right ring to it, at just the right time.

It's Not a Fair Marketplace

Social scientists sometimes use the analogy of a marketplace when thinking about how different ideas will catch on. There is something to this analogy: ideas compete for attention and acceptance, and better ideas naturally have an advantage. Almost nobody believes today that the sun revolves around the Earth, even though that idea once appeared irresistible and was a central tenet of Christianity for more than a thousand years.

The heliocentric view, placing the sun at the center of the solar system, was proposed as early as the third century BCE, but it lost out to the geocentric theories of Aristotle and Ptolemy. Aristotle was considered the foremost authority on almost all scientific matters in premodern Europe, and Ptolemy's work perfected the system and proved of practical value—for example, when using astronomical charts.

Eventually, more-accurate ideas can prevail, particularly when backed up by a coherent scientific methodology. If there are predictions that can be checked by others, that helps too. Yet this can take a while. Ptolemy's system was criticized by Muslim scholars starting about 1000 CE, but they never fully abandoned the idea that the Earth was at the center of everything. Heliocentrism in its modern form began to be developed by Nicolaus Copernicus in the early 1500s; it was significantly advanced by Johannes Kepler at the start of the 1600s and by Galileo Galilei shortly after. It then took decades for these ideas and their implications to spread through European scientific circles. Newton's *Principia*, which built on and extended Galileo's and Kepler's ideas, was published in 1687. In 1822 even the Catholic Church accepted that the Earth revolves around the sun.

However, the marketplace for ideas is an imperfect frame for technology choices, which are at the heart of this book. To many people, the word *market* implies a level playing field in which different ideas try to outcompete each other primarily on their merits. This is not how it happens most of the time.

As the evolutionary biologist Richard Dawkins emphasized, bad but catchy ideas can sometimes succeed spectacularly—think of conspiracy theories or crazy fads among investors. There is also a natural "rich-gets-richer" phenomenon when it comes to ideas: as we have already mentioned, the more an idea is repeated and the more one hears it from many different sources, the more plausible and compelling it seems.

Even more problematic for the marketplace-for-ideas notion is that an idea's validity in the eyes of people depends on the prevailing distribution of power in society. It is not just the self-confidence and the social networks that powerful people have for propagating their ideas. It is also whether your voice is amplified by existing organizations and institutions, and whether you have the authority to counter objections. You may have an idea about how to develop a technology or well-reasoned concerns about unintended consequences to which we should pay more attention. But if you do not have the social means to explain why this is a better technological path and the social status to make others listen, your idea will not go very far. This is what we capture with the second dimension of persuasion power: agenda setting.

Agenda Setting

Whoever asks the questions, sets the priorities, and rules options in and out has formidable powers to frame public discussion and convince others. Humans have an impressive ability to use collective knowledge, and this is what makes technology so important for society. But our powers to reason and our brains are also limited. We think through coarse categories and sometimes make false generalizations. We often rely on fast rules of thumb and simple heuristics to make decisions. We have myriad biases, such as a tendency to find evidence for what we already believe ("confirmation bias") or thinking that rare events are more common than they really are.

Particularly important for our discussion is that when it comes to complex choices, we tend to consider only a few options. That is natural, for it is impossible for us to consider all feasible choices and pay equal attention to everybody who may have an opinion. As it is, our brain already consumes 20 percent of our energy, and it would have probably been hard for it to become much more sophisticated and powerful during the process of evolution. Even when it comes to the decision of which crackers and which cheese to buy, if we paid attention to all options, we would have to consider more than one million options (more than 1,000 times 1,000, since more than a thousand types of crackers and cheese each are readily available). We typically do not need to consider so many choices because we can use shortcuts and well-honed heuristics to make reasonably good decisions.

One of our most powerful heuristics is to learn from others. We observe and imitate. Indeed, this social aspect of intelligence is a huge asset when it comes to building collective knowledge because it enables an efficient process of learning and decision making. But it also creates various vulnerabilities and weaknesses for the powerful to exploit. Sometimes what we learn is not what is good for us but is what others want us to believe.

In fact, we tend to learn from and listen to those who are more eminent in society. This too is natural: we could not feasibly pay attention to the experiences and advice of thousands of people. Concentrating on those who have proven that they know what they are doing is a good heuristic.

But who is competent? Those who are successful in the task at hand are obvious candidates. Yet we often do not observe who is good at which specific task. A reasonable heuristic is to pay greater attention to people who have more prestige. Indeed, we almost instinctively believe that the ideas and recommendations of those who have status are worthier of attention.

Our willingness to follow social status and prestige and imitate successful individuals is so deep in our psyche that it appears ingrained. You can even see it in the imitative behavior of children as young as 12 months.

Psychologists have long studied how children imitate—and, in fact, overimitate—adult behavior. In one experiment, an adult demonstrated how to get a toy out of a plastic puzzle box with two openings, one on top and one in front. The experimenter first opened the top, then the front, and finally reached from the front to get the toy. The first step was completely unnecessary. All the same, when children were asked to perform the task themselves, they faithfully repeated the first unnecessary step. Perhaps they did not understand that this was an unnecessary step? That was not the case at all. When asked about it at the end of the experiment, they knew full well that unlocking the top was "silly and unnecessary." But they still imitated it. Why?

The answer seems to be related to social status. The adult is the expert and has the status conferred by this position. Hence, children are inclined to suspend disbelief and imitate what he or she is doing. If the adult is doing it, even if it seems unnecessary and silly, there must be a reason for it. Indeed, older children are more likely to engage in this type of overimitation when they get better at social cues and relations, and this means getting better at recognizing social status and following what they perceive as expertise.

In similar experiments, chimpanzees skipped the first step and directly opened the front of the puzzle box. This is not because chimpanzees are smarter but presumably because they are not as predisposed as humans are to respect, accept, and imitate (apparent) human expertise.

Another ingenious experiment dug a little deeper into this type of behavior. The researchers had preschoolers watch videos in which different models use the same object in one of two different ways. They also had bystanders, played by confederates of researchers, who could also be seen observing the models. Pre-schoolers were much more likely to pay attention to whoever was being watched by the bystanders. When given a choice later, the preschoolers were much more likely to follow the choices made by the more-watched model.

The preschoolers were imitating not just to learn how and what to use, but they were following the other learners, something that the authors interpret as a prestige cue, a marker of who has prestige and is perceived as having the right expertise. It seems like it is instinctive for us to heed the views and practices of people who we think are successful; even more tellingly, we judge who is successful by seeing who is being obeyed and followed by others—back to social status again!

Respecting social status and imitating successful people has clear evolutionary logic, for these are the people who are likely to have thrived because they have made correct choices. But the snag is also obvious. Our tendency to pay more attention to those with high status and prestige generates powerful feed-backs: those who have other sources of social power will have high status, and we will tend to listen to them more, conferring on them greater persuasion power as well.

In other words, we are such good imitators that it is difficult for us not to absorb information embedded in the ideas and visions we encounter, which are often those offered by power-ful agenda setters. Experiments confirm this conclusion as well, showing that even when people see irrelevant information that is labeled as not being reliable, they have a hard time resisting tak-ing it seriously. This is exactly what the researchers found in the puzzle box experiment: when the children were told that opening the top lock was unnecessary, they still stuck to their imitation behavior. A similar phenomenon was found on social media sites for news items containing misinformation. Many participants could not discount misinformation even when it was clearly flagged as unreliable, and their perceptions were still affected by what they saw.

It is this instinct that agenda setting exploits: if you can set the agenda, you must be worthy of status, and you will be listened to.

The Bankers' Agenda

In the run-up to the global financial crisis of 2007–2008, the executives who ran large global banks had plenty of agenda-setting power. They were viewed as highly successful by an American culture that puts great weight on material wealth. As risk taking and profit margins in the industry grew, finance executives became wealthier, pushing up their prestige even higher.

When things went badly, the same firms suffered losses that were so large that they would face bankruptcy. This is when the "too-big-to-fail" card was played. Policy makers, who were previously persuaded that big and highly leveraged was beautiful in finance, were now convinced that allowing these gargantuan firms to fail would cause an even greater economic disaster.

When he was asked by a journalist why he robbed banks, Willie Sutton, an infamous criminal of the Great Depression era, is reputed to have said, "That's where the money is." In modern times, titans of finance assiduously build persuasion power because that's where the money is now.

During the 2007–2008 economic crisis, the heads of big banks were perceived as having considerable expertise because they controlled an important sector of the economy and were fawned over by the media and politicians as highly talented people who were richly rewarded for their specialized knowledge. This status and the persuasion power that followed meant that just over a dozen bankers became the ones framing the choices facing the US economy: either bail out the banks' shareholders, creditors, and all their executives on favorable terms, or let these firms fail and force economic ruin.

This framing left out realistic options, such as keeping the banks as intact legal entities by providing financial support while at the same time not allowing shareholders and executives to profit. The framing also precluded the option of firing or prosecuting bankers who had broken the law—for example, by deceiving customers and contributing to the financial meltdown in the first place. It ignored obvious policy actions that could have provided greater assistance to home owners in distress—because the prevailing view was that their bankruptcy would not cause system-wide risks, and it would be bad for banks if borrowers could cut their mortgage payments!

It even left out the option of temporarily withholding the lavish bonuses of the traders and executives in the very institutions that triggered the crisis and received government bailouts. The insurance company AIG was saved by a government support of $182 billion in the fall of 2008, yet it was allowed to pay nearly half a billion dollars in bonuses, including to people who had wrecked the company. In the middle of the deepest recession since the 1930s, nine financial firms that were among the largest recipients of bailout money paid five thousand employee bonuses of more than $1 million per person—supposedly because this was needed to retain "talent."

Wall Street's broader social network helped in its agenda setting because it encompassed many of the other people who had a say regarding what should be on the agenda. The revolving door between the financial sector and officialdom played a role, too. When your friends and former colleagues are asking you to see the world in a particular way, you pay attention.

Of course, agenda setting is intertwined with ideas. If you have a compelling idea, you are more likely to set the agenda, and the more you are successful in setting the agenda, the more plausible and powerful your idea becomes. The big-finance-is-good rhetoric became irresistible because the bankers and those who agreed with them had come to formulate the story, ask the questions, and interpret the evidence.

Ideas and Interests

Wall Street's machinations in the run-up to and during the 2007–2008 financial crisis may create the impression that agenda-setting power matters because it allows a group or individuals to protect their bottom line. Ideas do, of course, have a way of supporting the economic or political interests of the powerful people propagating them. But the influence of agenda setting goes far beyond selfish interest. In fact, if you tell others to follow what is blatantly good for you, they will balk, seeing it as a crude attempt to get what you want. For an idea to be successful, you need to articulate a broader viewpoint that transcends your interests or, at the very least, appears to do so.

There is another reason why powerful ideas are often not the openly selfish ones. You will be a much better advocate for an idea if you passionately believe in it, and this becomes more likely if you can convince yourself that this is not just a selfish ploy, but in the name of progress. It was thus much more important for the success of this vision that bureaucrats, policy makers, and journalists who had much less direct material interests became strong proponents of the big-finance-is-good rhetoric.

However, this dynamic also implies that ideas may diverge from interests. Once you have a set of ideas you believe in, these concepts shape the way you look at facts and weigh different trade-offs. In this way, you will start being driven by ideas even independently from your interests. Viewpoints that are passionately held have a way of becoming more dominant, even infectious.

It was not Lesseps's economic interests that made him push for a particular design of the Panama Canal, to be built at sea level with harsh conditions for workers. Nor was his almost magical belief in "men of genius" to come up with technological solutions rooted in selfish calculations. Lesseps was genuinely convinced that this was the right way to use the available scientific knowledge and technology for the common good, and he was able to persuade others because he had been hugely successful in the past and had the ear of many people in France.

Likewise, what was overwhelmingly dominant during the global financial crisis was not just the interests of the people who ran big banks (even if those were served quite well, thank you). It was a vision that these prominent bankers themselves completely believed in (weren't they fabulously wealthy, after all?). As Lloyd Blankfein, head of the investment bank Goldman Sachs, put it in 2009, he and his colleagues were doing "God's work." It was this combination of past success and a narrative of working for the common good that was so captivating to journalists, lawmakers, and the public. Anyone who questioned this approach was met with righteous indignation.

So far, we have explained how ideas can spread and become dominant, alongside the role of agenda setting, which confers a special position to those who can frame the debate.

Who can do so? The answer is those with high social standing. Because those with social power have greater ability to set the agenda, we see a circle that can turn vicious: the more power and status you have, the easier it is for you to set the agenda, and when you set the agenda, you obtain even more status and power. Nevertheless, the rules of the game matter greatly as well, and they can amplify or limit inequality in the power to persuade.

When the Rules of the Game Keep You Down

The aftermath of the US Civil War illustrates the central role of agenda-setting power, rooted in the ability of some groups to be at the table. There was a committed contingent of abolitionists in the North who believed that the war should transform the political, economic, and social life in the South, and thought this would be good riddance. As one of the leading abolitionists, Samuel Gridley Howe, argued in the run-up to the Civil War, "We have entered upon a struggle, which ought not to be allowed to end until the Slave Power is completely subjugated, and *emancipation made certain*." (Italics in original.)

The Emancipation Proclamation opened a new phase in American history on New Year's Day in 1863. The Thirteenth Amendment, abolishing slavery, followed at the end of 1865. The Fourteenth Amendment, ratified in 1868, granted citizenship and equal protection to all previously enslaved people. Recognizing that this was not a change that could be made at the stroke of a pen, federal troops were stationed in the South to implement these changes. The Fifteenth Amendment followed in 1870, granting Black American men the right to vote. It was now a crime to deny the vote based on "race, color, or previous condition of servitude."

At first, this looked like the ideal of equal rights for everyone, including in the political sphere. This was the era of Reconstruction in the South, when Black Americans made notable economic and political gains. They would not have to put up with low wages and daily coercion in plantations, they could open businesses with much less intimidation, and they would no longer be barred from sending their children to schools. Black Americans jumped at the chance for economic empowerment and political engagement. Before the Civil War, almost all southern states prohibited the instruction of slaves, and over 90 percent of the region's adult Black population was illiterate in 1860. This changed after 1865.

As part of this broader push for more opportunity, by 1870, Black Americans had raised and spent over $1 million on education. Black farmers wanted their own land, along with control over what they would plant and how they would live. For those in towns and cities, as well as in rural areas, there was a push for better working conditions and higher wages, and Black Americans began organizing strikes and signed collective petitions demanding better working conditions and higher wages. Even in rural areas, the labor market for Blacks started being transformed, with collective bargains over contract terms and wage schedules.

This improvement in economic conditions was backed by political representation. Between 1869 and 1891, every session of the Virginia General Assembly had at least one Black member. There were fifty-two Black Americans in the state legislature of North Carolina and forty-seven in South Carolina. Even more tellingly, between 1869 and 1876 the US had its first two Black senators (both elected from Mississippi) and fifteen Black representatives (elected from South Carolina, North Carolina, Louisiana, Mississippi, Georgia, and Alabama).

Yet it all came crashing down. As early as the second half of the 1870s, the political and economic rights of Black Americans were being curtailed. In the words of historian Vann Woodward, "The South's adoption of extreme racism was due not so much to a conversion as it was to a relaxation of the opposition." And there was a major relaxation of the opposition after the contested election of 1876 led to the Hayes-Tilden Compromise, which put Republican Rutherford Hayes into the White House, but only because he agreed to end Reconstruction and withdraw federal troops from the South.

Soon thereafter, Reconstruction gave way to the phase known as Redemption, in which southern White leadership pledged to "redeem" the South from federal interference and the emancipation of Blacks. This White elite succeeded in turning back the clock, and the South became what one of the most influential Black intellectuals of the early twentieth century, W. E. B. Du Bois, aptly characterized as "simply an armed camp for intimidating black folk."

This armed camp was of course about coercion of Black Americans in the South, including extrajudicial lynchings and other killings and the use of local law enforcement for repression. But this coercion power was rooted in and complemented by southern racists' success in persuading the rest of the nation that it was acceptable for Blacks to be systematically disadvantaged, discriminated against, and forcibly repressed. Southern White persuasion power was particularly important in making the rest of the country accept segregation and the systemic discrimination against Blacks that came to be known as the "Jim Crow" laws.

How could everything go so wrong? This question has many answers, obviously. But the most important ones were related to the lack of sufficient social power and agenda setting to propagate ideas of full economic and social equality.

It did not help that Black Americans were not given a full chance for economic empowerment. As a leading antislavery politician of the era, Congressman George Washington Julian, observed in March 1864, when proposing land reform for the South, "Of what avail would be an act of Congress totally abolishing slavery, or an amendment of the Constitution forever prohibiting it, if the old agricultural basis of aristocratic power shall remain? Real liberty must ever be an outlaw where one man only in three hundred or five hundred is an owner of the soil." Unfortunately, that old agricultural basis of power remained effectively unchallenged.

President Lincoln had understood that access to economic resources was critical for the advancement of Black Americans' freedom and supported the decision of General William Sherman to distribute "forty acres and a mule" to some freedmen. But after Lincoln's assassination, his pro-slavery successor, Andrew Johnson, revoked Sherman's orders, and freed people never received the resources necessary for any kind of economic independence. Even in the heyday of Reconstruction, Black Americans remained dependent on economic decisions made by White elites. Worse, the plantation system, which had until then relied on slave labor, was not uprooted. Many planters kept their large landholdings and continued to rely on low-wage Black Americans still locked in coercive employment relationships.

Equally important in the failure of Reconstruction was the fact that Black Americans never achieved true political representation. They were never fully represented. Even when there were Black politicians in Washington, they were far from the true seat of power, such as the important congressional committees and the back rooms where deals were made. As a result, they could not set the agenda and steer the pivotal debates. In any case, their national office holding soon came to an end as Reconstruction lost its momentum and started being unwound.

Black Americans fought and died in the Civil War, and they were the ones to suffer the consequences of slavery and Jim Crow. Nevertheless, because the key decisions that would determine their livelihood and political future were in the hands of others, what was given to them could be and was taken away when political calculations or coalitions changed—for example, when Andrew Johnson became president or in the Hayes-Tilden Compromise.

Black Americans knew what they wanted and how they could achieve it, as they demonstrated during the early phases of Reconstruction. Yet because they did not have effective political representation and the ability to influence the agenda, they did not shape the narrative of the nation. When politics and priorities shifted in the corridors of national power, they had no recourse to counter the fallout that this implied for their future.

Toward the end of the nineteenth century, as the United States engaged in overseas imperial expansion in the Philippines, Puerto Rico, Cuba, and Panama, there was a resurgence of racist thinking across the country. In a milestone verdict, the Supreme Court's *Plessy v. Ferguson* decision in 1896 concluded that "legislation is powerless to eradicate racial instincts" and allowed the constitutionality of "separate but equal" practices in the Jim Crow South. This was the tip of a much uglier iceberg. In October 1901 the editors of the *Atlantic Monthly* (a publication that supported equal rights) summarized this mood change among people in the North:

> Whatever blessings our acquisition of foreign territory may bring in the future, its influence upon equal rights in the United States has already proved malign. It has strengthened the hands of the enemies of negro progress, and has postponed further than ever the realization of perfect equality of political privilege. If the stronger and cleverer race is free to impose its will upon "newcaught, sullen peoples" on the other side of the globe, why not in South Carolina and Mississippi?

Writing in the same issue of the magazine was one of the most influential historians of the era, William A. Dunning. Dunning was a northerner, born in New Jersey, educated at Columbia University, and on the faculty at Columbia for his entire career. Yet he and his many students were highly critical of Reconstruction, which they argued had allowed "carpetbaggers" (northern interlopers) to control the votes of freedmen, aided and abetted by "scalawags" (southern Whites). The so-called Dunning School was a mainstay of the conventional wisdom in the first half of the twentieth century, in the North just as much as in the South, influencing depictions of American history in print and film, including in the 1915 movie by D. W. Griffith, *The Birth of a Nation.* The movie became one of the most influential films in history and deeply influenced social and political views with its unfavorable portrayal of Black Americans and its justification for racism and Ku Klux Klan violence.

How can you defend yourself against such racism if the majority group will not listen to your views? And the majority will not listen unless you have some ability to set the agenda.

A Matter of Institutions

We cannot understand how things went so badly wrong for Black Americans after Reconstruction without recognizing the role of economic and political power and the underlying economic and political institutions.

Economic and political institutions shape who has the best opportunities to persuade others. The rules of the political system determine who is fully represented and who has political power, and thus who will be at the table. If you are the king or the president, in many political systems you will have ample influence on the agenda—sometimes you can even directly dictate it. Likewise, economic institutions influence who has the resources and the economic networks to mobilize support and, when necessary, pay politicians and journalists.

Persuasion power is more potent if you have a compelling idea to sell. But, as we have seen, that too depends partly on institutions. For example, if you are rich or politically powerful, you will command social status, which then makes you more persuasive.

Social status is conferred by society's norms and institutions. Is it financial success or good deeds that matter? Are we impressed by those who have inherited family wealth or those who have earned it themselves? By those who claim to speak for and to the gods? Do we think bankers are to be respected and placed on a pedestal or treated as rather ordinary businesspeople, as was the case in the United States during the 1950s?

Social status also reinforces other power inequities: the greater your status, the more you can use it to gain an economic advantage, become politically more vocal and influential, and in some societies even gain more coercion power.

Institutions and ideas coevolve. Today, many around the world cherish democracy because the idea of democracy has spread and we accept it as a good form of government, with evidence supporting that it leads to good economic outcomes and a fairer distribution of opportunities. If trust in democratic institutions collapses, democracies around the world would soon follow. In fact, research shows that as democracies perform better in terms of delivering economic growth, public services, and stability, support for democracy grows considerably. People expect better from democracy, and when democracy delivers, it tends to flourish. But once democracy fails to live up to expectations, it ceases to become such an attractive prospect.

The impact of political institutions on ideas is even stronger. Better ideas and those backed by science or well-established fact have an advantage. But often things are not cut and dried, and it will be ideas that monopolize the agenda and, even more perniciously, those that can sideline counterarguments that will have an advantage. Political and economic power matter because they decide who has a voice and who can set the agenda, and because they place different people with distinct visions at the decision-making table. Once you are welcome in all high-status forums, your persuasion power grows, and you can start reshaping political and economic power.

History also matters: once you are at the table, debating important matters and influencing the agenda, you tend to stay there. All the same, as the aftermath of the US Civil War amply demonstrates, people remake these arrangements, especially during critical moments, when power balances shift and new thinking and options suddenly start being viewed as feasible or even inevitable.

History is not destiny. People have "agency"—they can make social, political, and economic choices that break its vicious circles. The power to persuade is no more preordained than is history; we can also refashion whose opinions are valued and listened to and who sets the agenda.

The Power to Persuade Corrupts Absolutely

Even if we are likely to end up with the vision of the powerful, can we at least hope that their vision could be sufficiently inclusive and sufficiently open, especially because they often appeal to the common good in justifying their designs? Perhaps they will act responsibly, so we do not have to suffer the consequences of self-centered visions applied zealously despite the costs that they impose on scores of others. This is likely to be wishful thinking; as a British historian and politician, Lord Acton, famously remarked in 1887,

> Power tends to corrupt and absolute power corrupts absolutely. Great men are almost always bad men, even when they exercise influence and not authority: still more when you superadd the tendency or the certainty of corruption by authority. There is no worse heresy than that the office sanctifies the holder of it.

Lord Acton was arguing with a prominent bishop about kings and popes, and there is no shortage of examples, historical or modern, of rulers with absolute power misbehaving absolutely.

But his aphorism applies just as aptly to persuasion power, including the power to persuade oneself. Put simply, the socially powerful often convince themselves that it is their ideas (and often their interests) that matter and find ways of justifying neglecting the rest. You will recognize this in Lesseps's ability to rationalize coercion against workers in Egypt and ignore the evidence that malaria and yellow fever were killing thousands in Panama.

There is perhaps no better evidence for this type of corruption than the work of social psychologist Dacher Keltner. In experiments spanning the last two decades, Keltner and his collaborators have amassed a huge amount of data that the more powerful people become, the more likely they are to act selfishly and ignore the consequences of their actions on others.

In a series of studies, Keltner and colleagues looked at the traffic behavior of drivers with expensive cars relative to those with inexpensive ones. They observed that more than 30 percent of the time, the more expensive cars crossed an intersection before it was their turn, cutting off other vehicles. In contrast, the same likelihood was about 5 percent for drivers of inexpensive cars. The contrast was sharper when it came to behavior toward pedestrians attempting to cross at a crosswalk (in this instance, the pedestrians were part of the research team, moving toward the crosswalk as the car was approaching). Drivers of the most expensive cars cut off pedestrians over 45 percent of the time, while the drivers of the least expensive cars almost never did so.

In lab experiments, Keltner and his team also found that richer and higher-social-status individuals were more likely to cheat, by unrightfully taking or claiming something. The rich were more likely to report greedy attitudes as well. This was not just true in their self-reports but also when the researchers designed experiments in which they could track whether subjects cheated or engaged in other unethical behaviors.

Even more strikingly, the researchers discovered that cheating could be triggered in lab settings simply by making subjects feel more high status—for instance, by encouraging them to compare themselves to people with less money.

How can powerful people engage in such selfish, unethical behavior? Keltner's research suggests that the answer may be related to self-persuasion—about what is and is not acceptable and what is in the common good. The rich and the prominent convince themselves that they are simply taking their just deserts, or even that being greedy is not beyond the pale. As the unscrupulous investor Gordon Gekko in the 1987 movie *Wall Street* put it, "Greed is right, greed works." Interestingly, Keltner and his collaborators also saw that other non-rich people can be nudged to behave more like the rich when they are given statements expressing positive attitudes toward greed.

We argued above that in the modern world the power to persuade is the most important source of social power. But with such persuasiveness, you tend to convince yourself that you are correct, and you become less sensitive to others' wishes, interests, and plights.

Choosing Vision and Technology

Social power matters in every aspect of our lives. It becomes particularly consequential for the direction of progress. Even when couched in appeals to the common good, new technologies do not benefit everybody automatically. Often, it is those whose vision dominates the trajectory of innovation who benefit most.

We have defined vision as the way in which people come to think about how they can turn knowledge into new technologies targeted at solving a specific set of problems. As in chapters 1 and 2, technology here means something broader than just the application of scientific knowledge to generate new products or production techniques. Working out what to do with steam power and deciding what type of canal to build are technological choices. And so is how to organize agriculture and who to coerce in the process. Visions of technology thus permeate almost every aspect of our economy and society.

What is true of social power in general becomes especially central when we turn to visions of technology. It is easy to ignore others when you have a compelling narrative about how to enhance our species' dominion over nature. Those who do not agree with this viewpoint and those who suffer can be cast aside, with no more than lip service paid to their suffering. When a vision becomes overconfident, these problems are magnified. Now those who stand in the way or argue that there might be alternative paths can be viewed as unimportant or out of touch, if not downright misguided. They can just be crushed. The vision justifies everything.

This of course does not mean that there is no way of reining in selfishness and hubristic visions. But it does very much mean that we cannot expect this type of responsible behavior to emerge automatically. As Lord Acton pointed out, we cannot count on social responsibility among those who hold great power. We can count on it even less among those who have forceful visions and dreams of shaping the future. The cards are further stacked against responsibility because the power to persuade corrupts and makes the powerful less likely to understand or care about others' woes.

We need to reshape the future by creating countervailing forces, particularly by ensuring that there is a diverse set of voices, interests, and perspectives as a counterweight to the dominant vision. By building institutions that provide access to a broader range of people and create pathways for diverse ideas to influence the agenda, we can break the monopoly over agenda setting that some individuals would otherwise enjoy.

It is equally about (social) norms—what society finds acceptable and what it refuses to consider and reacts against. It is about the pressure that ordinary people can put on elites and visionaries, and it is about their willingness to have their own opinions rather than be entrapped by dominant visions.

We must also find ways of curtailing selfish, overconfident visions, and this too is about institutions and norms. Hubris is much less powerful when it is not the only voice at the table. It becomes enfeebled when it is confronted with effective counterarguments that cannot be brushed aside. It (hopefully) starts to fade when it is recognized and mocked.

What's Democracy Got to Do with It?

Although there is no surefire way of achieving these objectives, democratic political institutions are crucial. Debates about the pros and cons of democracy go back at least to Plato and Aristotle, neither of whom was very keen on this political system, fearing the cacophony that it might engender. These fears and the all-too-frequent concerns about the resilience of democracy in the popular press today notwithstanding, the evidence is clear that democracy is good for economic growth, for delivering public services, and for reducing inequality in education, health, and opportunities. For example, research shows that countries that have democratized increased their GDP per capita by about 20 percent over the next two or three decades that followed democratization, and this was often accompanied with greater investments in education and health.

Why do democracies do better than dictatorships or monarchies? Unsurprisingly, there is no single answer. Some dictatorships are really badly managed, and most nondemocratic regimes tend to favor firms and individuals that are politically connected, often granting them monopolies and allowing resource expropriation for the benefit of elites. Democracies tend not only to break down oligarchies but also to constrain rulers and inculcate law-abiding behavior. They generate more opportunities for the less well-off and allow a more equal distribution of social power. They are often pretty good at resolving internal disputes through peaceful means. (Yes, democratic institutions have not been doing too well lately in the US and much of the rest of the world, and we will return to why that may be in Chapter 10.)

There is also another reason for democratic success: cacophonous voices may be the greatest strength of democracy. When it is hard for a single viewpoint to dominate political and social choices, there are more likely to be opposing forces and perspectives that undercut selfish visions imposed on people, regardless of whether they want them or benefit from them.

This democratic advantage is related to an idea proposed more than two hundred years ago by a French philosopher, the Marquis de Condorcet. Condorcet made the case for democracy using what he called a "jury theorem." According to his theorem, a jury—for example, consisting of twelve people with different viewpoints—is more likely to reach a good decision than would a single individual. Everyone will bring their own perspective and biases, which may vary from issue to issue. If we appoint one of them as decision maker or ruler, that individual may make bad decisions. However, if we put several people with different perspectives in the room and the ultimate decision aggregates their viewpoints, under plausible conditions this is likely to lead to better decisions. Democracy, when it works well, operates like a very large jury.

Our argument for democracy is a little different, though related. The democratic advantage may not be just the aggregation of separate views, but rather the encouraging of diverse perspectives to engage with and counterbalance each other. The strength of democracy is thus in the deliberation among different viewpoints, as well as in the disagreements that this often generates. Hence, as noted in Chapter 1, a major implication of our approach is that diversity is not a "nice to have" feature; its presence is necessary to counteract and contain the overconfident visions of elites. Such diversity is also the essence of democracy's strength.

This argument is almost diametrically opposed to a commonly held view among political elites in many Western democracies, which is based on the idea of "delegation to the technocrats." This viewpoint, which has gained a strong following in recent decades, maintains that important policy decisions, such as monetary policy, taxation, bailouts, climate mitigation, and AI regulation, should be decided by technocratic experts. It is better for the public not to get too involved in the details of such governmental affairs.

Yet it was exactly this technocratic approach that led to the policies that first encouraged Wall Street bankers and then—on incredibly generous terms—bailed out and absolved them during the 2007–2008 financial crisis. Tellingly, most of the key decisions before, during, and after the crisis were made behind closed doors. Viewed in this light, the technocratic approach to democracy can easily get trapped by a specific vision, such as the big-finance-is-good view that most policy makers bought into during the early 2000s.

In our assessment, a major part of democracy's real advantage is avoiding the tyranny of narrow visions. To make this happen, we should cherish and strengthen the diversity of voices in democracy. Ordinary people, sidelined by the technocratic consensus, seem to understand this. In surveys, support for democracy goes together with a disdain for overbearing experts, and those who believe in democracy do not want to cede political voice in favor of the experts and their priorities.

Such diversity is often maligned by experts who argue that regular people cannot provide valuable inputs into highly technical matters. We are not advocating that there should be a set of citizens from all backgrounds deciding the laws of thermodynamics or the best way to design speech-recognition algorithms. Rather, different technology choices—for example, on algorithms, financial products, and how we use the laws of physics—tend to have distinct social and economic consequences, and everybody should have a say on whether we find these consequences desirable or even acceptable.

When a company decides to develop face-recognition technology to track the faces in a crowd, to better market products to them or to make sure that people do not participate in protests, their engineers are best placed to decide *how* to design the software. But it should be society at large that should have a voice in *whether* such software should be designed and deployed. Listening to diverse voices requires that these consequences are made clearer and that nonexperts can speak about what they want to see happen.

In sum, democracy is an essential pillar of what we view as the institutional foundations of an inclusive vision. This is partly because of the more equal distribution of social power and the better laws that democracy typically provides. But equally, it is about ensuring a framework in which ordinary people become well informed and politically active, and in which norms and social pressure bring diverse perspectives and opinions to the table, prevent monopolies over agenda setting, and cultivate countervailing powers.

Vision Is Power; Power Is Vision

Progress has a way of leaving many people behind unless its direction is charted in a more inclusive way. Because this direction governs who wins and who loses, there is often a struggle over it, and social power determines whose favorite direction prevails.

We have argued in this chapter that in modern societies it is the power to persuade—even more so than economic, political, and coercion powers—that is critical in these decisions. Lesseps's social power did not come from tanks or cannons. Nor was he particularly rich or the holder of any political office. Rather, Lesseps had the power to persuade.

Persuasion is especially important when it comes to technology choices, and the technological visions of those who can convince others are more likely to emerge as dominant.

We also explored where the power to persuade comes from. Ideas and charisma of course matter. But there are more-systemic forces shaping persuasion power as well. Those with the ability to set the agenda, typically high-status people with access to the corridors of power, are more likely to be persuasive. Social status and access are both shaped by a society's institutions and norms; they determine whether there is room at the table for diverse voices and interests when the most important decisions are made.

Our approach emphasizes that such diversity is critical because it is the most surefire way of building countervailing powers and containing overconfident and selfish visions. All of these considerations are general, but once more, they become particularly important in the context of technology.

We further saw how persuasion power generates strong self-reinforcing dynamics: the more people listen to you, the more status you gain and the more successful you become economically and politically. You are thus enabled to propagate your ideas more forcefully, amplifying your power to persuade and further boost your economic and political resources.

This feedback is even more important when it comes to technology choices. The technological landscape not only determines who prospers and who languishes, but it also critically influences who holds social power. Those enriched by new technologies, or whose prestige and voice are magnified, become more powerful. Technological choices are themselves defined by dominant visions and tend to reinforce the power and status of those whose vision is shaping technology's trajectory.

This self-reinforcing dynamic is a type of vicious circle. Students of history and political economy have highlighted such dynamics, documenting the pathways that make the rich politically more influential and how this additional political power enables them to become richer. The same is true of the new vision oligarchy that has come to dominate the future of modern technology.

You may think that it is much better to be controlled by the power to persuade rather than the power to repress. In many ways, that is right. But there are two senses in which persuasion power may be equally pernicious in the modern context. For one, those with the power to persuade also persuade themselves to ignore those who will suffer because of these choices and the collateral damage they produce (because the persuaders are on the right side of history and working for the common good). In addition, biased choices propagated by the power to persuade are less evident than those that are supported by violence, so they may be easier to ignore and potentially harder to correct.

This is a vision trap. Once a vision becomes dominant, its shackles are difficult to throw off because people tend to believe its teachings. And, of course, things are much worse when vision gets out of control, encouraging overconfidence and blinding everyone to its costs.

People outside of the tech sector and away from contemporary corridors of power understandably feel frustrated, but in truth they are not helpless against this vision trap. People can support alternative stories, build more inclusive institutions, and strengthen other sources of social powers that weaken the trap.

Because technology is highly malleable, there is no scarcity of compelling stories that can support alternative paths for technology. There are always many technological choices, with very different consequences, and if we get stuck with a single idea or a narrow vision, it is very often not because we are short of options. Rather, it is because those setting the agenda and commanding social power have imposed it on us. Correcting this situation is partly about changing the narrative: dissecting the driving vision, revealing the costs of the current path, and giving airtime and attention to alternative futures of technology.

Ordinary people can also work toward building democratic institutions to broaden agenda-setting power. When different groups are entitled to be at the table, when economic inequalities and thus social status differences are limited, and when diversity and inclusion are enshrined in laws and rules, it becomes harder for the viewpoints of a few people to hijack the future of technology.

Indeed, we will see in later chapters that institutional and societal pressures have at least sometimes pushed visions and the direction of progress in a more inclusive direction. What we are proposing has been done and can be done again.

Before we turn to applying these ideas in the current context, in the next three chapters we discuss the complex and sometimes impoverishing role of technological change, first in preindustrial agriculture and then during the early stages of industrialization. In both cases we will see that in the name of the common good, narrow visions drove innovations and the application of new techniques. Gains accrued to those controlling technology, often harming rather than benefiting most of the population. Only when robust countervailing powers developed did a different direction of progress, more favorable to sharing prosperity, start to emerge.

4

Cultivating Misery

And Babylon, so often destroyed.
Who rebuilt it so many times? In which of the houses
Of gold-gleaming Lima did the construction workers live?

—BERTOLT BRECHT,
"Questions of a Worker Who Reads," 1935

The poor in these parishes may say, and with truth, *Parliament may be tender of property: all I know is, I had a cow, and an act of Parliament has taken it from me.*

—ARTHUR YOUNG, *An Inquiry into the Propriety of Applying Wastes to the Better Maintenance and Support of the Poor,* 1801 (italics in original)

Italian scholar Francesco Petrarca (better known as Petrarch) famously argued that the era following the collapse of the Western Roman Empire in the year 476 was a time "surrounded by darkness and dense gloom." Petrarch was referring to the paucity of advances in poetry and art, but his pronouncements came to define how generations of historians and social commentators thought of the eight centuries that followed the glory of the Roman Empire. Conventional wisdom long held there was essentially no progress of any kind, including technological breakthroughs, until the Renaissance began to turn things around starting in the 1300s.

We now know that this view was wrong. There was significant technological change and improvement in economic productivity in Europe during the Middle Ages. Practical innovations included:

- better rotation of crops across different fields
- greater use of legumes to feed animals and add nitrogen to the soil
- the heavy wheeled plow, pulled by six or eight oxen
- increased use of horses for plowing and transportation
- better harnesses, stirrups, saddles, and horseshoes
- more use of animal manure as fertilizer
- widespread adoption of the wheelbarrow
- early fireplaces and chimneys, which greatly improved indoor air quality
- mechanical clocks
- the basket wine press
- good mirrors
- the spinning wheel
- improved looms
- improved use of iron and steel
- expanded access to coal
- scaled-up mining of all sorts
- better barges and sailing ships
- advances in stained-glass windows
- the very first eyeglasses

Yet there was also something quite dark about this era. The lives of people working the soil remained hard, and peasants' standard of living may even have declined in parts of Europe.

The technology and the economy progressed in a way that proved harmful to most of the population.

Perhaps the most defining technology of the Middle Ages was the mill, whose rising importance is well illustrated by the English experience after the Norman Conquest of 1066. At the end of the eleventh century, there were about 6,000 water-powered mills in England, which worked out to just about one mill per 350 people. Over the next 200 years, the number of waterwheels doubled, and their productivity increased significantly.

The earliest water mills involved a small wheel that rotated in the horizontal plane below a grindstone, to which it was connected by a vertical axle. Later, more-efficient designs introduced a larger vertical wheel, mounted outside the mill and connected by gears to the grinding mechanism. The improvements were striking. Even a small vertical waterwheel, operated by five to ten people, could generate two or three horsepower, the equivalent of thirty to sixty workers doing the work by hand—more than a threefold increase in productivity. The larger vertical mills of the later medieval period boosted output per worker to as much as twenty times the level that hand milling could achieve.

Waterwheels could not be adopted everywhere: they needed a sufficient flow of water running down a steep enough gradient. Starting in the 1100s, windmills extended the reach of mechanical power, greatly expanding milling grain for bread and ale and fulling (preparing) cloth for wool processing. Windmills boosted economic activity in flat parts of the country with rich soils, such as East Anglia.

From 1000 to 1300, water mills and windmills and other advances in agricultural technology roughly doubled yields per hectare. These innovations also helped kick-start English woolen cloth textiles, which later played a pivotal role in industrialization. Although it is difficult to determine exact numbers, agricultural productivity per person is estimated to have increased by 15 percent between 1100 and 1300.

You might think that these technical and productivity advances would lead to higher real incomes. Alas, the productivity bandwagon—productivity increases that lift wages and workers' living standards—did not materialize in the medieval economy. Except for those belonging to a small elite, there were no sustained improvements in living standards and some episodes of deterioration. For most people, better agricultural technology during the Middle Ages deepened their poverty.

The rural population of England did not have a comfortable existence in the early eleventh century. Peasants worked hard and achieved little more consumption than the bare minimum necessary for survival. The available evidence suggests that these people were squeezed even further over the next two centuries. The Normans reorganized agriculture, strengthened the feudal system, and intensified implicit and explicit taxation. Farmers had to hand over more of their agricultural output to their social superiors. Over time, feudal lords imposed more-onerous labor requirements as well. In some parts of the country, peasants spent twice as many hours per year in the lord's field as had been the norm before the conquest.

Although food production was growing and peasants were working harder, malnutrition worsened, and consumption levels dropped toward the threshold below which subsistence becomes impossible. Life expectancy remained low and may have deteriorated to just twenty-five years at birth.

Then things got much worse in the early 1300s, with a string of famines, culminating in the Black Death in the middle of the century, which wiped out between one-third and one-half of the English population. This virulent bubonic plague was bound to kill many people, but it was the combination of the bacterial infection and chronic malnutrition that was responsible for the staggering death toll.

If not to the peasantry, where did all the additional output that came from water mills and windmills, the horseshoes, the loom, the wheelbarrow, and the advances in metallurgy go? Some of it was used for feeding more mouths. England's population increased from around 2.2 million in 1100 to about 5 million in 1300. But as the population rose, so did the size of the agricultural workforce and the level of agricultural production.

Overall, higher productivity and lower consumption levels for most of the population brought a huge increase in the "surplus" of the English economy, meaning the amount of output, mostly food, wood, and cloth generated above the minimum level necessary for the survival and reproduction of the population. This surplus was extracted and enjoyed by a small elite. Even under the most expansive definition, this elite, including the king's retinue, nobles, and high clergy, made up no more than 5 percent of the population. But it still captured most of the agricultural surplus in medieval England.

Some of the food surplus went to support the newly burgeoning urban centers, whose population increased from two hundred thousand in 1100 to about a million in 1300. Urban living standards appear to have improved, in stark contrast with what happened in more-rural areas. A wider variety of goods, including luxuries, became available to city inhabitants. London's expansion reflected this growing opulence; its population more than tripled, to around eighty thousand.

Most of the surplus was eaten up not by urban centers but by the large religious hierarchy, which built cathedrals, monasteries, and churches. Estimates suggest that by 1300, bishops, abbots, and other clerics together owned one-third of all agricultural land.

The church's construction boom was truly spectacular. After 1100, cathedrals were established in twenty-six towns, and eight thousand new churches were built. Some were enormous projects. Cathedrals were stone constructions at a time when most people lived in ramshackle houses. Most were designed by superstar architects and some took centuries to complete, with hundreds of workers, including skilled artisans and a great deal of unskilled physical labor quarrying stone and carrying materials.

Construction was expensive, costing between £500 and £1,000 per year, approximately 500 times the annual income of an unskilled worker at the time. Some of this money was raised through voluntary donations, but a significant portion was funded by periodic levies and taxes on the rural population.

In the 1200s there was competition to see which community could build the tallest structure. Abbot Suger of Saint-Denis in France, which was in the grips of a similar cathedral boom, represented the prevailing view, arguing that these glorious buildings should be equipped with every imaginable ornament, preferably in gold:

> Those who criticize us claim that this celebration [of the Holy Eucharist] needs only a holy soul, a pure mind and faithful intention. We are certainly in complete agreement that these are what matter above all else. But we believe that outward ornaments and sacred chalices should serve nowhere so much as in our worship, and this with all inward purity and all outward nobility.

Estimates from France suggest that as much as 20 percent of total output may have been spent on religious building construction between 1100 and 1250. This number is so high that, if true, it implies that roughly all production beyond what was needed to feed people went into church building.

The number of monasteries expanded as well. In 1535 there were between 810 and 820 religious houses, "great and small," in England and Wales. Almost all of these were founded after 940, and most first enter the records between 1100 and 1272. One monastery held more than seven thousand acres of arable land, while another owned more than thirteen thousand sheep. Additionally, thirty towns, known as monastic boroughs, were under the control of monkish orders, which meant that the church hierarchy also lived off the revenue from these towns.

Monasteries had a voracious appetite. They were expensive to build and operate. The annual income of Westminster Abbey in the late 1200s was £1,200, mostly derived from agriculture. Some of these agricultural empires were truly sprawling. The monastery of Bury Saint Edmunds, one of the richest, owned the rights to the income of more than sixty-five churches.

To make matters worse, monasteries were exempt from tax. As their land holdings and control over economic resources grew, less was left for the king and the nobility. Compared to the one-third of agricultural land the church controlled, the king held one-sixth of all land (by value) in 1086. But by 1300, he was receiving only 2 percent of the total land income in England.

Some monarchs tried to rectify this imbalance. Edward I enacted the Statute of Mortmain in 1279, attempting to close a tax loophole by prohibiting the donation of further land to religious organizations without royal permission. These measures were not effective, however, because the ecclesiastical courts, which were under the ultimate control of the bishops and abbots, helped devise legal work-arounds. Monarchs were not strong enough to wrest revenue away from the medieval church.

A Society of Orders

Why did the peasants put up with their lot, accepting lower consumption, longer work hours, and worsening health even as the economy was becoming more productive? Of course, part of it was that the nobility specialized in the control of the means of violence in medieval society and was not shy in using them when the need arose.

But coercion could only go so far. As the Peasants' Revolt of 1381 demonstrates, when the ordinary people became angry, it was not easy to put them down. Triggered by efforts to collect unpaid poll taxes in the southeast of England, the rebellion quickly grew, and rebels started articulating demands to reduce taxation, abolish serfdom, and reform the law courts that were so consistently biased against them. According to Thomas Walsingham, a contemporary chronicler, "Crowds of them assembled and began to clamour for liberty, planning to become the equals of their lords and no longer bound by servitude to any master." Henry Knighton, another observer at the time, summed up the events: "No longer restricting themselves to their original grievance [regarding a poll tax and how it was being collected] and not satisfied by minor crimes, they now planned much more radical and merciless evils: they determined not to give way until all the nobles and magnates of the realm had been completely destroyed."

The rebels attacked London and broke into the Tower of London, where the king, Richard II, was taking refuge. The revolt ended because the king agreed to the rebels' demands, including the abolition of serfdom. It was only after the king gathered a much larger force and reneged on his promises that the rebels were defeated and as many as fifteen hundred were tracked down, caught, and executed, often savagely—for example, by being drawn and quartered.

Most of the time, discontent never boiled up to such levels, because the peasantry was persuaded to acquiesce. Medieval society is often described as a "society of orders," consisting of those who fought, those who prayed, and those who did all the work. Those who prayed were crucial in persuading those who labored to accept this hierarchy.

There is some nostalgia for monasteries in the modern imagination. Monks are credited with transmitting to us many classic writings from Greco-Roman times, including Aristotle's works, or even with saving Western civilization. They are associated with various productive activities, and monasteries today sell products ranging from hot sauce to dog biscuits, fudge, honey, and even (until recently) printer ink. Belgian monasteries are world renowned for their ale (including what some regard as the world's best beer, Westvleteren 12, from the Trappist Abbey of Saint-Sixtus). One monastic order in the Middle Ages, the Cistercians, is famous for clearing land to plant crops, for exporting wool, and, at least initially, for not wanting to benefit from the work of others. Other orders insisted on poverty as a lifestyle choice for their members.

However, most medieval monasteries were not into production or fighting poverty, but in the business of prayer. In these turbulent times, when the population was deeply religious, prayer was tightly linked to persuasion. Priests and the religious orders gave advice to people and justified the existing hierarchy, and more importantly they propagated a vision of how society and production should be organized.

The clergy's ability to persuade was amplified by their authority as God's emissaries. The church's teaching could not be questioned. Any publicly expressed skepticism would quickly lead to excommunication. The laws also favored the church, as well as the secular elite, and empowered local courts, run by the feudal elite, or ecclesiastic courts, which were under the control of the church hierarchy.

The question of whether clerical authority was superior to secular authority remained contentious throughout the Middle Ages. Archbishop of Canterbury Thomas Becket famously locked horns with Henry II over the issue. When the king insisted that serious crimes committed by clerics must be heard in the royal courts, Becket responded: "This will certainly not be done, for laymen cannot be judges of [church] clerks and whatever this or any other member of the clergy has committed should be judged in a church court." Becket was former lord chancellor and royal confidant to the king, and he saw himself as standing up for liberty—or one form of liberty—against tyranny. The king saw this stance as a betrayal, and his fury eventually resulted in Becket's killing.

But this royal use of force backfired, in the sense that it only increased the church's persuasion power and ability to stand up to the king. Becket came to be regarded as a martyr, and Henry II had to pay public penance at his tomb. The tomb remained an important shrine until 1536, when, influenced by the Protestant Reformation and wanting to get remarried, Henry VIII turned against the Catholic Church.

A Broken Bandwagon

This unequal distribution of social power in medieval Europe explains why the elite could live comfortably while the peasantry was miserably poor. But how and why did new technologies further impoverish much of the population?

The answer to this question is intimately linked to the socially biased nature of technology. How technology is used is always intertwined with the vision and interests of those who hold power.

The most important part of the production landscape of medieval England was reorganized following the Norman Conquest. The Normans intensified lords' dominance over peasants, and this context determined wages, the nature of agricultural work, and the way that new technology was adopted. Mills represented a significant investment, and in an economy where landowners had grown larger and politically stronger, it was natural that they would be the ones making these investments, and in a way that further strengthened their hand against the peasants.

Feudal lords operated large tracts of land themselves, with substantial control over their tenants and everyone else living in their manors. This control was critical because rural residents were required to perform unpaid, essentially coerced, labor on the lord's property. The exact terms of this work—how long it should last and how much it would coincide with the harvesting season—were often negotiated, but local courts, controlled by the lords, made the decisions when there were disagreements.

Mills, horses, and fertilizers raised productivity, for more agricultural output could now be produced using the same amount of labor and land. But the productivity bandwagon was nowhere to be seen. To understand why not, let us revisit the economics of the productivity bandwagon from Chapter 1.

Although mills save labor in various tasks such as grinding corn, they also increase the marginal productivity of workers. According to the productivity bandwagon perspective, employers should hire more people to work in the mills, and competition for workers should push wages up. But as we have seen, the institutional context matters enormously. Greater demand for workers leads to higher wages only when employers compete to attract labor in a well-functioning, noncoercive labor market.

There was no such labor market in medieval Europe and little competition between mills. As a result, wages and obligations were often determined by what the lords could get away with. The lords also decided how much the peasantry should pay to have access to the mills and set some of the other taxes and the dues that they owed. Thanks to their greater social power under Norman feudalism, lords of the manor could tighten the screws quite a bit.

But why would the introduction of new machines and the resulting higher productivity lead to more squeezing of peasants and worse living standards? Imagine a setting in which new technologies raise productivity but lords cannot (or do not want to) hire additional workers. They still would like to have more work hours to go with their more productive technology. How to achieve this? One way, often ignored in standard accounts, is to increase coercion and squeeze more labor out of existing workers. Then productivity gains benefit landowners but directly harm workers, who now suffer both greater coercion and longer work hours (and possibly even lower wages).

This is what happened after mills were introduced in medieval England. As new machines were deployed and productivity rose, feudal lords exploited the peasantry more intensively. The working hours of laborers rose, with less time left to tend to their own crops, and their real incomes and household consumption fell.

The distribution of social power and the vision of the age also defined how new technologies were developed and adopted. Critical decisions included where new mills would be built and who would control them. In England's society of orders, it was viewed as just and natural that both lords and monasteries operated mills. The same people also had the authority and power to ensure that no competition would appear. This enabled the lord's mill to process all the grain and cloth in the local economy at prices set by the lord. In some instances, feudal lords even managed to ban home milling. This path of technology adoption exacerbated economic and power inequality.

The Synergy Between Coercion and Persuasion

We can see the role of the dominant vision of medieval society, backed up by the coercive power of the religious and secular elites, in biasing the path of technology adoption in the story of Herbert the Dean's attempt to build a windmill in 1191. The abbot of Bury Saint Edmunds, one of the richest and most powerful monasteries, was not pleased by this entrepreneurship and demanded that the windmill be demolished immediately because it would compete with his monastery's mills. According to Jocelin of Brakelond, who worked for the abbot, "Hearing this, the Dean came and said that he had the right to do this on his free fief, and that free benefit of the wind ought not to be denied to any man; he said he also wished to grind his own corn there and not the corn of others, lest perchance he might be thought to do this to the detriment of neighboring mills."

The abbot was furious: "I thank you as I should thank you if you had cut off both my feet. By God's face, I will never eat bread till that building be thrown down." In the abbot's interpretation of customary law, if a mill existed, he could not prevent the dean's neighbors from using it, and this would constitute competition for the monastery's own mills. However, under the same interpretation the dean did not have the right to build a windmill without the abbot's permission.

Although such arguments could in principle be contested, in practice there was no way for the dean to challenge them because all matters related to the monastery's rights were decided in ecclesiastical court, which would rule in favor of the powerful abbot. The dean hastily pulled his mill down just before the bailiffs arrived.

Over time, the church's control over new technologies intensified. By the thirteenth century, the monastery of Saint Albans in Hertfordshire spent £100 upgrading its mills and then insisted that tenants bring all their corn and cloth to these mills. Even though tenants lacked access to other mills, they refused to comply. Hand processing their cloth at home was preferable to paying the monastery's high fees.

But even this small amount of independence ran up against the monastery's design of being the sole beneficiary of new technologies. In 1274 the abbot attempted to confiscate cloth from tenants' houses, which resulted in physical confrontations between tenants and monks. Unsurprisingly, when the tenants protested in the King's Court, the ruling went against them. Their cloth had to be processed by the abbot's mills, and they would have to pay the fees the monastery set.

In 1326 there was an even more violent confrontation with Saint Albans over whether tenants were allowed to grind their grain at home using hand mills. The monastery was besieged twice, and when the abbot ultimately prevailed, he seized all home millstones and used them to pave a courtyard in the monastery. Fifty years later, as part of the Peasants' Revolt, farmers stormed the monastery and broke up the courtyard, "the symbol of their humiliation."

Overall, the medieval economy was not bereft of technological progress and major reorganizations. But it was a dark age for English peasants because the Norman feudal system ensured that higher productivity would accrue to the nobility and the religious elite. Worse, the reorganization of agriculture paved the way to greater surplus extraction and more-onerous obligations from the peasantry, whose living standards declined further. New technology served to further favor the elite and intensify the peasants' misery.

These difficult times for ordinary people were the result of the religious and aristocratic elite structuring technology and the economy to make it hard for most of the population to prosper. Day-to-day sway over the population through persuasion power rested on a strong bedrock of religious belief reinforced by court action and coercion.

A Malthusian Trap

An alternative interpretation of stagnant living standards during the Middle Ages is rooted in the ideas of Reverend Thomas Malthus. Writing at the end of the eighteenth century, Malthus argued that the poor were feckless. If you gave them enough land for a cow, they would just have more children. As a result, "Population, when unchecked, increases in a geometrical ratio. Subsistence increases only in an arithmetical ratio. A slight acquaintance with numbers will shew the immensity of the first power in comparison of the second." Because there was a limit on available land, an increase in the population would increase agricultural output by less; consequently, any potential improvement in living standards for the poor would not last and would be quickly eaten up by more mouths to feed.

This uncharitable view, which blames the poor for their misery, does not fit the facts. If there is any kind of Malthusian "trap," it is the trap of thinking that there is an inexorable law of Malthusian dynamics.

The poverty of the peasantry cannot be understood without recognizing how they were coerced—and how political and social power shaped who benefited from the direction of progress. During the thousands of years before the Industrial Revolution, technology and productivity were not stagnant, even if they did not improve as steadily and rapidly as they did after the middle of the eighteenth century.

Who benefited from new technologies and productivity increases depended on the institutional context and the type of technology. During many critical periods, such as those discussed in this chapter, technology followed the vision of a powerful elite, and productivity growth did not translate into any meaningful improvements in the lives of the majority of the population.

But the hold of the elite over the economy ebbed and flowed, and not all productivity increases were directly under their control the way the new mills were. When yields in the lands worked by peasants rose and the lords were not dominant enough to grab the additional surplus, living conditions for the poor improved.

After the Black Death, for example, many English lords, facing untilled fields and a shortage of labor, attempted to extract more from their servile laborers without paying more. King Edward III and his advisers were alarmed about the demand for higher compensation from workmen and pushed through legislation intended to curb these wage demands. The Statute of Labourers of 1351 was adopted as part of this effort and began, "Because a great part of the people and especially of the workmen and servants has now died in that pestilence, some, seeing the straights of the Masters and the scarcity of servants, are not willing to serve unless they receive excessive wages." It stipulated harsh punishments, including imprisonment, to any workman leaving their service. It was particularly important that higher wages should not be used to attract laborers away from their fields, so the statute decreed, "Let no one, moreover, pay or permit to be paid to anyone more wages, livery, mead or salary than was customary as has been said. . . ."

These royal commands and laws were to no avail, however. Labor shortage swung the pendulum in favor of peasants, who could refuse their lords' demands, ask for higher wages, refuse to pay fines, and, if necessary, walk away to other manors or into towns. In the words of Knighton, the workmen were "so arrogant and obstinate that they did not heed the king's mandate, but if anyone wanted to have them, he had to give them what they asked."

The outcome was higher wages, as John Gower, a contemporary poet and commentator, described: "And on the other hand it may be seen that whatever the work may be the labourer is so expensive that whoever wants anything done must pay five or six shillings for what formerly cost two."

A House of Commons petition of 1376 put the blame squarely on how labor shortages empowered servants and laborers, who "as soon as their masters accuse them of bad service, or wish to pay them for their labour according to the form of the statutes . . . take flight and suddenly leave their employment in district." The problem was that "they are taken into service immediately in new places, at such dear wages that example and encouragement is afforded to all servants to depart into fresh places. . . ."

The shortage of labor did not only increase wages. The power balance between lords and peasants shifted across rural England, and lords started reporting lack of respect from their inferiors. Knighton described "the elation of the inferior people in dress and accoutrements in these days, so that one person cannot be discerned from another, in splendor or dress or belongings." Or, as Gower put it, "Servants are now masters and masters are servants."

In other parts of Europe where the dominance of rural elites persisted, there was no similar erosion of feudal obligations and no similar evidence that wages rose. In eastern and central Europe, for example, the peasantry was even more harshly treated and thus less able to articulate demands, even in the midst of labor shortage, and there were fewer towns where people could easily escape to. Prospects for peasant empowerment remained weaker.

In England, however, the power of local elites eroded over the next century and a half. Consequently, as a famous account of the period explains, "The lord of the manor was forced to offer good conditions or see all his villeins [peasants] vanish." In these social circumstances, real wages drifted up for a while.

The dissolution of monasteries under Henry VIII and the subsequent reorganization of agriculture was another step that altered the balance of power in rural England. The slow growth in the real incomes of the English peasantry before the beginning of the industrial era was a consequence of this type of drift.

In the course of the Middle Ages as a whole, there were periods when higher yields increased fertility and population outgrew the capacity of the land to feed people, sometimes leading to famine and demographic collapse. But Malthus was wrong in thinking that this was the only possible result. By the time he was formulating his theories at the end of the eighteenth century, English real incomes, not just population, had been on an upward trajectory for centuries, with no sign of inescapable famines or plagues. Similar trends are visible in other European countries during this time period, including Italian city-states, France, and the areas that today make up Belgium and the Netherlands.

Even more damning to Malthusian accounts, we have seen that the surplus generated by new technologies during the medieval era was eaten up not by the excessively fertile poor but by the aristocracy and the church in the form of luxuries and ostentatious cathedrals. Some of it also contributed to higher living standards in the largest cities, such as London.

It is not just evidence from medieval Europe that strongly refutes the idea of a Malthusian trap. Ancient Greece, led by the city-state of Athens, experienced fairly rapid growth in output per capita and living standards between the ninth and fourth centuries BCE. During this almost five-hundred-year period, house sizes increased, floor plans improved, household goods multiplied, consumption per person grew, and various other indicators of life quality got better. Even though population expanded, there was little evidence of Malthusian dynamics setting in. This era of

Greek economic growth and prosperity was terminated only by political instability and invasion.

There was also growth of output per capita and prosperity during the Roman Republic, starting sometime around the fifth century BCE. This period of prosperity continued all the way into the first century of the Roman Empire and most likely came to an end because of political instability and the damage wrought by authoritarian rulers during Rome's imperial phase.

Extended periods of preindustrial economic growth, with no sign of Malthusian dynamics, were not confined to Europe. There is archaeological and sometimes even documentary evidence suggesting similar long episodes of growth in China, in the Andean and Central American civilizations before European colonization, in the Indus Valley, and in parts of Africa.

The historical evidence strongly suggests that the Malthusian trap was not a law of nature, and its existence looks highly contingent on particular political and economic systems. In the case of medieval Europe, it was the society of orders, with its inequities, coercion, and distorted path of technology, that created poverty and lack of progress for most people.

Original Agricultural Sin

Socially biased technology choices were not confined to medieval Europe and have been a mainstay of preindustrial history. They began as early as, if not earlier than, agriculture itself.

Humans started experimenting with plant and animal domestication a long time ago. Dogs were already cohabitating with Homo sapiens more than fifteen thousand years back. Even as they continued to forage—hunt, fish, and gather—humans selectively encouraged the growth of some plants and animals and began influencing their ecosystem.

Then around twelve thousand years ago, a process of transitioning to settled, permanent agriculture based on fully domesticated plants and species started. We now know that this process took place, almost surely independently, in at least seven places around the world. The crops that were at the center stage of this transition varied from place to place: two types of wheat (einkorn and emmer) and barley in the Fertile Crescent, part of what is now called the Middle East; two types of millet (foxtail and broomcorn) in northern China; rice in southern China; squash, beans, and maize in Mesoamerica; tubers (potatoes and yams) in South America; and a variety of quinoa in what is now the eastern United States. Several crops were domesticated in Africa, south of the Sahara; Ethiopia domesticated coffee, which deserves special commendation and should probably count double.

Absent written records, no one knows exactly what happened and when. Theories about timing and causation remain hotly contested. Some scholars claim that the warming of the planet created abundance, which in turn caused settlements and agriculture. Other experts maintain the opposite, that necessity was the mother of innovation and episodic scarcity was the main force pushing humans to boost yields through domestication. Some claim that permanent villages came first, followed by the emergence of social hierarchy. Others point to signs of hierarchy in goods found in graves that predate settlements by thousands of years. Some join the famous archaeologist Gordon Childe, who coined the term *Neolithic Revolution* to describe this transition, seeing it as foundational to the advancement of technology and humanity. Yet others follow Jean-Jacques Rousseau and maintain that settling down to till fields full-time was human society's "original sin," paving the way to poverty and social inequality.

The likely reality is that there was a great deal of diversity. Humans experimented with various crops and many ways of domesticating animals. Early cultivars included legumes (peas, vetch, chickpeas, and their relatives), yams, potatoes, and various vegetables and fruits. Figs may have been among the first plants to be cultivated.

We also know that farming did not spread rapidly and that many communities continued foraging even as agriculture became well established in nearby places. For example, recent DNA evidence shows that indigenous European hunter-gatherers did not adopt cultivation for thousands of years and that farming ultimately came to Europe because Middle Eastern farmers moved there.

In the process of these social and economic changes, many different types of societies emerged. For example, in Göbekli Tepe, now in central Turkey, we have archaeological records of settlements dating back to 11,500 years before the present, involved in both agriculture and foraging for more than a thousand years. Remains of grave goods and rich artwork suggest a significant degree of hierarchy and economic inequality in this early civilization.

In another famous site, Çatalhöyük, less than 450 miles to the west of Göbekli Tepe, we have a slightly later civilization with very different features. Çatalhöyük, which also lasted more than a thousand years, appears to have had a fairly egalitarian social structure, with little inequality in grave goods, no evidence of clear hierarchy, and very similar houses for all inhabitants (especially on the east mound, where the settlements existed for a long time). The people of Çatalhöyük seem to have achieved a healthy diet combining cultivated crops, wild plants, and hunted animals.

Around 7,000 years before the present, a very different picture starts taking shape throughout the Fertile Crescent: permanent agriculture, often based on a single crop, becomes the only game in town. Economic inequality intensifies, and a very clear social hierarchy emerges with elites at the top consuming a lot and doing none of the production. Around this time, the historical record becomes clearer as well, for writing emerges. Although this record is written by the elite and their scribes, the opulence they achieved and the huge power they commanded over the rest of their societies are apparent.

The Egyptian elite, around whom pyramids and tombs were built, seem to have enjoyed relatively good health. They certainly had access to medical services, such as they were, and at least some of their mummies suggest long and healthy lives. In contrast, peasants suffered from, among other conditions, schistosomiasis, a parasitic disease spread through water; tuberculosis;

and hernias. The ruling elite traveled in comfort and do not seem to have worked hard. Anyone not willing to pay the required taxes to support such lifestyles could expect to beaten with wooden sticks.

The Pain of Grain

The early diversity notwithstanding, grains ended up on top in most places where settled agriculture took root. Wheat, barley, rice, and maize are all members of the grass family—small, hard, dry seeds known to botanists as caryopses. These cereal grains, as they are commonly known, share some appealing characteristics. They have low moisture content and are durable once harvested and thus easy to store. Most importantly, they have high energy density (calories per kilogram), making them attractive to transport, which is crucial if they are going to feed populations that are far from the cultivation sites. These grains can also be handled at scale if you have the labor force to sow, maintain, and harvest them. In contrast, tubers and legumes are harder to store, can rot easily, and have much lower calories per volume (about one fifth of what cereal grains provide).

Looked at from the perspective of achieving large-scale production and deriving significant energy from agriculture, the introduction of cereal grains is an exemplar of technological progress. It was this suite of crops and production methods that enabled the emergence of dense settlements, cities, and then ultimately larger states. But, once again, the way this technology was applied had very unequal consequences.

In the Fertile Crescent prior to 5,000 years ago, there is no indication of any town having more than 8,000 inhabitants. At that time, Uruk (in southern Iraq), however, dramatically breaks the record, with 45,000 inhabitants. Over the next two millennia, the size of the largest cities creeps upward: 4,000 years ago, there were 60,000 people living in both Ur (Iraq again) and Memphis (the capital of a unified Egypt); 3,200 years ago, we see Thebes (Egypt) with around 80,000 inhabitants; and 2,500 years ago, Babylon reaches a population of 150,000.

In all these places, the evidence is clear that a centralized elite benefited greatly from new technologies. Most other members of these societies did not.

We do not know the living conditions of early agriculturalists with any certainty. But under the auspices of early centralized states, most people engaged in full-time grain cultivation seem to have been decidedly worse off than their foraging ancestors. Existing estimates indicate that foragers worked somewhere around five hours per day, ate a wide variety of plants and plenty of meat, and had healthy lives, achieving levels of life expectancy at birth that ranged from twenty-one to thirty-seven years. Infant mortality rates were high, but people who reached the age of forty-five could be expected to live another fourteen to twenty-six years.

Settled grain cultivators worked probably twice as much, more than ten hours per day. The work became much harder as well, especially after grains emerged as the main crop. There is plenty of evidence suggesting that their diets deteriorated, compared with the less settled lifestyle. As a result, farmers were shorter by four or five inches, on average, than the foragers and had significantly more skeletal damage and much worse dental problems. Farmers also suffered more from infectious disease and died younger than their foraging cousins. Their life expectancy at birth is estimated to have been around nineteen years.

Full-time farming was particularly tough on women; their skeletons show signs of arthritis from all the work involved in grinding grain. Mortality rates in childbirth were also significantly higher among farmers, and these societies became distinctly more male dominated.

Why did people adopt, or at the very least acquiesce to, a technology that involved much backbreaking work, unhealthy lives, so little consumption for themselves, and such a steep hierarchy? Of course, nobody alive twelve thousand years ago could have foreseen the type of society that would emerge from settled agriculture. Nevertheless, just as in the medieval period, technological and organizational choices in early civilizations favored the elite and impoverished most people. In the Neolithic case, new technologies evolved over a much longer span of time—thousands of years, rather than hundreds in the medieval experience—and the dominant elite often emerged slowly. All the same, in both cases a political system that placed disproportionate power in the hands of that elite was critical. Coercion played a role, of course, but the persuasion power of religious and political leaders was often the decisive factor.

Slavery became more common than it had been during the earliest days of agriculture, and there were significant numbers of slaves in civilizations ranging from ancient Egypt to Greece. There was also plenty of coercion for everyone else, when necessary. But, as in the Middle Ages, this was not how people were controlled daily. Coercion was often in the background, while persuasion was center stage.

Pyramid Scheme

Take the pyramids, which are a symbol of the opulence of the pharaohs. Building pyramids cannot be considered as investing in public infrastructure that boosted the material well-being of ordinary Egyptians over time, although it did create a lot of jobs. To build the Great Pyramid of Khufu at Giza approximately 4,500 years ago, a rotating workforce that numbered 25,000 per extended shift toiled for around 20 years. This was a much bigger construction project than any single medieval cathedral. For more than 2,000 years, each Egyptian ruler aspired to build his or her own pyramid.

At one time it was commonly assumed that these workers must have been coerced by ruthless overseers. We now know that this is not what happened. The people who built the pyramids were paid decent wages, many of them were skilled craftsmen, and they were fed well—for example, with beef, the most expensive meat available. They were most likely convinced to work hard through a combination of tangible rewards and persuasion.

Fascinating records exist for some of the work, including details of how one Giza work gang, the escort team of "The Uraeus of Khufu Is Its Prow," spent its time. There is no mention in these day-to-day accounts of punishment or coercion. Rather, the surviving fragments represent the kind of skilled labor and hard work associated with building medieval cathedrals: stone needs to be transported from a quarry to the Nile, then along the river by boat, and then hauled to the construction site. There is no mention of slavery, although some modern experts take the view that there were likely forced labor obligations for ordinary workers, similar to those that existed in feudal times and that Lesseps used to build the Suez Canal during the 1860s.

In the pharaonic period, skilled Egyptian artisans could be fed and paid because surplus food was squeezed out of the agricultural labor force. The technology of grain production enabled a large volume of crops in the fertile Nile Valley, which could then be transported to cities. But it was also thanks to the willingness of ordinary peasants to provide a huge amount of labor for so little reward. And this was in turn because they were persuaded by the authority and glory of the pharaoh as well as, of course, by his ability to crush opposition if necessary.

No one knows exactly what motivated ancient people; we cannot see inside the minds of farmers who lived two thousand or seven thousand years ago, and they did not leave written records of their aspirations or plight. It seems likely that organized religion helped convince them that this life was appropriate or indeed their unavoidable fate. Centralized farming cosmologies are quite clear that there is a hierarchy, with gods at the top, kings and priests in the middle, and peasants firmly at the bottom. The reward for not complaining varies across belief systems, but in general it is some form of deferred compensation. The gods have assigned you this role, so shut up and get back to work in the fields.

In the Egyptian belief system, helping the rulers to a better equipped afterlife was a major motivating idea. Ordinary people might not expect any improvement; servants would remain servants, and so on. But the gods approved of people who provided service, built pyramids, and handed over food that helped the rulers to greater glory and a bigger mausoleum. The really unlucky traveled with their masters directly to the afterlife; there is evidence in some pyramids that courtiers and other staff may have been ritually murdered at the time that their ruler was interred.

The Egyptian ruling elite lived in towns and comprised some combination of priestly hierarchy and "divine kings," claiming legitimacy or even direct descent from the gods. This pattern is not unique to Egypt. Temples and other monuments appear in most early civilizations and typically for the same reasons as the medieval church constructed cathedrals—to legitimize the rule of the elite by honoring their deity and to maintain people's faith.

One Kind of Modernization

Neither grain monoculture nor the highly hierarchical social organization that extracted most of the surplus from farmers was preordained or dictated by the nature of the relevant crops. These were choices. Other societies, often in similar ecological conditions, specialized in different types of agriculture, including tubers and legumes. In early Çatalhöyük, grains appear to have been combined with a rich array of wild plants, and meat consumption came from domesticated sheep and goats, as well as undomesticated animals, such as aurochs, foxes, badgers, and hares. In Egypt, before grain monoculture became established, emmer wheat and barley were cultivated at the same time that waterfowl, antelopes, wild pigs, crocodiles, and elephants were hunted.

Even cereal cultivation did not always produce inequality and hierarchy, as the more egalitarian Indus Valley and Mesoamerican civilizations illustrate. Rice farming in Southeast Asia took place in the context of less hierarchical societies for thousands of years, and the onset of greater social and economic inequality appears to coincide with the introduction of new agricultural and military technologies during the Bronze Age. The complex of large-scale grain cultivation, high level of surplus extraction, and top-down control was typically a result of political and technological decisions made by elites, when they were powerful enough and could persuade the rest to go along with it.

For Neolithic times and the age of the Egyptian pharaohs, we can form only rough guesses of how new technologies were selected and used, and what kinds of arguments were offered to convince people to adopt them and cast existing arrangements aside. In the eighteenth century, however, we can see more clearly how a new vision of agricultural modernization emerged in England. What comes clearly into view is how those who stood to gain got their way by linking arguments for their preferred technology choice with what they claimed to be the common good.

By the mid-1700s, English agriculture had changed a great deal. Serfdom and most of the vestiges of feudalism had faded away. There were no lords who could directly command the local economy and compel others to work in their fields or process grain in their mills. Henry VIII had dissolved the monasteries and sold off their land in the mid-1500s. The rural elites were now the landowning gentry, with several hundred acres or more and an increasingly keen eye for how to modernize agriculture and boost the surplus that they captured.

The process of agricultural transformation had been ongoing for centuries, and more use of fertilizers and improved harvesting technologies had increased productivity, pushing up output per hectare between 5 and 45 percent, depending on the crop, over the preceding five hundred years. Economic and social change likely accelerated from the middle of the sixteenth century onward. As the hold of landowners and monasteries weakened, productivity gains also started spreading to peasants. From around 1600, we see real wages creeping upward more steadily, bringing improved nutrition and slightly better health to the peasantry.

As population increased, so did the demand for agricultural produce. Higher agricultural yields became a topic of national policy debates. To be sure, there were parts of the English rural economy that needed to be modernized. Much of the land was now private property, operated by the gentry, their tenants, or smallholders. But in some parts of the country a significant amount was "common land," over which members of the local community had the informal customary rights to graze cattle, collect firewood, and hunt. There were also unfenced, open fields that were being farmed. As land became more valuable, an increasing number of landowners wanted to "enclose" these lands, which meant removing peasants' customary rights to use them. Enclosures involved turning informally shared commons into formal private property, protected by law, typically as an extension of existing estates.

Enclosures of various kinds had been going on since the fifteenth century in an ad hoc way. In many parts of the country, landowners could achieve this by convincing the local population to acquiesce to enclosures, in return for monetary or other compensation. Yet in the eyes of the British elite of the late eighteenth century, there was much need for further modernization, especially by expanding their land holdings. About a third of all agricultural land was still held as common land and could potentially be turned into their private property.

Although the rhetoric was couched in terms of productivity increases and what was good for the country, the proposed modernization was far from neutral. It meant taking access to land away from peasants and expanding commercial agriculture. The vision of the age came to see the customary rights of landless peasants as a vestige of the past that needed to be modernized. If peasants did not want to relinquish these customary rights, then they had to be compelled to do so.

In 1773 Parliament passed the Enclosure Act, making it easier for large landowners to push through the reorganization of land they desired. Parliamentarians enacted this new law because they believed, or wanted to believe, that enclosures would be in the national interest.

Arthur Young, a farmer and influential writer, had a distinctive voice in these arguments. In his early work, Young had emphasized the importance of new agricultural techniques, including fertilizers, more scientific rotation of crops, and better plows for harvesting. Consolidated landholdings would make these technologies more effective and easier to implement.

But what about the resistance of the peasantry to enclosures? To understand Arthur Young's perspective on this, we must first recognize the context in which he was situated and the broader vision guiding technology and agricultural reorganization. Britain was still a hierarchical society. Its democracy was by the elite and for the elite, with less than 10 percent of the adult male population having the vote. Worse, this elite did not think much of their less privileged compatriots.

Malthus's writings were indicative of the mood of the time and the worldview of well-off people. Malthus thought that it was more humane not to let the living standards of the poor increase too much in the first place, lest they end up back in misery as they had more children. He also argued that "a man who is born into a world already possessed, if he cannot get subsistence from his parents on whom he has a just demand, and if the society do not want his labour, has no claim of *right* to the smallest portion of food, and, in fact, has no business to be where he is. At nature's mighty feast there is no vacant cover for him" (italics in original).

Young, like many of his contemporaries among the upper and middle classes, started out with similar notions. In 1771, almost three decades before Malthus's argument was published, Young wrote: "If you talk of the interests of trade and manufactures, everyone but an idiot knows that the lower classes must be kept poor, or they will never be industrious."

Combining this skeptical view of the lower classes and his belief in the imperative to apply better technologies in agriculture, Young became an outspoken voice for further enclosures. He was appointed as a key adviser to the Board of Agriculture, in which position he drafted authoritative reports on the state of British agriculture and opportunities for improvement.

Young thus became a spokesperson for the agricultural establishment, consistently listened to by ministers and cited in parliamentary debates. As an expert, he wrote forcefully in support of enclosures in 1767: "The universal benefit resulting from enclosures, I consider as fully proved; indeed so clearly, as to admit no longer of any doubt, amongst sensible and unprejudiced people: those who argue now against it are merely contemptible cavillers." Seen through this lens, it was acceptable to strip the poor and uneducated from their customary rights and common lands because the new arrangements would allow the deployment of modern technology, hence improving efficiency and producing more food.

An increasing number of major landowners were keen to have public support and parliamentary approval for what they wanted to do, and Young became a useful ally. Here was a careful assessment of what needed to be done in the national interest, and if this perspective said that casting aside traditional rights and compelling the holdouts was necessary for progress, this was a price British society would have to pay.

By the early 1800s, however, the collateral damage of enclosures was becoming clear, at least to those who wanted to see it. That thousands were being forcefully pushed into deeper poverty was fine with Malthus. Perhaps surprisingly, Young's reaction to these developments was quite different.

Though infused by the prejudices of his time, Young was an empiricist at heart. As he continued to travel and observe first-hand what was going on as enclosures came into effect, his empirical findings came increasingly into conflict with his views.

Even more remarkably, at this point Young changed his position on enclosures. He continued to believe that consolidation of open fields and common lands would result in efficiency gains. But he recognized that much more was at stake. The way in which common property was being abolished had a major impact on who won and who lost from the change of agricultural technology. By 1800, Young had completely reversed his recommendations: "What is it to the poor man to be told that the Houses of Parliament are extremely tender of property, while the father of the family is forced to sell his cow and his land?"

He argued that there were different paths of reorganizing agriculture; land could be consolidated without trampling the rights of ordinary people and taking the means of subsistence from them. There was no need to completely expropriate the rural population. From here, he went further and articulated the case that providing means of subsistence to the rural poor, such as a cow or goats, was not an impediment to progress. They could better support their families and perhaps have greater commitment to the community and even more sympathy for the status quo.

Young may have even understood a subtler economic truth: once expropriated, poor peasants would become a more reliable source of cheap labor to landowners—perhaps one of the reasons why so many landowners were keen on expropriating them. Conversely, protecting their basic assets might be a way of ensuring higher wages in the rural economy.

When advocating for enclosures, Young was a highly regarded expert, celebrated by the British establishment. Once he had his turnaround, all that changed, and he was no longer welcome to publish whatever he wanted on behalf of the Board of Agriculture. His aristocratic boss at the board made it clear that any anti-enclosure views were not welcome in official circles.

The history of the enclosure movement is a granular illustration of the way that persuasion and economic self-interest shape who benefits from technological change and who does not. The vision of British upper classes on what constituted progress and how to achieve it was critical for the reorganization of agriculture. This vision, as usual, overlapped quite a bit with their self-interest—taking land away from the poor with no or little compensation was clearly beneficial for those doing the taking.

A vision that articulates a common interest is powerful even when—especially when—there are losers as well as winners from new technologies because it enables those doing the reorganization and technology adoption to convince the rest.

There are often many constituencies to be persuaded. It was difficult to convince the poor peasants whose customary rights were being taken away. More feasible and more essential was the persuasion of the urban public and those with political power, such as the parliamentarians. Young's scientific assessment of the necessity of the rapid rollout of enclosures played a significant role in this process. Predictably, landowners knew what conclusions they wanted to hear, and they embraced Young when he voiced these opinions and silenced him when he changed his mind.

Technological choice was critical as well. Even when couched in the language of progress and national interest, there were many intricate choices about the implementation of new technologies, and these decisions determined how much the elite benefited and how much hardship the peasants suffered. Totally expropriating the customary rights of poor peasants was a choice. We now know that it was not one that was dictated by the inexorable path of progress. Common lands and open fields could have existed for longer while British agriculture was being modernized. In fact, the available evidence suggests that these forms of land tenure were not inconsistent with new technologies and yield increases.

In the seventeenth century, open-field farmers had been at the forefront of those adopting peas and beans, and in the eighteenth century they kept up with the adoption of clover and turnips. There was more drainage installed on enclosed soil, but even in areas where that made a difference, output per hectare was higher only by about 5 percent in 1800. On arable land with lighter soils, which drain well naturally, and on land used for pasture, the yields for open-field farmers were within 10 percent of what enclosed farmers achieved. Output per worker was also only slightly higher for farmers working enclosed land.

The reorganization of agriculture set the tone for the next several decades of British economic development and determined who gained from it. People with property did well, including through parliamentary action where necessary. Those without property did not.

Technological modernization in agriculture became an excuse for expropriating the rural poor. Did this expropriation help with productivity improvements that were so sorely needed in late eighteenth-century Britain? There is no consensus on this question, with estimates ranging from no productivity gains to significant increases in yields. But there is no doubt that inequality increased and that those who had their lands enclosed lost out.

None of this was inevitable. The encroaching of customary rights and intensification of rural poverty were choices made and imposed on people in the name of technological progress and national interest. And Young's assessment stands: productivity gains could have been achieved without driving landless peasants into further misery.

The Savage Gin

What the history of enclosures makes abundantly clear is that technological reorganization of production, even when proclaimed in the interest of progress and the common good, has a way of further pushing down the already disempowered. A pair of historical episodes, from two very different economic systems and continents, are emblematic of its savage implications. In nineteenth-century America, we can see the implications of the transformative technology of the cotton gin.

In American economic history, Eli Whitney appears alongside Thomas Edison as one of the most creative technological entrepreneurs enabling transformative progress. Whitney invented an improved cotton gin in 1793 that quickly removed the seeds from upland cotton. In Whitney's own assessment, "One man and a horse will do more than fifty men with the old machines."

The early American cotton industry was based on a long-staple variety that did not do well when planted in areas away from the East Coast. An alternative, upland cotton, grew well in other environments. But its sticky green seeds were more tightly attached to the fiber and could not be easily removed by existing gins. Whitney's gin was a breakthrough in separating the seeds, and it greatly expanded the area where upland cotton could be cultivated. More cotton cultivation meant increased demand for slave labor in the "lower South," first across the interior of South Carolina and Georgia, and eventually into Alabama, Louisiana, Mississippi, Arkansas, and Texas. Cotton became king in these thinly settled areas, where Europeans and Native Americans had previously grown subsistence crops.

Cotton production in the South increased from 1.5 million pounds in 1790 to 36.5 million pounds in 1800 and 167.5 million pounds in 1820. By mid-century, the South provided three-fifths of America's exports, almost all of it cotton. Roughly three-quarters of the world's cotton was grown in the American South at that time.

With such a transformative change raising productivity so spectacularly, would one perhaps be justified to speak of the national interest and the common good? Perhaps this time farm workers also benefited? Perhaps the productivity bandwagon worked? Once again, not at all.

Although landowners in the South and many other southerners involved in the processing, production, and trade in the cotton supply chain benefited handsomely, the actual workers doing the production were pushed more deeply into exploitation. Even worse than in the medieval era, greater demand for labor, under conditions of coercion, translated not into higher wages but into harsher treatment so that the last ounce of effort could be squeezed out of the slaves.

Southern planters pursued various innovations to increase yields, including the use of new cotton varieties. But when human rights are weak or nonexistent as in medieval Europe or on southern plantations, improving technology can easily lead to more intense exploitation of labor.

In 1780, just after independence, there were about 558,000 slaves in the United States. The slave trade became illegal starting on January 1, 1808, when there were about 908,000 enslaved people in the country. The importing of slaves from outside the US shrank to near zero, but the number of enslaved grew to 1.5 million in 1820 and 3.2 million in 1850. In 1850, 1.8 million slaves worked in cotton production.

Between 1790 and 1820, 250,000 slaves were forced to move to the Deep South. Overall, around a million slaves were moved to plantations that had been made productive by gin technology. The enslaved population of Georgia doubled in the 1790s. In four "upcountry" counties of South Carolina, slaves rose as a percent of the population from 18.4 percent in 1790 to 39.5 percent in 1820 and 61.1 percent in 1860.

Judge Johnson of Savannah, Georgia, praised Whitney's contribution this way: "Individuals who were depressed with poverty and sunk in idleness, have suddenly risen to wealth and respectability. Our debts have been paid off, our capitals increased, and our lands trebled in value." By "our," the judge was of course referring only to White people.

The lives of enslaved people in cultivating tobacco, the dominant slave-grown crop of the eighteenth century in Virginia, were obviously not good. Nevertheless, the journey to the Deep South was unusually brutal, and their circumstances became much worse on cotton fields. Compared with tobacco, cotton plantations were larger, with the work "regimented and relentless." One slave recalled being driven harder when cotton prices rose: "When the price rises in the English market, even but half a farthing a pound, the poor slaves immediately feel the effects, for they are harder driven, and the whip is kept more constantly going."

As in medieval England, the institutional context was key about how progress took place and who benefited from it. In the US South it was always shaped by coercion. Violence and mistreatment of Black Americans intensified after the cotton gin opened a broad area across the South for cultivation. An already harsh system of slavery was about to become much worse.

Improved productivity most definitely did not mean higher wages or better treatment of Black workers. Account books were developed to record exactly how much had been extracted from slaves and to help plan how to squeeze more output from them. Harsh punishments, forms of torture in many cases, were routine, along with violence in all its forms, including sexual assault and rape.

As we argued in Chapter 3, slavery in the South was enabled in large part because White people in the North were persuaded to go along. This is where the vision of progress in late eighteenth-century America was pivotal. There had long been racist ideas, based on the notion that there was a natural hierarchy with Whites on top. But now new ones were added to the mix to make the plantation system acceptable to the whole country.

The doctrine of "positive good" was made famous by James Henry Hammond, a congressman who became governor of South Carolina, and further developed by John C. Calhoun, senator and vice president of the United States from 1825 to 1832. Their position was a direct response to those arguing that slavery was immoral. On the contrary, according to Hammond in his 1836 speech on the floor of the House of Representatives,

But [slavery] is no evil. On the contrary, I believe it to be the greatest of all the great blessings which a kind Providence has bestowed upon our glorious region. For without it, our fertile soil and our fructifying climate would have been given to us in vain. As it is, the history of the short period during which we have enjoyed it has rendered our Southern country proverbial for its wealth, its genius, its manners.

He continued, making the threat of violence plain if the US moved toward emancipation for slaves:

The moment this House undertakes to legislate upon this subject, it dissolves the Union. Should it be my fortune to have a seat upon this floor, I will abandon it the instant the first decisive step is taken, looking towards legislation on this subject. I will go home to preach, and if I can, to practise disunion, and civil war, if needs be. A revolution must ensue, and this Republic sink in blood.

And then came the claim that slaves were happy:

As a class, I say it boldly, there is not a happier, more contented race upon the face of the earth. I have been born and brought up in the midst of them, and so far as my knowledge and experience extend, I should say they have every reason to be happy. Lightly tasked, well clothed, well fed—far better than the free laborers of any country in the world, our own and those perhaps of the other States of this confederacy alone excepted—their lives and persons protected by the law, all their sufferings alleviated by the kindest and most interested care, and their domestic affections cherished and maintained—at least so far as I have known, with conscientious delicacy.

Hammond's speech became a standard refrain, with elements repeated many times over the decades: slavery was a southern issue, in which others should not interfere; it was essential to the prosperity of White people, particularly in the cotton industry;

and enslaved people were happy. And if the North insisted on pressing, the South would fight to defend the system.

A Technological Harvest of Sorrow

At first blush, nineteenth-century America may appear to have little in common with Bolshevik Russia. Look deeper, and there are uncanny parallels.

The cotton sector in the US flourished thanks to new knowledge, such as improved gins and other innovations, at the expense of Black slaves laboring on large plantations. The Soviet economy grew rapidly starting in the 1920s, with greater use of machinery, including tractors and combine harvesters, applied to grain fields. However, growth came at the expense of millions of small-scale farmers.

In the Soviet case, coercion was justified as a means of achieving what the leadership regarded as an ideal type of society. Lenin articulated this notion in 1920 when he said, "Communism is Soviet power plus the electrification of the whole country."

Communist leaders discerned early on that there was a great deal to learn from large-scale factory operations, including Frederick Taylor's "scientific management" methods and the assembly-line production of Henry Ford's car factories. In the early 1930s, about ten thousand Americans with specific skills, including engineers, teachers, metalworkers, carpenters, and miners, went to the Soviet Union to help install and apply industrial technology.

Although building industry was the primary objective, experience during the New Economic Policy of the 1920s indicated that bringing more people to work in factories needed to be supported by a sufficiently high and stable supply of grain. This grain was necessary not just to feed the growing urban population but also as a key source of export revenues, needed to finance the import of foreign industrial and agricultural machinery.

In the early 1920s, Leon Trotsky argued that forced collectivization of agriculture was the way forward for the Soviet Union. Nikolai Bukharin and Joseph Stalin opposed Trotsky, maintaining that industrialization would be possible while retaining small-scale farmers. Following Lenin's death, Trotsky's star faded; he was first exiled internally and then expelled from the Soviet Union in 1929.

At this point, Stalin made an about-face, pushing Bukharin aside and going all in on collectivization. Small-scale farmers, the kulaks, were becoming prosperous and should be viewed as a major anticommunist force. Stalin was also deeply suspicious of Ukrainians, some of whom had sided with the anticommunist rebels in the civil war.

Stalin believed that collectivization had to be combined with mechanization, and he regarded the United States as a role model. Agriculture in the US Midwest, which had similar soil and climate conditions to parts of the Soviet Union, was in the midst of rapid mechanization, with spectacular gains in productivity. Stalin needed grain exports to buy tractors, harvesters, and other equipment from the West, so the US experience of mechanization was an inspiring model.

By the early 1930s, collectivization and consolidation of smaller landholdings into bigger fields were proceeding full speed, and Soviet agriculture was becoming much more mechanized. In the 1920s, grain required 20.8 worker-days per hectare. This had fallen to 10.6 days in 1937, primarily because of the use of tractors and combine harvesters.

But the process of collectivization was massively disruptive, resulting in famine and the destruction of livestock. The output available for consumption (total production minus what is needed for seed and to feed animals) fell 21 percent between 1928 and 1932. There was some rebound, but total agricultural output increased by only 10 percent between 1928 and 1940—and much of that was a result of irrigation in Soviet-controlled parts of Central Asia that were boosting cotton production.

According to a careful recent estimate, total farm output at the end of the 1930s would have been 29–46 percent higher without collectivization, mostly because livestock production would have been higher. But grain "sales," as enforced transfers to the state were euphemistically called, were 89 percent higher in 1939 compared with 1928. Farmers were squeezed and squeezed hard.

The human toll was staggering. With a starting population of around 150 million, there were between 4 million and 9 million "excess deaths" caused by collectivization and forced food deliveries. The worst year was 1933, but the prior years also show elevated mortality. Living standards may have improved in urban areas, and construction and factory workers were increasingly well fed. Just as in medieval England and the US South, there was no sign of productivity gains raising the real incomes or improving the lives of agricultural workers.

Of course, Stalin's vision was not that of a medieval abbot or southern plantation owner. Rather than religion or the interests of the wealthy elite, technological progress in the Soviet Union was for the ultimate good of the proletariat, and the Communist Party knew best what that ultimate good was.

Indeed, technological progress was now doing the bidding of the Soviet leadership, whose grip on power would have been hard to maintain without some increase in economic output. All the same, whether the elite were feudal lords in medieval Europe, plantation owners in the US, or Communist Party bosses in Russia, technology was socially biased, and its application in the name of progress left devastation on its way.

None of this could have been achieved without the intensification of coercion. Millions of peasants put up with harsh exploitation because the alternative was being shot or sent to even more savage conditions in Siberia. During and after the collectivization of agriculture, a reign of terror spread across the Soviet Union. About a million people were executed or died in prison in 1937–1938 alone. About 17–18 million were sent to gulag labor camps between 1930 and 1956, a number that does not include all forced removals or the irreparable damage to family members.

But again, control was not just about coercion. As soon as Stalin decided to collectivize agriculture, the Communist propaganda machinery sprang into action and started marketing this strategy as progress. The most important constituency was members of the party, who had to be convinced so that the leadership's grip on power would continue and their plans could be implemented.

Stalin used all the means of propaganda available to him and, for both domestic and foreign consumption, presented collectivization as a triumph: "The successes of our collective-farm policy are due, among other things, to the fact that it rests on the *voluntary character* of the collective-farm movement and on *taking into account the diversity of conditions* in the various regions of the U.S.S.R. Collective farms must not be established by force. That would be foolish and reactionary" (italics in original).

The Soviet collectivization episode makes it clear, once more, that the specific way in which technology was applied to agriculture was not just biased but also a choice. There were many ways of organizing agriculture, and the Soviets themselves had experimented, with some success, with the smallholder model during the New Economic Policy under Lenin.

As in earlier episodes discussed in this chapter, the elite chose the path for agricultural technology based on their own vision. Millions of ordinary people paid the price.

Social Bias of Modernization

We live in an age obsessed with technology and the progress it will deliver. As we have seen, some prominent visionaries imagine today to be the best of times, whereas others argue that even more spectacular advances are around the corner, with boundless abundance, extended lifespans, or even colonization of new planets.

Technological changes have always been with us, along with influential people making decisions about what needs to be done and by whom. Over the past twelve thousand years, agricultural technology has advanced repeatedly and sometimes in dramatic ways. There have been times when, as productivity rose, ordinary people also benefited. But there was nothing automatic about these improvements trickling down to benefit the greater number of people. Shared benefits appeared only when landowning and religious elites were not dominant enough to impose their vision and extract all the surplus from new technologies.

In many of the defining episodes of agricultural transitions, the benefits were much more narrowly shared. These were times in which elites initiated a process of rapid transformation, often in the name of progress. Yet rapid change typically coincided not with any obvious notion of the common good but with gains for those spearheading new technologies. These transitions often brought little benefit to the rest.

Exactly how the common good was formulated differed across eras. In medieval times, the goal was a well-ordered society. In late eighteenth-century England, a growing population needed to be fed while keeping the price of food down. In the Soviet Union during the 1920s, the Bolshevik leaders argued about how best to build their version of socialism.

In all these periods, growth of agricultural productivity primarily benefited the elite. The people in charge, whether landowners or government officials, decided what machines to use and how to organize planting, harvesting, and other tasks. Moreover, despite demonstrable productivity gains, most people were consistently left behind. Workers in the fields failed to benefit from agricultural modernization; they continued to toil for longer hours, lived under harsher conditions, and at best did not experience any improvement in their material well-being.

For anyone who believes that the productivity benefits necessarily trickle down through society and improve wages and working conditions, these formative episodes are hard to explain. But once you recognize that technology's advances look after the interests of those who are powerful and whose vision guides its trajectory, everything makes a lot more sense.

Large-scale grain agriculture, mills monopolized by lords and abbots, the cotton gin intensifying slavery, and Soviet collectivization were specific technology choices, in each case clearly in the interests of a dominant elite. Predictably, what followed looks nothing like the productivity bandwagon: as productivity rose, powerful people extracted more effort from agricultural labor by pressing them to work longer hours and hand over more of their production. This is the shared pattern across medieval England, the US South, and Soviet Russia. The situation during enclosures in late eighteenth-century Britain was a little different, but the rural poor again lost out, this time because they were stripped of their customary rights, including their ability to collect firewood, hunt, and graze animals on common land.

We know less about what transpired during the millennia that followed the Neolithic Revolution. But by the time fully settled agriculture emerged, about seven thousand years ago, the pattern appears quite similar to what we have seen in more recent history. Across all of the well-known grain-centered ancient civilizations, most of the population appears to have been worse off than their ancestors had been in foraging. In contrast, the people in charge under settled agriculture were better off.

None of this can be considered as inexorable consequences of progress. Centralized, despotic states did not arise everywhere, and agriculture did not necessitate that an elite specialized in coercion and religious persuasion should extract most of the surplus. New technologies such as mills did not need to be under the tight monopoly of local elites. Nor did the modernization of agriculture require the expropriation of land from already poor peasants. In almost all cases, there were alternative paths available, and some societies made different choices.

Those alternate paths notwithstanding, the long history of agricultural technology exhibits a decided bias in favor of elites, especially when they could combine coercion and religious persuasion. This history suggests that we should always carefully examine ideas about what is or is not progress, particularly when powerful people are keen to sell us on a specific vision.

Naturally, agriculture is very different from manufacturing, and the production of physical goods is distinct from digital technologies or the potential future of artificial intelligence. Perhaps we can be more hopeful today? Perhaps the technologies of our age are inherently more inclusive? Surely the people in charge today are more enlightened than any pharaoh, southern planter, or Bolshevik?

In the next two chapters we will see that experience during industrialization was indeed different, but not because steam engines or the people in charge had a more natural tendency to be inclusive. Rather, industrialization brought large numbers of people together in factories and urban centers, created new aspirations among workers, and began to allow the development of counterbalancing forces of a kind that agricultural society had not experienced.

The first phase of industrialization was arguably even more socially biased and created even more dramatic inequities than agricultural modernization. It was only later that the rise of countervailing powers caused a dramatic course correction that, after many stops and starts, redirected much of the Western world onto a new path for technological changes and institutional developments that bolstered shared prosperity.

Unfortunately, as we will see from Chapter 8 onward, four decades of digital-technology deployment have undermined the sharing mechanisms that developed earlier in the twentieth century. And with the arrival of artificial intelligence, our future begins to look disconcertingly like our agricultural past.

5

A Middling Sort of Revolution

Necessity, which is allowed to be the mother of invention, has so violently agitated the wits of men at this time that it seems not at all improper, by way of distinction, to call it the Projecting Age.

—DANIEL DEFOE, *An Essay upon Projects*, 1697

The triumph of the industrial arts will advance the cause of civilization more rapidly than its warmest advocates could have hoped, and contribute to the permanent prosperity and strength of the country, far more than the most splendid victories of successful war. The influences thus engendered, the arts thus developed, will long continue to shed their beneficent effects over countries more extensive than those which the sceptre of England rules.

—CHARLES BABBAGE, *The Exposition of 1851: Views of the Industry, the Science, and the Government of England*, 1851

On Thursday, June 12, 1851, a group of agricultural laborers from Surrey, in the south of England, donned their best clothing and boarded a train bound for London. Their day out in the capital was not intended for idle sightseeing. Instead, their trip was subsidized by local people of means to provide a glimpse of the future.

In the enormous Crystal Palace, specially constructed in London's Hyde Park, the Great Exhibition presented legendary diamonds, dramatic sculptures, and rare minerals. However, the stars of the show were the new industrial machines. As the agricultural workers wandered the halls, it was as if they had landed on a different planet.

Almost every dimension of industrial production was on display. The entire cotton-production process, now mechanized from spinning yarn to weaving cloth, was prominent. So was a vast array of "moving machinery" powered by steam. There were 976 items under Class 5, "Machines for Direct Use, Including Carriages, Railways and Marine Mechanism," and 631 items under Class 6, "Manufacturing Machines and Tools." Perhaps the most impressive visual demonstration of the new industrial world was provided by a machine that could fold an unprecedented 240 envelopes per hour.

The machines were from Europe, the United States, and most of all the United Kingdom; this was a display of patriotic achievement, after all. There were 13,000 exhibitors, including 2,007 exhibitors from London, 192 from Manchester, 156 from Sheffield, 134 from Leeds, 57 from Bradford, and 46 from the Staffordshire Potteries.

Economic historian T. S. Ashton famously summed up the century leading up to the exhibition: "'About 1760 a wave of gadgets swept over England.' So, not inaptly, a schoolboy began his answer to a question on the industrial revolution. It was not only gadgets, however, but innovations of various kinds—in agriculture, transport, manufacture, trade, and finance—that surged up with a suddenness for which it is difficult to find a parallel at any other time or place." The steam engine allowed a leap forward in human control over nature, and in the lifetime of many visitors to the Great Exhibition, the technologies used in mining, cotton, and transportation had been transformed.

For almost all of human history, the food-production capacity of economies increased roughly in line with population. In good years, most people had enough to eat, with some margin for safety. In bad years, because of famine, war, or other disruptions, many would starve. The average growth rate of output per capita over long periods of time was barely above zero. Despite myriad medieval innovations that we discussed in Chapter 4, the quality of life of a European peasant circa 1700 was not much different from that of an Egyptian peasant two thousand or seven thousand years previously. According to the best available estimates, GDP per capita (in real, price-adjusted terms) was almost the same in 1000 CE as it had been a thousand years earlier.

The modern demographic history of our species can be divided into three phases. The first is a gradual population increase from about 100 million in 400 BCE to 610 million in 1700 CE. For most societies over most of that time, the well-to-do elites constituted no more than 10 percent of the population; everyone else lived on not much more than the bare minimum necessary for survival.

The second phase witnessed an acceleration, with world population increasing to 900 million in 1800. Industry began to develop in Britain, but growth rates were still slow, and skeptics could find many reasons why this would prove hard to sustain. Other countries were even slower in adopting new technologies. The average annual growth rate (per capita) from 1000 to 1820 was just 0.14 percent for Western Europe as a whole and 0.05 percent for the entire world.

Then came the third, completely unprecedented phase, already evident by 1820, beginning with output per person more than doubling in the following century across Western Europe. Growth rates for output per capita among the larger European economies ranged from 0.81 percent in Spain to 1.13 percent in France per annum from 1820 to 1913.

Pre-industrial economic growth was a little more rapid in England, enabling the country to pull ahead of the previous technology leaders, such as Italy and France, though still trailing that epoch's powerhouse, the Netherlands. English national production per capita doubled from 1500 to 1700. Growth in Britain, as it came to be known after the unification of England and Scotland in 1707, picked up pace thereafter, raising national output by another 50 percent over the following 120 years as Britain became the most productive country in the world. Over the subsequent 100 years, output per person accelerated and reached an average annual growth rate of approximately 1 percent, which meant British output per person more than doubled between 1820 and 1913.

Behind these statistics lies a simple fact: useful knowledge expanded dramatically during the nineteenth century, including for all aspects of engineering. Railway networks enabled the transport of larger quantities of goods at cheaper prices, and allowed people to travel as never before. Ships became bigger, and freight costs for long-distance sea travel fell. Elevators made it possible to live and work in taller buildings. By the end of the century, electricity had started transforming not just lighting and the organization of factories but all aspects of urban power systems. It had also created the basis for telegraph, telephones, and radio, and later all sorts of household appliances.

Big breakthroughs in medicine and public health significantly lowered the burden of disease and consequently reduced the morbidity and mortality associated with living in crowded cities. Epidemics were increasingly brought under control. Lower infant mortality meant that more children survived into adulthood, and, together with lower maternal mortality, this significantly raised life expectancy. The population of industrializing countries increased sharply.

It was not only practical innovations in engineering and methods of production. There was also a transformation in the relationship between science and industry. What had previously seemed smart but rather theoretical now became of fundamental importance for industry. By 1900, the leading economies of the world had substantial industrial sectors. The largest firms had research and development departments, aiming to turn scientific knowledge into the next wave of products. Progress became synonymous with invention, and both seemed unstoppable.

What drove this broad-based surge in the invention of useful things? We will see in this chapter that a large part of the answer is a new vision.

The machinery on display in the Crystal Palace was not produced by a narrow elite or a class of top-level scientists but was the work of an emergent entrepreneurial class, originating primarily from the Midlands and the north of England. Almost all these entrepreneur inventors were "new" people, in the sense that they were not born into nobility or riches. Rather, they strove from modest beginnings to acquire wealth through success in business and technological ingenuity.

In this chapter we argue that it was first and foremost the rise and emboldening of this new class of entrepreneurs and inventors—the essence of Daniel Defoe's Projecting Age—that was responsible for the British industrial revolution. Chapter 6 then explores how this new vision for progress failed to benefit everyone and how this situation started to change later in the nineteenth century.

Coals from Newcastle

Perhaps nobody epitomizes this new Projecting Age better than George Stephenson. Born in 1781 to illiterate, poor parents in Northumberland, Stephenson did not go to school and started reading and writing only after he was eighteen. However, by the early decades of the nineteenth century, Stephenson was recognized not just as a leading engineer but also as a visionary innovator shaping the direction of industrial technology.

In March 1825, Stephenson was called to testify before a parliamentary committee. At issue was a proposed railway between Liverpool and Manchester, connecting a major port with the heart of the burgeoning cotton industry. Because any potential route would involve the compulsory purchase of land, an act of Parliament was needed. Backers of the railroad company had enlisted Stephenson to survey the route.

Opposition to the new railway line was strong. It came from local landowners who did not want to cede their property rights, and even more powerfully from the owners of the lucrative canals that ran along the same route and would face stiff competition from railways. The duke of Bridgewater, one such owner, was reported to have earned well over 10 percent per year on his canal (an impressive rate of return at the time).

At the parliamentary hearing, Stephenson's suggested route was ripped to pieces by Edward Alderson, a distinguished lawyer hired by the canal interests. Stephenson's work had been sloppy: one of his proposed bridges had a height three feet below the maximum flood level of the river it would cross; some of his cost estimates were obviously rough guesses; and he was vague on important details, such as how exactly the baseline for the survey had been determined. Alderson summed up with the elegant language of a top Cambridge graduate and future prominent judge, calling the railway plan "the most absurd scheme that ever entered into the head of man to conceive." He continued: "I say he [Stephenson] never had a plan—I believe he never had one—I do not believe he is capable of making one. . . . He is either ignorant or something else which I will not mention."

Stephenson struggled to reply. He lacked the kind of privileged education that prepares one to respond to such rebukes with effective rejoinders, and still spoke with a strong Northumbrian accent that people from the south of England found hard to understand. Overstretched and understaffed, Stephenson had hired a weak team to do the survey, had failed to supervise them properly, and was caught unaware by Alderson's aggressive questioning.

However, whatever else Stephenson may have been, he was certainly not ignorant. By the early 1800s, Stephenson was known throughout the Tyneside coalfields, in the northeast of England, as a reliable mining engineer who earned a decent living helping pit operators sort out technical problems.

In 1811 he had his breakthrough. A rudimentary steam engine was failing to pump water effectively out of a new mine, High Pit, rendering it useless and even dangerous. All the respectable local specialists had been consulted, but to no avail. Stephenson wandered up to the engine house one evening and took a close look at the problem. He confidently predicted that he could greatly improve the water-pumping capability of the engine, providing he was allowed to hire his own workers. Two days later, the pit was pumped dry. The rest is history—railway history.

In 1812 Stephenson was placed in charge of all machinery for collieries owned by a group of rich landowners known as the Grand Allies. In 1813 he became an independent consulting engineer, still helping the Grand Allies but increasingly building and deploying his own steam engines. The most powerful of these engines could draw 1,000 gallons of water per minute from 50 fathoms (300 feet). He also built underground haulage systems that pulled coal across a network of rails using stationary engines.

The idea of moving coal from the pit to the market via rail was well established. Since the late seventeenth century there had been "wagon ways," along which horses pulled wagons on rails usually of wood, but sometimes iron. As the demand for coal in urban areas grew, a group of merchants based in Darlington resolved to build an improved set of rails to connect pits with navigable waterways. Their concept was to allow all types of appropriate vehicles run by approved operators paying fees, much like a toll road.

Stephenson's vision was different, and ultimately much bigger. Despite his modest background, haphazard education, and difficulties expressing himself when faced with hostile Cambridge lawyers, Stephenson's ambition was boundless. He believed in technology as a practical way of solving problems and had the self-confidence to ignore the limited thinking of the prevailing social hierarchy.

On the same day the Stockton and Darlington Railway Act became law, April 19, 1821, George Stephenson called upon Edward Pearse, a prominent Quaker merchant in Darlington and leading supporter of the proposed new line. At that moment, there were three main approaches in the mix for this railway and other similar projects: continue to use horses; install stationary engines, which would pull wagons up the hills and let gravity do the rest; and build locomotives that would run on rails.

The traditionalists preferred to stick with horses. Though cumbersome, this approach worked. Some more-forward-looking engineers with impressive credentials recommended stationary engines, which were already used to pull carts underground. An improvement but a modest one.

Stephenson's view, that steam engines with metal wheels would easily generate enough traction on iron rails, was quite different from the established wisdom, which maintained that smooth rails would not provide a powerful engine with enough friction to accelerate and decelerate safely. It would be more like skating on ice. Stephenson's understanding was based on experience in mines. He proceeded to persuade Pearse that steam engines on iron rails should become a significant part of the solution.

Not that Stephenson had a locomotive in hand or had solved the practical problems standing in the way of producing working engines for railroads. Existing low-pressure or "atmospheric" steam engines, of the kind that Thomas Newcomen had first built, James Watt later significantly improved, and George Stephenson himself had fixed at High Pit, were too bulky and did not generate enough power. More powerful high-pressure engines existed but had never been demonstrated to work consistently at scale, let alone pull heavy coal wagons up and down hills every day.

Building a high-pressure steam engine that was light enough to move itself was a spectacular challenge; early models leaked, were underpowered, or even blew up with tragic consequences. Wrought iron was too brittle for the rails. Engines and wagons needed some form of suspension system.

Still, Stephenson and his colleagues gradually managed to improve on existing engine designs and to demonstrate that a locomotive could run safely at what was then an extraordinary speed: six miles per hour over a thirty-mile route. The official opening of the line and the running of Stephenson's train was treated as a great event, drawing national attention, with a stream of international visitors soon to follow.

However, the Stockton and Darlington rail line had some serious design flaws that soon became evident, including building only one line with "passing loops" at various points. Rules about who should give way to whom were frequently violated. Drunk operators of horse-drawn coal wagons further complicated matters. Derailments and fisticuffs were frequent. Allowing multiple parties to operate on the same rails was not a workable solution. But Stephenson learned the painful lessons well and determined to operate future rail services differently.

Stephenson's ambition and technical know-how were not his only assets. His enthusiasm for steam locomotives was infectious. It was this enthusiasm that had brought Edward Pearse on board, as early as July 1821, making him conclude that "if the railway be established and succeeds, as it is to convey not only goods but passengers, we shall have the whole of Yorkshire and next the whole of the United Kingdom following with railways."

Over the next five years, Stephenson continued to improve his engines, the rails they ran on, and the operation of an integrated system. He always preferred to hire his own men, most of whom were coal-field engineers with minimal or no formal education. They were a group of tinkerers, working their way carefully through dangerous terrain, literally and metaphorically.

Boilers exploded, heavy equipment was dropped, and engine brakes failed. Calamity was never far from early railways. Stephenson's brother and brother-in-law both died in industrial accidents during these early years.

Despite these setbacks, Stephenson's reputation as a problem solver grew. And the devastating cross-examination by Alderson was not enough to prevent the Liverpool and Manchester line from receiving parliamentary approval in 1826. After some further twists and turns, Stephenson was placed in charge of the entire project and was given the authority to design and build the first modern railway line.

Operations began in September 1830. All trains running on the dual track were owned and operated by the railway company, which also demanded serious commitment from its workers. In return, in a regional labor market where the prevailing wage was one pound per week, the railway paid double that.

Early engine drivers and their firemen, who stood next to them on locomotive engines, needed to be highly skilled. The first trains had no brakes; the only way to stop them was by adjusting a series of valves in the right order to put the wheels into reverse. In the early days, there was only one driver in the country who could do so in the dark (others required a fireman to hold a light in just the right way).

The people who sold railway tickets needed to be incorruptible because they handled considerable amounts of cash. Workers who managed any aspect of safety, human or machinery, needed to show up on time and follow the rules. It helped to provide railway cottages for employees, as well as smart uniforms to wear. But paying premium wages was also a major part of the new industrial math—and the most important way that higher productivity was shared with workers.

Stephenson and his success epitomize what happened with railways and more broadly across other sectors. Practical men, born to scant resources, were able to propose, fund, and implement useful innovations. Each of those innovations consisted of small adjustments that, taken individually, increased productivity by boosting the efficiency of machines in some fashion.

One outcome was the introduction of a new transportation system across which productivity was dramatically increased and entirely new possibilities emerged. Railways reduced the cost of coal in urban areas, as intended. But the true impact was much bigger. They significantly expanded passenger travel over both short and long distances. They stimulated further improvements in metalworking, paving the way to the next stage of British industrialization in the second half of the nineteenth century. They were also foundational to the later advances in industrial machinery.

Railways revolutionized the transport of material, goods, and services as well. Milk and other food products could be brought daily to big cities, enabling these products to be drawn from a wider area, for they no longer needed be produced by small-scale farms located within walking or cartable distance. How people moved around the country and thought about distance also changed profoundly, paving the way to such things as suburbs and days out at the seaside, which were unimaginable for most people before railways.

George Stephenson also gives us a clue about the deeper causes of the early British lead in the adoption of railways and everything else in the early Industrial Revolution, including big factories, rapidly expanding cities, and new ways of organizing trade and finance.

People like Stephenson were a new breed. The Middle Ages, as we have seen, was a time of rigid hierarchy, where everyone had their place. The scope for upward social mobility was limited. But by the mid-1700s, the "middling sort" of people—from modest origins but viewing themselves firmly in the middle class—could dream big and rise fast in Britain. Three things were remarkable about this. The first was that they aspired to rise in a way that may reasonably be considered unprecedented for people of modest social standing in preindustrial Europe. The second was that those ambitions so often centered around technology, how it could solve practical problems and make them rich and famous. They also acquired a range of mechanical skills to put these dreams into practice. The third, and the most remarkable one, was that British society let them realize these dreams.

What enabled them to have such ambitions and the temerity to try to put them into practice was a deep set of social and institutional changes that British (and earlier English) society had undergone over the preceding centuries. The same institutional changes ensured that the rising middle class was hard to resist.

Before discussing how this mind-set created the Projecting Age, it is useful to think about the centrality of technology. Was the focus on technology because of the earlier Scientific Revolution, which altered how people, especially intellectuals, thought about nature? We will see that the answer is no, for the most part.

Science at the Starting Gate

In 1816 Sir Humphry Davy received a great honor for his scientific work, the Royal Society's Rumford medal. One of the country's leading chemists, based at the Royal Institution in London, Davy had investigated the cause of mining disasters and, based on careful lab experiments, determined that a new kind of "safety lamp" would reduce the chance of fatal explosions. There was national acclaim, which was personally pleasing. Davy also welcomed the confirmation that applying science could improve people's lives.

He was therefore mortified to find that someone else with no scientific education claimed to have invented an equally effective safety lamp at the same time as, or perhaps even before, Davy's innovation. That other innovator was none other than George Stephenson.

Davy, though of humble origin, was very much a product of the Scientific Revolution, standing on the shoulders of Robert Boyle (1627–1691), Robert Hooke (1635–1703), and Isaac Newton (1643–1727), all of whom had been leading lights in the Royal Society of London for Improving Natural Knowledge, founded in November 1660. Davy was a pioneer in the study of the properties of gases, including nitrous oxide. He had also demonstrated how batteries could be used to generate an electric arc, which was a crucial step toward understanding the properties of electricity and artificial lighting.

By 1816, Davy was not lacking in self-confidence. He jumped to the conclusion that Stephenson's work must have been the result of plagiarism and wrote to Stephenson's prominent supporters, the Grand Allies, demanding that they acknowledge that their coal-mining protégé could not possibly be at the frontier of innovation: "The Public Scientific Bodies to which I belong must take Cognizance of this indirect attack on my Scientific fame, my honour and varacity [sic]."

The Grand Allies were not impressed by Davy's claims. Precisely when and how Stephenson had built and tried out his lamp was well documented by people they trusted. William Losh, one of the allies, dismissed the idea that London-based organizations could somehow determine what was or was not original: "Satisfied as I am with my conduct on this subject I must say that I am wholly indifferent as to the cognizance which may be taken of it by the 'Public Scientific Bodies' to which you belong."

Another of Stephenson's supporters, the earl of Strathmore, was even more scathing in his response to Davy, articulating how he saw, and why he would help, people like Stephenson: "I can never allow any meritorious Individual to be cried down because he happens to be placed in an obscure situation—on the contrary, that very circumstance will operate in me as an additional stimulus to endeavour to protect him against all overbearing efforts."

The safety-lamp controversy illustrates not just how far Britain had moved from its medieval society of orders by this time but also the contrast between two approaches to innovation. The first, represented by Davy, was based on what we now regard as modern scientific methods and was advancing rapidly. By the early decades of the nineteenth century, it had become largely "evidence based"—for example, requiring hypotheses to be tested in labs or other controlled settings, and be replicable. The second, epitomized by Stephenson, did not care about publications or impressing scientists but instead focused on solving practical problems. Even if this approach was indirectly influenced by the scientific knowledge of the era, it was all about practical knowledge, often acquired while adjusting machines to see what improved performance.

A vivid demonstration of this point is provided by the Rainhill Trials, organized by the Liverpool and Manchester Railway in 1829 to determine what kind of locomotive it should use. As chief engineer of the Liverpool and Manchester line, Stephenson was in charge of designing and building the main routes, figuring out where bridges and tunnels should be and what kind of gradient and corners to allow, and solving the difficult problem of how to cross a treacherous marshy area. The directors of the Liverpool and Manchester line had accepted steam-powered locomotives with metal wheels, running on iron rails, with a line of track in each direction. No horse-drawn wagons with drunk drivers would be allowed.

The directors decided on an open competition to determine who would supply the locomotives. The competition was going to be carried out in public, with clearly specified criteria. By this point, the principles of steam engines, advanced by James Watt in 1776, were out in the public domain for all to build upon. Watt had worked to prevent the development of high-pressure engines, assiduously defending his patents on earlier engine models in court and arguably slowing down the rate of innovation by others. But the patents expired in 1800, removing the remaining barriers to the application of this knowledge by others.

The Rainhill Trials were a combination instant Nobel Prize and reality show. The prize money itself was significant (500 pounds), but it was obvious that the market to be established was immense, not just in Britain but across Europe and America, and surely soon around the world. Every potential inventor and distinguished scientist must have paused to take note.

This was, arguably, the most compelling engineering moment in human history to date. Henry Booth, a Liverpool corn merchant and key backer of the rail line, was impressed by the range of entrants: "Communications were received from all classes of persons, each recommending an improved power or an improved carriage; from professors of philosophy, down to the humblest mechanic, all were zealous in their proffers of assistance: England, America, and Continental Europe were alike tributary."

Like the judges in any good bake-off, the directors had clear views on what they wanted to see: a locomotive having four or six wheels, with manageable boiler pressure, running on a gauge of 56.5 inches, and not costing more than 550 pounds per engine. This machine would need to pull three tons for each one ton of locomotive weight over 70 miles at an average speed of at least 10 miles per hour. The trials were to be conducted along a flat piece of track known as the Rainhill Level, with difficult gradients at both ends.

Preliminary assessments ruled out most of the entrants for simply failing to meet the specified criteria. There were five finalists.

One of these, *Cycloped*, was likely a joke that also made the point that technology had moved beyond a point of no return. In this machine, a horse walked on a treadmill, which rotated the wheels. No steam was involved, and the result was quick disqualification. The final showdown was therefore between four steam-powered locomotives, one of which (*Perseverance*) could not get above 6 miles per hour. Another (*Novelty*) suffered debilitating boiler leaks, and a third (*Sans Pareil*) cracked a cylinder.

The winner was *Rocket*, designed and built by George Stephenson and his son, Robert.

The contribution of the Royal Society, its members, or the scientific establishment at large to these competitions was essentially zero. No members of the scientific establishment played any role in the design of the engines, in the work on how the metal parts were cast and put together, and in the way in which steam was generated or smoke was handled.

The attitude of the practical innovators of this era is exemplified by Stephenson's plans for his son's education. He put great effort to ensure that Robert had the best possible opportunities to acquire all fields of knowledge necessary to become an excellent engineer. This meant attending good schools, but only up to a point. Robert left school at sixteen. He was immediately thrown into practical work with his father and others engaged in engineering to solve real-world problems in mining, surveying, and engine building.

Even more importantly, scientific advances, by themselves, cannot explain why the Industrial Revolution was British. The Scientific Revolution was a thoroughly pan-European affair. Boyle, Hooke, and Newton were English, but many of the most innovative thinkers of this revolution, such as Johannes Kepler, Nicolaus Copernicus, Galileo Galilei, Tycho Brahe, and René Descartes, never set foot in Britain. They communicated among themselves and with their English peers in Latin, underscoring the Europe-wide nature of this enterprise.

Equally, Europe was not even unique in experiencing an extended period of scientific breakthroughs. China was far ahead of Europe in science in 1500, and arguably had a lead as late as 1700. The Song Dynasty (960–1279) was a particularly creative time. The major technological breakthroughs that first took place in China include gunpowder, the water clock, the compass, spinning, smelting, and advances in astronomy. In fact, almost all the big European innovations of the Middle Ages and early Industrial Revolution can be plausibly traced back, directly or indirectly, to China. Chinese technologies that were adopted relatively early by Europeans include the wheelbarrow, movable-type printing, and clocks. Also important were ideas that later propelled the Industrial Revolution, including Chinese machines for mechanized spinning, iron smelting, and canal locks. The Chinese also made extensive use of paper money, which was for a time used for both local and long-distance trade.

True, Chinese authorities did not encourage scientific inquiries after the Song Dynasty, and the shared vision of rigorous, empirical science that took root in Europe starting in the seventeenth century had no equivalent in China. Nevertheless, the absence of Chinese industrialization until the twentieth century shows that scientific advances by themselves were not enough to kick-start the Industrial Revolution.

This assessment is not meant to downplay the role of science in industrialization. The Scientific Revolution provided three critical contributions. First, science prepared the ground for the mechanical skills of the ambitious entrepreneurs and tinkerers of the age. Some of the most important scientific breakthroughs—for example, those involving iron and steel—became part of the practical knowledge of the era and thus contributed to the base of useful facts upon which entrepreneurs built in designing new machines and production techniques.

Second, as we elaborate more in Chapter 6, starting around the 1850s scientific methods and knowledge became much more important for industrial innovation because of advances in electromagnetism and electricity, and then later with a growing focus on new materials and chemical processes. For example, the development of the chemical industry was tightly linked to scientific discovery, with the invention of the spectroscope in 1859 a leading example. More broadly, the telegraph (1830s), the Bessemer process for making steel (1856), the telephone (1875), and electric light (commercialized in 1880) arose much more directly from scientific investigations.

Third, the reason why so many ambitious young men such as George Stephenson were drawn to technology was because they grew up at a time that had been shaped by the Age of Discovery. This era, starting in the mid-fifteenth century, saw major advances in maritime technologies and Europeans' expansion into parts of the world that they had had little contact with previously. The Scientific Revolution was very much bundled in people's minds with this process of discovering and potentially shaping the physical and social environment. Europeans could now sail ships over previously hostile waters, subjugate other populations, and expand their dominion over nature.

If not science directly, what were the primary factors that helped Britain launch the Industrial Revolution?

Why Britain?

Detailed economic histories have established the basic pattern of formative events for industry. There was a sustained rise in the cotton textile sector from the early 1700s, with northern entrepreneurs playing a key role. New machinery greatly increased the productivity of first spinning and then weaving.

At the same time, artisans active in other sectors, such as iron making and pottery, figured out how to introduce other machines to improve quality while also boosting production per worker. A notable step forward occurred with the switch from waterpower to steam as the energy source to pump water out of mines. From the start of the nineteenth century, steam became the main energy source for factories. From the 1820s, putting steam engines on wheels enabled much faster and cheaper transportation over long distances. New ways to raise finance emerged during the nineteenth century, making it easier to trade across long distances, build large factories, and fund a global railway construction boom.

All these elements are hard to dispute, and the basic time line for the rise of an industrial sector is not in question. But what explains why this occurred in Britain before anywhere else? And why starting in the eighteenth century?

Ever since the term *industrial revolution* was coined in the late nineteenth century, a wide variety of thinkers have put forward explanations for "why Britain was first." Theories can be usefully grouped into five main buckets: geography, culture (including religion and innate entrepreneurship), natural resources, economic

factors, and government policies. Some of these are quite ingenious, but all the leading contenders leave important unanswered
questions.

One view is that something about Britain's geography was
particularly conducive to economic development. But this seems
strange as a general proposition, given that England and other
parts of the British Isles were an economic backwater at least
until the sixteenth century. For thousands of years, most European prosperity remained concentrated around the Mediterranean basin. Even when the Age of Discovery opened trade routes
through the Atlantic, Britain remained significantly behind Spain,
Portugal, and the Netherlands in benefiting from new colonial
opportunities.

As we discussed in Chapter 4, from the Norman Conquest
in 1066 until the early 1500s, England was a feudal system. The
king was strong, and the barons were periodically troublesome,
particularly when control of the throne was in question. The peasants were often pressed down hard. People who lived in a few
towns acquired some additional rights over the years but nothing
close to what was achieved in the leading cities of Italy during
the Renaissance (from the 1330s to about 1600). English backwardness was reflected in the arts, which compared rather poorly
to other parts of Western Europe as well as to China. England
produced little of lasting value during the entire medieval period.

Did Britain's status as an island confer some advantages? Perhaps, in terms of reducing the number of invasions over the years.
But foreign invasion or instability was not a major issue for the
technologically most advanced part of the world, China, from
the 1650s to the middle of the nineteenth century, until the Tai-
ping Rebellion and the opium wars. Moreover, other European
nations, including Spain in the Reconquista period (700–1492) or
Italy during the Renaissance, had no trouble combining participation in military conflict with generating prosperity. France and
Spain faced no major invasion threat during the 1600s and 1700s,
and the Netherlands was forged by the need to keep the Spanish
and French at bay.

The British eventually built a formidable navy, but it was not overwhelmingly stronger than its rivals until well into the industrial age. British naval forces were substantially smaller than the Spanish fleet in the 1500s, defeated repeatedly by the Dutch in the 1600s, and outmaneuvered with great consequence by the French during the American Revolution in the 1770s. In 1588 the English survived a powerful Spanish fleet, the Armada, sent by Spanish monarch Philip II to invade, not because of any superiority of their naval technology or strategy but mostly by sheer luck: bad weather and a series of mistakes doomed the Spanish effort.

Britain has rivers suitable for waterwheels, and moving goods over inland waterways was initially much easier and cheaper than using roads. Some of Britain's rivers could readily be connected to each other and to the sea by canal, and this was useful at the end of the eighteenth century (hence the opposition from the Duke of Bridgewater and other canal interests to the development of early railways).

Nevertheless, other countries, including Germany, Austria, and Hungary, contain impressive amounts of navigable water, and France had a notable push to build canals that long predated the British investments in such infrastructure. Besides, the canal-based transportation phase was relatively short-lived in British industrialization. Most of the Industrial Revolution moved by rail, and British railway pioneers were only too eager to sell engines, wagons, and all relevant accessories to anyone interested in buying in Europe or elsewhere. Transferring technology proved easy, whether it was by leasing, copying, or improving designs. By the 1830s, for example, Matthias Baldwin was building locomotives in Pennsylvania, and by the 1840s, his engines were arguably better suited to long-distance haulage under American conditions than any imported designs.

It has become fashionable in some quarters to argue for another aspect of geography. Industrial development is claimed to be easier at some latitudes, in part because these are intrinsically healthier. But Britain had no discernible advantage in terms of public health in the preindustrial phase. Infant mortality was high and life expectancy at birth quite low. There was also an inability to deal with serious waves of disease, a point made painfully clear by the experience of the Black Death, which wiped out between a third and a half of England's population in the 1300s.

Could there have been some other advantage to be in a "lucky latitude"? As we discussed in Chapter 4, the Near East and the Eastern Mediterranean were early to adopt what has been commonly regarded as "civilization," which means that people living in those places have been writing things down and living under the authority of a state for longer than anyone else. But those social and political systems hardly proved themselves conducive to sustained economic growth.

Even when industrial technologies became widely available during the 1800s, the original Fertile Crescent area did not rush to adopt new machinery or build big factories. Nor did other places that had early civilizations, such as Greece or southern Italy. If there was any special advantage that ancient history conferred for eighteenth-century industrialization, it would be strange that Britain would be its recipient. It is a long way from the Fertile Crescent to Birmingham.

Additionally, most of these geographic features do not set Britain apart from China. China has powerful rivers in its heartland and a long coast. A large part of the country is in the lucky latitudes. Yet it did not turn any of its amazing scientific advances into industrial technology.

If not geography, could it have been culture that set England and then Britain apart? Was there some deep cultural advantage across the broad swath of British people, in terms of their attitudes about risk, enterprise, community, or something else? Such an explanation is again hard to square with the fact that before 1500 or 1600, English society does not seem to have much cultural advantage when compared with neighboring parts of Western Europe.

It is true that in the late sixteenth century, most of the country shifted from Catholicism to Protestantism. In the early 1600s, Galileo's astronomical work was hindered by Catholic dogma and by an Italian church hierarchy that was determined to preserve its monopoly on interpreting scripture. Working at the far end of the same century, Isaac Newton and his English contemporaries still had to step carefully when it came to religion, even if they did not face the same personal dangers or blockages imposed by the remnants of medieval theocracies.

However, there were plenty of other European countries that turned Protestant without adopting industrial technologies early, including Scandinavia, Germany, and what became the Czech Republic. France, a predominantly Catholic country, was at least on par with Britain in terms of general scientific knowledge in the eighteenth century. France was also among the fastest adopters of industrial technologies in the early nineteenth century. Catholic Bavaria became an innovation and industrial powerhouse in the 1800s, a position it still occupies today. One place in northwestern Europe that adopted early textile technology ahead of Britain was the predominantly Catholic Bruges, now in Belgium. Bruges had the most skilled European spinners and weavers during the thirteenth century.

It is also unlikely that religious minorities, such as Quakers or other nonconformist Protestant sects in the north of England, played a defining role. Although such religious beliefs influenced the outlook and ambitions of some people, most other countries that experienced the Reformation had a similar mix of groups but did not industrialize until later.

Perhaps it was the luck of having a few extraordinary entrepreneurs who made early breakthroughs? Individuals were important, but this transformation was much more than just about a handful of people. In the textile industry, for example, at least three hundred men made significant contributions to the development of modern manufacturing techniques during the 1700s. More broadly, the Industrial Revolution involved investments made by thousands of people, and more likely tens of thousands if we include all the relevant decision makers and investors during the eighteenth and early nineteenth centuries.

Natural resources were not the defining factor in British industrialization either. One of the most influential alternative views puts greater weight on the availability of coal. Britain did benefit from decent-quality iron ore available close by coal deposits in the north and midlands of England. But this does not explain the critical early phase of the British industrial revolution, led by water-powered textile factories. One study estimated how developed the British economy would have been in 1800 if James Watt's steam engine had never been invented. The conclusion: the level of development achieved by January 1, 1801, would have been reached by February 1, 1801—a delay of only one month!

Coal and iron became much more crucial in the second phase of the Industrial Revolution, after about 1830. But the most essential raw material for the first part of the industrial phase was cotton, which does not grow in Britain or in most parts of Europe.

Another set of arguments emphasizes various economic factors that might have advantaged Britain. Most importantly, adoption of technologies that save on labor becomes much more attractive when wages are high because in this case greater cost reductions can be secured by the use of new technologies. By the mid-1700s, wages in some parts of Britain, in particular London, were higher than almost anywhere else in the world. But in this, Britain was not unique either. Wages were high in the Netherlands and commercially oriented parts of France as well.

In any case, labor costs were most likely a contributing factor rather than the major driver of British industrialization. Productivity increases in textiles, when they finally got underway, were truly spectacular—tenfold and then hundredfold increases in output per person. Relatively modest differences in wages between Britain and the Netherlands or France are unlikely to have been the critical determinant of when and whether these technologies would be adopted.

Moreover, the channel from wages to technology adoption applies when labor costs are high relative to productivity. Instead, if workers are more productive, then replacing them is not as attractive. Part of the reason why wages were high in eighteenth-century Britain was its highly skilled, well-trained artisans.

Could these artisanal or practical engineering skills of the workforce have been the trigger for the British industrial revolution? The mechanical knowledge of innovators like George Stephenson was important, but the general skills of the workforce do not appear to have been a critical factor. Workers with specialized skills and correspondingly high productivity in their trade were not widespread throughout the British economy. Literacy gives one indication of the country's general skill levels. Only 6 percent of English adults could sign their names in 1500, rising to 53 percent in 1800. The Dutch had higher literacy rates in both years, whereas Belgium was ahead in 1500 and just behind in 1800. France and Germany started at almost the same level as England; by 1800, they had fallen behind, at 37 percent and 35 percent, respectively.

Moreover, many of the iconic technologies of the age, rather than using artisanal skills honed over the centuries, were targeted at replacing them with machinery and the cheaper labor of unskilled men, women, and children. Most famously, skilled weavers were thrown out of their jobs by mechanized frames, triggering what later came to be known as the Luddite riots (on which we will have more to say in Chapter 6).

Agricultural productivity is also unlikely to have generated a decisive edge for Britain. Agricultural yields had risen during the preceding centuries, and this laid the scene for spectacular urban growth. But here too Britain was not exceptional. Agricultural productivity rose in many parts of Western Europe, including France, Germany, and the Netherlands, which witnessed rapid urban growth as well. Moreover, as we saw in Chapter 4, the extent of this growth was limited everywhere in medieval Europe and was unlikely to have been a trigger for industrialization. The fact that these gains were not broadly shared also meant that they did not generate widespread demand for textiles or luxury products in Britain.

Relatively high levels of artisanal skills, wages, and agricultural productivity do not distinguish Britain from China either. The historian Mark Elvin claimed that from the fourteenth century onward, China was in a "high-level equilibrium trap," precisely because it had high wages and productivity but still showed no inclination to industrialize.

The British population, and the demand for food and clothing, grew rapidly in the seventeenth and early eighteenth centuries. The population of England rose from 4.1 million in 1600 to 5.5 million in 1700. But the bigger growth in population came during industrialization. For example, from 1700 to 1841, when the first comprehensive census was held, population increased roughly threefold. This growth was in part a consequence of rising incomes and better nutrition. It was also enabled by the revolution in transport, which carried enough food to the cities.

Early financial innovation is not where we should look for the origin story of the Industrial Revolution either. Many more-consequential financial innovations had taken place earlier in Italy during the Renaissance and in the Netherlands, and had fueled the growth of Mediterranean and then Atlantic trade and voyages; the British Isles were at the time a financial backwater. By the early 1700s, London-based financiers were willing to fund long-distance trade, but they were hesitant to dip their toes into industrial waters, at least during the early years. Profits made in trade tended to be reinvested in trade. The establishment of the Bank of England was good for public finance and for credit used in overseas trade, although it was quite disconnected from industrial development. For the most part, those northern entrepreneurs financed their ventures with retained earnings, alongside loans from friends, family, and people within their own business community.

Similarly, the legal environment regulating finance and business contracts was cumbersome at least until the railway age. For example, the modern version of limited liability was not fully established in law until the 1850s. It is very hard to argue that Britain had some practical legal advantage that was not available to other European countries.

Overall, there is no indication that Britain had any inherent advantage in the availability of finance for new ventures that used machines. Compared to well-established continental practice, the commercial banking system remained rudimentary until at least the early years of the nineteenth century.

Could it be government policy that put Britain ahead? Following the Glorious Revolution of 1688, Britain had a strong Parliament, and the property rights of landowners and merchants were well protected. Yet the same was true in other countries, such as France, where a great deal of feudal privilege still protected traditional landowners, and merchants were also secure against expropriation.

The British government was keen to build its overseas empire and, over time, strengthened the navy with the rationale of supporting international trade. But this colonial empire was small in economic terms for a long time. Britain gained control over most of India only in the second half of the eighteenth century, shortly before it lost the North American colonies.

Estimates of profits from the slave trade and Caribbean plantation economies indicate that this form of human trafficking and exploitation did contribute resources to industrialization, but this direct effect is not large enough to explain what happened. In addition, while Britain was a major participant in the Atlantic slave trade, Portugal, Spain, France, the Netherlands, and Denmark were just as active, and some of these countries generated much larger profits over the centuries than did Britain.

There was no conscious British strategy or government policy supporting industrialization. In any case, such ideas were a long way from being plausible when no one understood the nature of what could be invented and how profound its effects could be. If any European country led the way with attempting to encourage the growth of industry, it was France when Jean-Baptiste Colbert was in charge of its economic policy during the 1600s.

Some people have argued that it was the opposite, the lack of government action described by economic philosopher Adam Smith as "laissez faire," that was important for British economic growth. Yet most other European countries did nothing to help—or prevent—industrialization either. When the French government adopted a semi-coherent industrialization strategy under Colbert, it gave a boost to France's industrial production, making it harder to believe that lack of any government policy could have been the British secret sauce. In any case, the age of laissez-faire in Britain follows the early, defining phase of industrialization, which was characterized by government policies that protected woolen textiles and then helped British exports.

A Nation of Upstarts

What set apart Britain from its peers was the outcome of a long process of social change that had created a nation of upstarts.

By the mid-nineteenth century, tens of thousands of middle-status Britons had formed the idea that they could rise substantially above their station through entrepreneurship and command of technologies. Other parts of Western Europe saw a similar process of social hierarchies loosening and ambitious men (and rarely women in those patriarchal times) wishing to gain in wealth or status. But nowhere else in the world at that time do we see so many middle-class people trying to pierce through the existing social hierarchy. It was these middling sort of men who were critical for the innovations and the introduction of new technologies throughout much of the eighteenth and nineteenth centuries in Britain.

By the early 1700s, the zeitgeist had become what Daniel Defoe identified as the Projecting Age. Middle-class Englishmen were looking for an opportunity to advance, whether that be through sound investment or get-rich-quick financial speculation. The South Sea Bubble, which burst in 1720, was an extreme case but also an exemplar of the fascination with new ventures, particularly on the part of small investors seeking profits.

It was in this context that innovators around what we now call industrial processes began to emerge. The most successful of the early entrants include Abraham Darby (pig iron in blast furnaces fueled by coke, 1709), Thomas Newcomen (steam engine, 1712), Richard Arkwright (spinning frame, 1769), Josiah Wedgwood (Etruria pottery works, 1769), and James Watt (much improved steam engine, 1776). These men could not, for the most part, read Latin and did not spend much time with scholarly works.

Darby was the son of a yeoman farmer. Newcomen was an ironmonger who sold tools to mines. Arkwright's parents were too poor to send him to school, and his first occupation was barber and wigmaker. Wedgwood was the eleventh child of a potter. Watt's father was a shipbuilder, which puts him into a higher social class than the others. But by the time of James Watt's schooling, his father was seeking work as an instrument maker, his previous business having collapsed.

These pioneers, like almost everyone else who shaped technology through at least 1850, were practical men without extensive formal education. Much like George Stephenson, they began small and were able to expand over decades as investors and customers began to appreciate their new offerings.

Of 226 people who founded large industrial enterprises during this period, precisely two came from the peerage and less than 10 percent had any connection to the upper classes. However, they were not people who started out at the very bottom of society. Most had fathers engaged in small-scale manufacture, some sort of craft work, or trade. And most of these industrialists had practical skills and engaged in the same kind of small-scale enterprise before creating what became larger businesses.

These men were all extremely ambitious—not what you'd expect from people born into modest means in a society of orders like the one of medieval Europe. More remarkably, they believed in technology, both as an engine of progress and as a means for their own social elevation. But what was most notable about them was that they succeeded.

How did they become so emboldened? What gave them the idea that they could do so, using the power of technology? And what ensured that their efforts were not blocked and/or somehow neutralized?

By the time these men came on the scene, a slow process of social and political change had eaten away some of the most stifling aspects of English social hierarchy, preparing the ground for this emboldening. Notions of individualism and vestiges of popular sovereignty dating back a thousand years may have played a role by providing the raw material for some of these changes. But what was most defining was a series of major institutional transformations that shaped this process of social change and convinced the aristocracy to accommodate these new people.

The Unraveling

In 1300 the idea of ascending from nothing to national prominence would not have occurred to most English people, and the notion that this could be done through invention would have seemed preposterous. In 1577 clergyman William Harrison would characterize the defining feature of society in his *Description of England* as "We in England divide our people commonlie into foure sorts," and he described these as gentlemen (including nobility); citizens in English towns; yeoman farmers; and, at the lowest level, laborers, poor husbandmen, artificers, and servants. More than a century later, when Gregory King drew up his famous *Ranks, Degrees, Titles and Qualifications*, he used roughly the same categories. Which group one fell into, be it in 1577 or 1688, determined one's status and power.

This stratified "society of degrees" was widely accepted and had deep historical roots. Following the Norman Conquest in 1066, England's new rulers had established a centralized feudal system with a great deal of power in the hands of the king. The goal of the monarch was the acquisition of territory through marriage and conquest. The military was based primarily on the feudal obligations of lords and lesser nobility to provide troops. Commercial endeavors were rarely seen as a priority.

But even by 1300, there was some erosion of this position, including the famous Magna Carta of 1215, which paved the way to the creation of the first Parliament and granted some rights to the Church and prominent nobles—while also paying lip service to the rights of people more broadly. All the same, when Elizabeth I ascended to the throne in 1558, English social hierarchy looked remarkably unchanged since the 1300s. And the country was still an economic laggard, far behind Renaissance Italy or the early textile industry present in what are now Belgium and the Netherlands.

Elizabeth's father, Henry VIII, had been a shock to the traditional system. Henry spearheaded political changes with far-reaching consequences. He confronted the Catholic Church and the ecclesiastic orders to marry Anne Boleyn, and he ultimately declared himself the head of the Church of England in 1534. Continuing down this path, he dissolved the monasteries and seized their considerable properties after 1536. At the start of this process, about 2 percent of the male population belonged to religious orders, which collectively owned a quarter of all land. This land was sold off, initiating another round of social changes: the holdings of some rich families increased significantly, and so did the number of people who owned at least some land.

By the end of Henry's rule, many of the foundations of the medieval society of orders were crumbling. But the fruits of this transformation can be more easily seen during Elizabeth I's long reign, between 1558 and 1603. A strong merchant class, especially in London and other port cities, was already evident in these decades and was becoming more assertive and active in overseas trade. The changes in the countryside may have been even more momentous. This is the period during which we see the emergence of the yeoman farmers and skilled artisans as both economic and social forces.

The social changes that were underway accelerated because of England's overseas expansion. The "discovery" of the Americas by Columbus in 1492 and the rounding of the Cape of Good Hope by Vasco da Gama in 1497 opened up new, lucrative opportunities for Europeans. England was a latecomer to the colonial adventures, and by the end of Elizabeth's reign, it had no significant colonies abroad and a navy that was barely strong enough to confront the Spanish or the Portuguese.

But England's weakness was also its strength in this instance. When Elizabeth decided to throw her lot into the colonial scramble, she turned to privateers, such as Francis Drake. These adventurers would equip their own ships and, authorized by a letter of marque, attempt to raid Spanish or Portuguese possessions or seize their shipping. If things went well, the monarch could expect a generous share of the take; Drake's successful circumnavigation of the world generated a large fortune for Elizabeth. If things went badly, there was at least some plausible deniability.

The Atlantic trade significantly altered the balance of political power in England by enriching and emboldening overseas merchants and their domestic allies. London and other ports became a strong source of political support for anyone opposed to high rates of taxation and the arbitrary power of kings. Merchant and overseas colonial interests became increasingly outspoken in political circles, and this mattered in an age of true political and social upheaval.

At the start of the seventeenth century, James I asserted that he had inherited the "divine right of kings," implying a view of society that would have been familiar to Norman monarchs or Egyptian pharaohs. The king, representing God on Earth, was entitled to rule in the same way that a father would over his family, and society should look up to him and obey him as well-mannered children. This attitude and associated high-handed actions by James and his son, Charles I, did not sit well with rural landowners and urban merchants, paving the way to the English Civil War, in 1642–1651.

The full implications of the Civil War could not have been understood by its participants. But there were moments when it became clear that something was really stirring in English society. The extent of political and social transformation is most evident in the ideas that a group of radical men, the Levellers, articulated.

The Levellers were a social protest movement in the early years of the Civil War, represented in the Parliamentary New Model Army. Their main demand was political rights for all ("one man, one vote"), as well as what we would now call human rights more broadly. Their demands came to a head in the so-called Putney debates of October–November 1647, when they confronted the army's leaders. Colonel Thomas Rainsborough, one of the most articulate Levellers, put it this way:

For really I think that the poorest he that is in England has a life to live as the greatest he; and therefore truly, sir, I think it's clear that every man that is to live under a government ought first by his own consent to put himself under that government; and I do think that the poorest man in England is not at all bound in a strict sense to that government that he has not had a voice to put himself under.

Rainsborough's vision was based on universal suffrage:

I do not find anything in the Law of God that a lord shall choose twenty burgesses, and a gentleman but two, or a poor man shall choose none. I find no such thing in the law of nature, nor in the law of nations. But I *do* find that all Englishmen must be subject to English laws; and I do verily believe that there is no man but will say that the foundation of all law lies in the people; and if it lie in the people, I am to seek for this exemption. (Italics in original)

Army leaders, including Oliver Cromwell and the then commander in chief, Lord Fairfax, pushed back. For them, political power had to be kept in the hands of people who owned land and property. After several rounds of vigorous debate, the Levellers lost out, and their ideas faded from the scene.

The Civil War ended in victory for the parliamentarians and was followed by a commonwealth that lasted until 1660. But in retrospect, we should view the next three decades as a continuation of the struggle to set limits on royal power—and which social groups would be allowed to fill the vacuum.

This culminated in the Glorious Revolution of 1688, but the word *revolution* should not deceive us; this was nothing like the French Revolution of 1789. There was no redistribution of property, no assertion of universal rights of the sort the Levellers favored, and no dramatic change in how the country was governed. Most significantly, the people who gained power thought that preserving property, and the rights of property owners, should be the central organizing principle of political life.

These social currents are not only crucial to understanding how English and then British society started changing rapidly, but they also explain some of its distinctive features.

We have thus arrived at some answers to the questions that we posed earlier. What was critical for the British industrial revolution was the entrepreneurship and innovativeness of a cadre of new men from relatively modest backgrounds. These men had practical skills and the ambition to be technologically inventive.

In principle, it could be feudal lords or local strongmen who innovate, but that rarely happens. Lords could order their peasants to innovate, but that is just as unlikely. Abbots could lead the way, applying the resources of their monasteries; this sometimes happened in medieval times, but not often. Hence, the ascent of a new group of people was crucial for industrial innovation. Most importantly, these men had to be resourceful and strive to rise up by becoming wealthy, and society had to let them do it. It was the decline of feudal society in Britain that enabled them to dream and dream big.

Feudalism declined in other parts of Europe, although its order was not challenged to the same extent as it was in Britain. There were peasant rebellions and new philosophical ideas in France, Germany, and Sweden. Yet these did not alter the basis of power as the English Civil War and the Glorious Revolution did, and the extent of economic and social change never reached the same proportions as the one in British society.

This explanation also provides the right perspective on China. Even if China had the scientific breakthroughs and some of the other prerequisites for industrialization, it did not have the right institutional structure to encourage new, innovative people to challenge established ways of organizing production and existing hierarchies. China was not exceptional in this regard; it was just like most of the rest of the world. A few scientific ideas developed around the fringes of organized society were not seen as—and indeed were not—threatening to its order. Moreover, these innovations might have military value, as with gunpowder, or they could help calculate when exactly religious holidays should fall, as with astronomy. But they would certainly not become the basis for an industrial revolution.

Even though there was a social revolution in Britain, it was not one that truly challenged the extant social hierarchy. It was a revolution within the system, and its ambitions were characterized by a fixation on property, in the sense that people who got rich should be taken seriously.

If you wanted to move up socially, you needed to acquire wealth. Conversely, if you could acquire wealth, there was no limit on how high you could rise. And, in the rapidly changing British economy of the eighteenth century, wealth was not tied just to land ownership. One could make money through trade or by building factories, and social status would follow. In this relatively fluid environment, it was natural for many ambitious men from modest origins to strive to succeed within a modified version of the existing order rather than try to overthrow the entire social edifice.

Thomas Turner's diary summarizes his contemporary middle-class aspirations in the mid-eighteenth century: "Oh, what a pleasure is business! How far preferable is an active busy life (when employed in some honest calling) to a supine and idle way of life, and happy are they whose fortune it is to be placed where commerce meets with encouragement and a person has the opportunity to push on trade with vigour."

It was not only commerce and production; developing new technology was a natural place for the dreams and ambitions of people from middle-class backgrounds in the Age of Discovery. Old truths and established ways were crumbling. As Francis Bacon had anticipated, command over nature was increasingly on people's minds.

New Does Not Mean Inclusive

British industry emerged through a revolution of vision. It was fueled and implemented by thousands of men (and some women) of humble origins, limited education, and little inherited wealth. Crucially, these men were rebels within the social order.

New people replacing an age-old hierarchy sounds like the stuff that could produce an inclusive vision, and if so, we should expect this vision to propel us toward shared prosperity. Unfortunately, this was most definitely not what happened in the short term.

In eighteenth-century and early nineteenth-century Britain, the working poor had no political representation and, aside from occasional demonstrations, no way to express themselves collectively. The emboldened middle class, in turn, aspired to rise within the existing system. They accepted its values, and many of them, including Richard Arkwright, bought estates in order to improve their social standing.

In the words of the contemporary commentator Soame Jenyns, "The merchant vies all the while with the first of our nobility in his houses, table, furniture, and equipage." Or as another contemporary, Philip Stanhope, the earl of Chesterfield, put it, "The middle class of people in this country [are] straining to imitate their betters."

These aspirants also adopted the Whig aristocracy's condescending view of the rural and urban poor, who were considered as the "meaner sort," a world apart from themselves, the aspiring middling sort, who could be incorporated into the system. Gregory King thought that these poor were "decreasing the wealth of the nation," not contributing to it. In the words of another contemporary, William Harrison, they had "neither voice nor authoritie in the common wealthe, but are to be ruled and not to rule other."

With this vision, it was entirely natural for this aspirant class to focus on accumulating wealth without worrying about improving the living standards of their employees and their broader community. Consequently, as we will see in Chapter 6, industrial entrepreneurs' choices of technology, organization, growth strategy, and wage policies enriched themselves while denying their workers the benefits of productivity increases—until the workers themselves had enough political and social power to change things.

6

Casualties of Progress

And so muscular force, or *mere* Labor, becomes daily more and more a drug in the market, shivers at the approach of winter, cringes lower and lower at the glance of a machine-lord or landlord, and vainly paces street after street, with weary limbs and sinking heart, in quest of "something to do."

—HORACE GREELEY, *The Crystal Palace and Its Lessons: A Lecture*, 1851 (italics in original)

In the industrial epoch alone has it become possible that the worker scarcely freed from feudal servitude could be used as mere material, a mere chattel; that he must let himself be crowded into a dwelling too bad for every other, which he for his hard-earned wages buys the right to let go utterly to ruin. This manufacture has achieved, which, without these workers, this poverty, this slavery could not have lived.

—FRIEDRICH ENGELS, *The Condition of the Working-Class in England in 1844*, 1845

The 1842 Report from the Royal Commission of Inquiry into Children's Employment was a shocker. For decades there had been growing disquiet about what was referred to as "the condition of England," including how children lived and worked. But with little systematic information available, there was a great deal of disagreement about exactly what young people did in coal mines and factories and whether this constituted an issue that needed to be addressed through legislation.

The royal commission conducted a careful, three-year investigation, including interviews with children, their family members, and employers in all parts of the country. The first report focused on mines, and lengthy appendices provided verbatim quotations.

Young children were doing hard work for long hours, deep underground. The testimony of David Pyrah, from Flockton in West Yorkshire, was typical:

> I am going on 11, I worked at one of Mr. Stansfield's pits. I was lamed at Christmas by a sleeper falling on me, and have been off work since. I went to work usually at 6, but at 4 on odd days. We came out at 6 or 7, sometimes at 3—whenever our work was done. We found it very hard work. The roads [height of the tunnel] were nearly a yard but at the face it was half a yard. I did not like it because it was very low and I had to work till night.

The smallest children would operate trapdoors ("trappers"). Once they were bigger, they could pull loads of coal along rails, bent over or even on their hands and knees ("hurriers"). William Pickard, general steward at the Denby mine, explained that children were valuable underground because they could fit into smaller spaces:

> We used trappers till lately and they used to go and begin as early as 6 years old. . . . They come at 8 or 9 years old to hurry. The thinnest [coal] bed we are working is only 10 inches. We cut the gates 26 inches high. The youngest children go there.

Girls were employed alongside boys. Sarah Gooder, eight years old, reported that she operated a trapdoor that was used to prevent the spread of dangerous gases:

I'm a trapper in the Gawber pit. It does not tire me but I have to trap without a light and I'm scared. I go at 4 and sometimes at half past 3 in the morning, and come out at 5 and half past. I never go to sleep. Sometimes I sing when I've light, but not in the dark; I dare not sing then, I don't like being in the pit.

The interview with Fanny Drake, age fifteen, from Overton, also in West Yorkshire, made the health implications of moving a coal cart underground abundantly clear:

I push with my head sometimes and it makes my head so sore that I cannot bear it touched; it is soft too. I often have headaches and colds and coughs and sore throats. I cannot read, I can say my letters.

Parents understood full well what their children were doing and admitted that this was because the families needed the money and other potential sources of employment were less attractive. As a Mrs. Day explained,

I have two girls in the pit: the youngest is 8 and the oldest will be 19 in May. If the girls don't go into the pits they will have to take a bowl and go begging.

Employers were candid, too. Employing children in this fashion was all about maintaining the profitability of mining operations. As Henry Briggs, co-owner of a mine in Flockton, put it,

We could not have horse roads or even higher roads when the coal seams are so thin, because it would be so expensive. If children were to be stopped from working in pits the best Flockton seams must cease to be mined because it would cost too much to increase the height of the gates.

Wood had been the primary fuel throughout the medieval period but was already replaced by coal by the 1600s. Coal has a higher energy density, calories per kilogram and per volume, compared with wood. It was also possible to move large amounts of coal by barge or sailing ship, further lowering the cost of transportation per unit of thermal energy provided.

By the mid-1700s, pits were being dug deeper underground. The shafts were not more than 50 meters in the late 1600s, but depths increased to 100 meters after 1700, 200 meters by 1765, and 300 meters after 1830. Machines also began to have an impact, first using waterwheels and windmills to lift coal and then with Newcomen steam engines pumping water out from mines after 1712. Later in the century there was a large mining sector, including in the Northeast, with coal moved on rails from the pithead, pulled by horses. Higher-efficiency steam engines were developed in part to help prevent flooding in deeper mines. Improving the transportation of coal by harnessing steam power on wheels was a major motivation for George Stephenson and other railway inventors of the early 1800s.

By the 1840s, coal mining was one of the best-established modern sectors in the country, using state-of-the-art mechanical equipment. More than two hundred thousand people worked in coal mining, with 20–40 percent of those employed in each mine being children.

Careful observers of working conditions at that time were under no illusions about the lives of children. In agriculture, for example, family members as young as six years old had always looked after animals and helped with other tasks, especially at harvest time. Children had also long assisted their parents with artisanal work, including spinning yarn.

However, children working long hours, semi-naked under incredibly unsanitary and dangerous conditions, had no historical parallel at this scale. By the mid-1850s, the conditions of working children showed no signs of improvement. If anything, they worsened as mines were dug deeper.

Coal mines were vividly horrific, but they were not so unusual. Working conditions in cotton and other factories, documented in the second report of the Royal Commission, were similarly draconian. And it was not just children who suffered. Workers did not see much, or any, improvement in their real incomes but ended up toiling longer hours and under harsher conditions than they used to before the age of the factories. Pollution and infectious diseases in dense cities with deficient infrastructure shortened lives and increased morbidity.

It became increasingly evident to the Victorians that although industrialization had made some people very rich, most workers lived shorter, less healthy, and more brutal lives than had been the case before industry began to develop. By the mid-1840s, authors and politicians on all sides of the political spectrum were asking: Why had industrialization worsened so many lives, and what could be done about it? Was there a way to encourage the growth of industry while also sharing the benefits more broadly?

There was an alternative way, and we will see in this chapter that Britain in the second half of the nineteenth century embarked upon it. Technology's bias against working people is always a choice, not an inevitable side effect of "progress." To reverse this bias, different choices needed to be made.

Much better outcomes for the majority of the population followed when technological change created new opportunities for working people and wages could not be kept low anymore. They became a reality after countervailing powers against factory owners and the wealthy elites started developing in workplaces and then in the political arena. These changes triggered improvements in public health and infrastructure, enabled workers to bargain for better conditions and higher pay, and contributed to a redirection of technological change. But we will also see that for people around the world, especially those in the European colonies who had no political voice, industrialization's effects were often grim.

Less Pay for More Work

The productivity bandwagon suggests that as technology advanced rapidly during the early phases of the Industrial Revolution, wages should have risen. Instead, real incomes of the majority stagnated. Work hours increased, and conditions significantly deteriorated, amounting to lower hourly wages as more and more labor was extracted from British workers.

Detailed studies have reconstructed the cost of food and other essentials, such as fuel and housing, and the general pattern is reasonably clear. At the end of the 1600s, most English people consumed a "subsistence basket" that was little different from what was available to rural residents in medieval times. At the center of the working person's diet was grain, both in the form of food (bread) and drink (ale). For the English, the grain was wheat, mostly grown domestically. Some vegetables were available on a seasonal basis, and a small amount of meat might be eaten once or twice per week. Similar consumption baskets can be constructed for other parts of Europe, as well as for India and China. Three broad patterns emerge from these data.

First, from around 1650 to 1750 there was a slow improvement in real incomes in England, most likely as a result of productivity growth in agriculture and the expansion of long-distance trade with Asia and the Americas, which raised incomes in London and port cities such as Bristol and Liverpool and modestly increased wages throughout the country. In consequence, by around 1750, wages in England were somewhat higher compared to southern Europe, India, and China. For example, average calorie consumption for unskilled laborers was about 20–30 percent higher than it had been in medieval times, and they enjoyed a slightly more nutritious diet and more meat than had people five hundred years earlier. Other parts of the world remained mired in the same poor nutrition found in the 1200s.

Second, starting around 1750, there was fairly rapid productivity growth, especially in textiles. The earliest spinning machines increased output per hour of work nearly 400 times. In India at this time, spinning a hundred pounds of raw cotton took 50,000 hours of labor. In England, using a spindle mule in 1790, the same amount of output needed just 1,000 hours of labor. By 1825, with improved machinery, the work required was down to 135 hours of labor.

But real incomes moved little, if at all. The spending power of an unskilled worker in the mid-1800s was about the same as it had been fifty or even one hundred years earlier. There was also

1. Ferdinand de Lesseps: "the great canal digger."

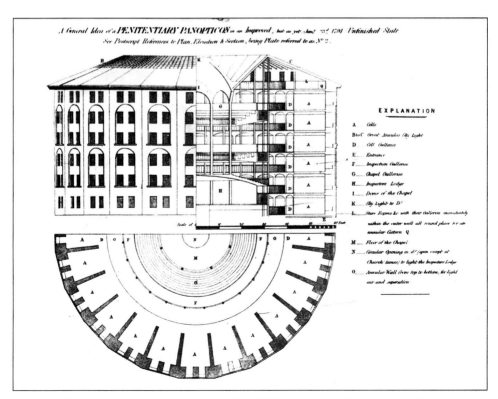

2. Jeremy Bentham's panopticon—proposed in 1791 for more "efficient" surveillance in prisons, schools, and factories.

3. The Suez Canal. According to Lesseps, "The name of the Prince who opens the great maritime canal will be blessed from century to century until the end of time."

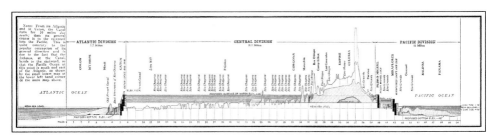

4. Lesseps's vision of a canal without locks at Panama was a complete failure, resulting in more than 20,000 deaths and financial ruin.

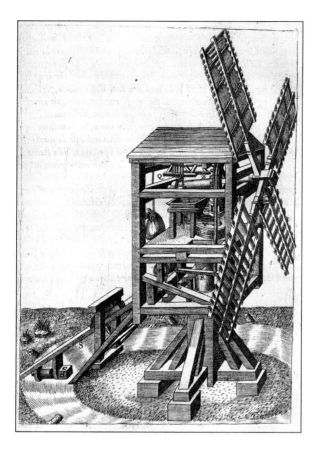

5. A major medieval technology that generated large productivity gains but few benefits for peasants.

6. Medieval productivity gains made possible monuments such as the Lincoln Cathedral, the tallest building in the world from 1311 to 1548.

7. Large textile factories, such as this water-powered cotton mill in Belper, Derbyshire, increased average productivity more than 100-fold. But conditions were unhealthy, workers had no autonomy, child labor was pervasive, and wages remained low.

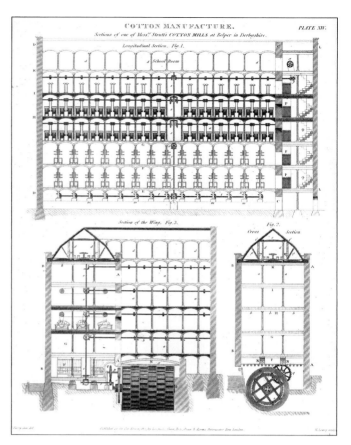

8. Inmates of All Saints Workhouse, Hertford, grinding labor, literally and metaphorically, as was typical for many recipients of "support" under the Poor Law.

9. Eli Whitney's gin boosted cotton production in the US South, paving the way for the expansion and intensification of slavery.

10. Eli Whitney was also a pioneer in the adoption of interchangeable parts in the US North, increasing the productivity of unskilled labor and reducing the need for skilled labor. This illustration shows machine gears designed by Charles Babbage, who was in pursuit of a "fully-automatic calculating machine."

11. George Stephenson's *Rocket* decisively won the Rainhill Trials in 1829 and became the basis for designs that swept the world.

12. *Archimedes,* built in the 1880s, waiting for passengers in Euston Station. Railways paid high wages and spearheaded the expansion of British industry.

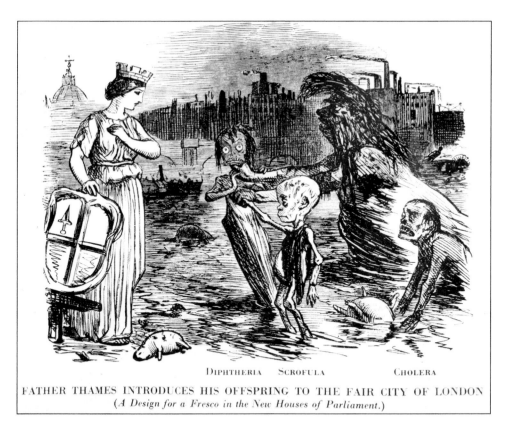

DIPHTHERIA   SCROFULA   CHOLERA

FATHER THAMES INTRODUCES HIS OFFSPRING TO THE FAIR CITY OF LONDON
(A Design for a Fresco in the New Houses of Parliament.)

13. Human waste and industrial effluent were dumped into the Thames, creating ideal breeding grounds for infectious diseases.

14. The London sewer system, designed by Joseph Bazalgette (top right), rivals the pyramids of ancient Egypt for imagination and application. Evaluated in terms of impact on public health, Bazalgette did better.

15. In this stylized engraving of a nineteenth-century dairy factory, every machine is connected by a belt to the same driveshaft.

16. According to Henry Ford, "The motor enabled machinery to be arranged according to the sequence of the work, and that alone has probably doubled the efficiency of industry. . . ." This is his Rouge Plant in 1919, with electricity throughout.

not much improvement in diet for most British workers during the first century of industrialization.

Third, although skilled workers enjoyed higher wages than others throughout this period, what it meant to be "skilled" changed a great deal. Men who worked looms in the early 1800s were considered skilled and commanded a premium wage. But as we will see later in this chapter, automation wiped out large categories of jobs that previously required artisanal skills, including the work carried out by male weavers. Those workers were then forced to seek employment as unskilled laborers, at a lower wage. At least through the mid-1800s, the wage gains of skilled industrial workers were precarious or even fleeting.

Equally important was the transformation of the British labor market throughout this period, with longer working hours and a very different organization of work. Indeed, as economic historian Jan de Vries has pointed out, the Industrial Revolution was very much an "industrious revolution," in the sense that first the British, and then everyone else, began to work much harder.

In the middle of the eighteenth century, the average working year involved about 2,760 hours, likely unchanged from 50 or 100 years earlier. By 1800, average hours had already risen to 3,115. Over the following 30 years, hours worked per year rose further to 3,366—an average of nearly 65 hours per week. However, longer hours did not mean higher incomes for most of the population.

Experts debate how much of this increase in work effort and hours was voluntary, taking place in response to better economic opportunities, and how much of it was imposed on workers. These are good questions to ask from our comfortable chairs in the twenty-first century, but most British people knew in the early 1800s that they had to work longer hours and in more arduous conditions than people fifty or a hundred years earlier. This was the only way to survive in the new manufacturing economy.

Plenty of items were made by skilled hands in small workshops before the Industrial Revolution took off. The late Middle Ages saw the growth of European book production, and making clocks also became a significant activity. After 1500, a sizable textile industry developed in England, centered on wool products, and coal and tin mining were well established by the 1600s.

In the "putting out system" for woolen textiles, much of the production took place in people's homes, where they could spin or weave at their own pace and were compensated according to a piece rate, depending on how much output they produced. This was hard work for little money. Nevertheless, workers had a considerable degree of autonomy in how and when they worked. Most people took advantage of this flexibility, adapting their work hours and style to their needs—for example, based on agricultural work they did on the side. They also took time off when they were tired or when they had had too much to drink the previous evening. Weavers typically did not work on Mondays, and sometimes even Tuesdays, and when necessary made it up by working Friday and Saturday nights. Most workers did not need to keep careful track of time or did not even have access to clocks.

Factory work changed all that. The modern imagery of early factories owes much to Adam Smith's vivid description of a pin factory in his classic book, *The Wealth of Nations*. Smith emphasized how the division of labor in factories improved efficiency by allowing each worker to focus on a very specific task in the process of pin making. But early factory organization had as much to do with worker discipline as the division of labor for technical efficiency. Factories imposed tight rules on when workers had to show up and when they could go home. They required significantly longer working hours and much more hierarchical decision making. Their organization was inspired by early modern militaries.

The drill code developed by Maurice of Nassau, a Dutch prince and the most influential tactician of the early 1600s, specified more than twenty separate steps involved in firing a musket. Perfecting a method dating back to the Romans, drill became the main way of organizing soldiers: small movements, subject to voice commands, that enabled infantry lines to pivot, form squares, reverse direction, and so on. With a few months of training, hundreds could learn to fight together closely, preserving cohesion under enemy fire or in the face of a cavalry charge. Using these methods, armies became larger. In the 1600s and early 1700s, they typically comprised tens of thousands of men. The New Model Army, which became the dominant force in the English Civil War during the 1640s, numbered over twenty thousand.

The English word *factory* is derived from a Latin root that means either an oil press or a mill. In the 1500s the term was used to designate an office or trading post that could be quite small. As a "building for making goods," the meaning can be traced back to the early 1600s. Starting about 1721, the word came to represent something quite new: a place where large numbers of people, many of them women and children, gathered to work with machines. Early textile factories employed as many as a thousand people and broke tasks down into simple components, emphasized repetitive motion, used strong discipline to keep everyone working together, and, of course, significantly reduced worker autonomy.

Richard Arkwright, one of the most successful innovators and factory owners of the era, built his first mills near coal-mining operations. This location was chosen not for easy access to fuel, for he was relying on water as his source of power. Instead, Arkwright's goal was to hire the family members of mine workers to work in his mills. Women and children were regarded as more dexterous and also more pliant than adult men would be in a highly regimented system. Water flowed all day and night, so the mill could work incessantly. Factories were expensive to build, and once this up-front capital cost was incurred, entrepreneurs wanted to use their equipment as intensely as possible, preferably around the clock and definitely long into the night.

The discipline in these new factories would have seemed familiar to Maurice of Nassau, although the extensive use of children would have been an eye-opener. All workers on a shift had to arrive at the same time. They needed to learn how to manage the machines, typically with a limited set of actions. Those actions needed to be precise; any deviation from the required pattern could disrupt production or damage the equipment. Even if Jeremy Bentham's panopticon, which we discussed in the Prologue, was not adopted widely, employees were closely supervised to ensure that they paid sufficient attention and followed orders.

Workers often complained about the conditions and especially resented the loss of autonomy in the hierarchical structure of factories. A folk ballad from Lancashire captured the feeling:

So, come all you cotton-weavers, you must rise up very
    soon,
For you must work in factories from morning until noon:
You mustn't walk in your garden for two or three hours
    a-day,
For you must stand at their command, and keep your
    shuttles in play.

There were numerous industrial accidents, with little regard for worker safety or compensation. A Manchester man whose son had been killed in such an accident stated that "I have had seven boys, but if I had 77 I should never send one to a cotton factory." It was not just the hard work, "from six in the morning till eight at night," but also working conditions, the discipline, and the hazards of the factory.

Because workers were not organized and lacked political power, employers could get away with paying low wages. Intensifying factory discipline, longer workdays, and tougher working conditions should be viewed in the same light. When employers are powerful and labor is not, productivity gains are not shared with workers—and profits are higher. Less pay for more work during this era was thus a consequence of the imbalance of power between capital and labor.

Employer efforts to keep wages low and extract as much work as possible were also helped by the harsh way that public policy treated poor people, including orphans, in Victorian Britain. In Arkwright's early factories, for instance, many of the employees were children from the local workhouse, placed there because their families could not support them. Legally regarded as "apprentices," these children were not allowed to leave their jobs, under penalty of law, and in any case could not leave if they wanted to eat. Consequently, they were hardly able to demand higher wages or better working conditions.

Large-scale construction projects in ancient Egypt and Rome had at their core skilled artisans, trained over many years. In contrast, British factories hired people without specialized trades, including of course women and children, many of whom did not acquire much in the way of new skills. Opening a trapdoor underground or pushing a coal cart with your head was not conducive to learning. If the children died or were hurt in a workplace accident, they could be easily replaced.

By 1800, the British cotton industry was the largest in the world, and great fortunes were being made. Arkwright became one of the richest men in England, famously lending five thousand pounds to the duchess of Devonshire to cover her gambling debts. The industrial middle class was rising fast, but the productivity bandwagon did not have many people on board.

Worse was yet to come.

The Luddites' Plight

On February 27, 1812, as the Industrial Revolution was literally and figuratively gathering steam, Lord Byron rose to address the House of Lords. Byron was a young man, famous already for his romantic poetry, and he spoke as eloquently as he wrote. But his topic that day was brutally real: the Frame Breaking Act, which proposed the death penalty for people who smashed newly invented textile machines, specifically for weaving cloth.

Turning raw cotton into clothing is an old business, but for two thousand years of recorded history, there had only been small improvements in production methods. Then came a wave of British inventions that, starting in the 1730s, mechanized spinning so that it could be done more cheaply in the large mills employing mostly unskilled workers.

Consequently, the real price of cotton yarn fell to about one-fifteenth of its previous level, at first great news for skilled artisans engaged in weaving. In response, cotton weaving expanded, although the boon to skilled weavers was short-lived. Subsequent waves of invention mechanized weaving, brought it inside factories, and reduced the need for skilled artisans there as well.

There was a wave of machine breaking in 1811–1812 by groups of textile workers calling themselves Luddites, after Ned Ludd, an apocryphal character who was supposed to have broken knitting frames in 1779. The Luddites were clear that they were not plunderers and thieves. A letter from Nottinghamshire Luddites stated that "plunder is not our object, the common necessaries of life is what we at present aim at." No matter, the government response was to propose capital punishment when the previous maximum penalty had been forced deportation to Australia.

Byron spoke with passion, anticipating two centuries of subsequent debate about technology and jobs:

> The rejected workmen, in the blindness of their ignorance, instead of rejoicing at these improvements in arts so beneficial to mankind, conceived themselves to be sacrificed to improvements in mechanism. In the foolishness of their hearts, they imagined that the maintenance and well doing of the industrious poor, were objects of greater consequence than the enrichment of a few individuals by any improvement in the implements of trade which threw the workmen out of employment, and rendered the labourer unworthy of his hire.

Byron did not spend long in politics, and he did not have much impact while he was active. This was unfortunate—he had a way with words:

> I have traversed the seat of war in the peninsula; I have been in some of the most oppressed provinces of Turkey; but never, under the most despotic of infidel governments, did I behold such squalid wretchedness as I have seen since my return, in the very heart of a christian country.

Industrialization was destroying good jobs, livelihoods, and lives. Subsequent decades proved that Byron was in no way exaggerating. In fact, he only saw part of the harm that would be done.

Horace Greeley, a prominent American newspaper editor, came to a similar conclusion after visiting the Great Exhibition of 1851 in London. He concluded that the cause of so much mid-century angst was machinery—automation—replacing workers:

> On every side the onward march of Invention is constant, rapid, inexorable. The human Reaper of thirty years ago, finds to-day a machine cutting grain twenty times as fast as ever he could; he gets three days' work as its waiter where he formerly had three weeks' steady harvesting: the work is as well done as of old, and far cheaper; but his share of the product is sadly diminished. The Planing Machine does the work of two hundred men admirably, and pays moderate wages to three or four; the Sewing Machine, of moderate cost, performs easily and cheaply the labors of forty seamstresses; but all the seamstresses in the world probably do not own the first machine.

As we explained in Chapter 1, machines can be used either to replace workers through automation or to increase worker marginal productivity. Examples of the latter include water mills and windmills, which took over some tasks previously done by hand but also increased the need for labor to process and handle the consequently cheaper grain and wool, including through the creation of new tasks.

Pure automation is different because it does not increase workers' contribution to output and hence does not create the need for additional workers. For this reason, automation tends to have more acute consequences for the distribution of income, creating big winners, such as owners of the machines, and many losers, including those displaced from their jobs. It is for this reason that the productivity bandwagon effect is weaker when there is a lot of automation going on.

Pervasive automation, especially in the textile industry, was one reason why the productivity bandwagon did not operate and wages did not increase, even as the British economy mechanized in the late eighteenth and early nineteenth centuries. As an early chronicler of the British factory system, Andrew Ure, observed in his 1835 book, *The Philosophy of Manufactures,*

In fact, the division, or rather adaptation of labour to the different talents of men, is little thought of in factory employment. On the contrary, wherever a process requires peculiar dexterity and steadiness of hand, it is withdrawn as soon as possible from the *cunning* workman, who is prone to irregularities of many kinds, and it is placed in charge of a peculiar mechanism, so self-regulating, that a child may superintend it. (Italics in original)

Having a child "superintend" the task was, unfortunately, not just a figure of speech.

The Luddites themselves seem to have understood not just what the machines of the age meant for them but also that this was a choice about how to use technology and for whose benefit. In the words of a Glasgow weaver,

The theorists in political economy attach more importance to the aggregate accumulation of wealth and power than to the manner of its diffusion, or its effects on the interior of society. The manufacturer possessed of capital, and the inventor of a new machine, study only how to turn them to their own profit and advantage.

Improvements in textile productivity did generate jobs in other sectors of the British economy—for example, in machine and tools manufacturing. Nevertheless, for decades this additional demand for labor was not enough to fuel wage growth. Moreover, any new job that skilled weavers could obtain was not commensurate with their skills and previous earnings. Luddites were right to worry about knitting frames decimating their livelihoods.

At this stage, British workers were not unionized and could not collectively bargain. Although the worst coercive practices from medieval times had already been eliminated, many workers toiled in a semi-coercive relationship with their bosses. The Statute of Labourers of 1351 was repealed only in 1863. The Statute of Artificers, enacted in 1562–1563, which similarly mandated compulsory service and prohibited workers from quitting their employers before the end of the contracted term, was still being used for prosecuting workers. A revised Master and Servant Act renewing the prohibition on breach of contract by workers was adopted by Parliament in 1823 and 1867. Between 1858 and 1867, there were ten thousand prosecutions under these acts. These cases typically started with the arrest of workers about whom there was a complaint. These laws were also used consistently against union organization, until they were fully repealed in 1875.

These conditions of the working class were entirely in line with the vision of the politically powerful segments of society. Their attitudes, and the implications thereof, are well illustrated by the 1832 Royal Commission into the Operation of the Poor Laws, convened to reform these laws dating back to Elizabethan times.

The old Poor Laws were already ungenerous and unforgiving to those in distress. But they were viewed by the new thinkers of the age as insufficiently motivating for the poor to get their act together and supply their labor. The commission consequently proposed organizing all poor relief in the context of workhouses so that the recipients of aid would continue to work. It also recommended tougher eligibility requirements and making poorhouses less hospitable so that people were motivated to choose work instead of relief.

The burden on taxpayers, primarily the aristocracy, the gentry, and the middle classes, was to be reduced as well. There was a political consensus, and the commission's recommendations were adopted in 1834, albeit in a watered-down version. The workhouse effectively created what one expert described as a "prison system to punish poverty."

In this environment, workers had little chance to receive higher wages or share in the profits of firms. Longer and less autonomous working days and stagnant real incomes were not the only fallout from early industrialization. The social bias of technology also had a broader impoverishing effect.

The Entrance to Hell Realized

Industrialization brought a great deal of pollution, particularly as the use of coal increased. The early boom in textiles was powered by water, but after 1800 coal became the fuel of choice for the increasingly ubiquitous steam engines. The largest waterwheels also powered factories, although these could be placed only where there was sufficient water flow. Steam engines meant that factories could be built anywhere—closer to ports, near coal, where workers were available, or all of the above.

With steam power, major industrial centers became a forest of chimneys, belching smoke all day and night. The first cotton mill was built in Manchester in the 1780s, and by 1825, there were 104 such operations. Reportedly, there were 110 steam engines in the city. According to one observer,

> A steam-engine of 100 horse-power, which has the strength of 880 men, gives a rapid motion to 50,000 spindles, for spinning fine cotton threads: each spindle forms a separate thread, and the whole number work together in an immense building, erected on purpose, and so adapted to receive the machines that no room is lost. Seven hundred and fifty people are sufficient to attend all the operations of such a cotton mill; and by the assistance of the steam-engine they will be enabled to spin as much thread as 200,000 persons could do without machinery, or one person can do as much as 266.

Pollution was out of control in the early phases of industrialization. It caused huge numbers of deaths and an unimaginable decline in the quality of life for most people. Friedrich Engels was scathing about the effect of pollution on the working class:

> The manner in which the great multitude of the poor is treated by society to-day is revolting. They are drawn into the large cities where they breathe a poorer atmosphere than in the country; they are relegated to districts which, by reason of the method of construction, are worse ventilated than any

others; they are deprived of all means of cleanliness, of water itself, since pipes are laid only when paid for, and the rivers so polluted that they are useless for such purposes; they are obliged to throw all offal and garbage, all dirty water, often all disgusting drainage and excrement into the streets, being without other means of disposing of them; they are thus compelled to infect the region of their own dwellings.

Sir Charles Napier, an experienced general, was posted to Manchester in 1839, commanding forces that were intended to keep the peace. Though not a radical like Engels, Napier was still appalled by the city's conditions, referring to it in his journal as "the entrance to hell realized!"

London's notorious fog, primarily caused by burning coal, created episodes of "acute pollution exposure" that were bad enough to account for one out of every two hundred deaths for more than a century.

Pollution was not the only reason lives became shorter and nastier in nineteenth-century Britain. Infectious diseases created an increasingly deadly threat for city dwellers. Although there had been some advances against established infectious diseases in the 1700s, particularly smallpox, crowded and fast-growing industrial cities created ideal breeding grounds for new epidemics. The first global epidemic of cholera broke out in 1817, followed by regular outbreaks until late in the century, when the importance of clean municipal water was fully understood.

Death rates in overcrowded industrial cities were rising sharply. In Birmingham, deaths per 1,000 in 1831 were 14.6, and 27.2 per 1,000 in 1841. Similar increases were recorded in Leeds, Bristol, Manchester, and Liverpool. In the new manufacturing towns, half of all children died before reaching age five.

Parts of Manchester had only thirty-three toilets for more than seven thousand people. Sunderland had one privy per seventy-six people. Most of these sanitary facilities were not connected to public sewers, resulting in urban cesspools that were seldom cleaned out. Anyway, most sewer systems could not handle the human waste that did find its way in.

In this environment, a very old disease, tuberculosis, reemerged as a scourge. Records show traces of tuberculosis in Egyptian mummies, and this disease has long haunted dense settlements. It became a major killer in the nineteenth century as crowding and unsanitary conditions reached unprecedented proportions in large cities. At its peak in mid-century, tuberculosis was responsible for about 60,000 deaths per year in England and Wales, during a time when annual total deaths were between 350,000 and 500,000. However, there was also evidence that most people suffered some form of tuberculosis during their life.

Highly contagious childhood diseases such as scarlet fever, measles, and diphtheria proved devastating well into the twentieth century, when effective vaccination programs were put in place. The presence of measles and tuberculosis, both respiratory diseases, worsened the effects of pollution, leading to higher mortality. Maternal mortality rates remained high throughout this period. Hospitals also spread infection until the importance of hand washing was properly understood, toward the end of the century.

The population of Manchester was just over twenty thousand in the early 1770s. By 1823, more than a hundred thousand people lived in the city, trying to fit into overcrowded accommodations, with filthy streets, not enough water, and soot everywhere.

Crowded and squalid living conditions, difficult lives, and cheap alcohol created another hazard: growing violence, including within families. No doubt there was domestic violence before industrialization, and treating children well, in terms of education, nutrition, and care, became the norm only in the twentieth century. Still, there was less abuse of alcohol when everyone was drinking weak ale. The British drinking of distilled spirits supposedly began only after the Battle of Ramillies in 1706. The drinking of gin took off in the 1700s, and by the mid-1800s, alcoholism was rife. As the price of tobacco came down, cigarettes also came within reach for the working class.

The educated British reaction was to argue there was a broader moral decline across the nation. Thomas Carlyle wrote influentially about this issue, coining the term "the condition of England" in 1839. A wave of social novels dealt with the evils of factory life, including works by Charles Dickens, Benjamin Disraeli, Elizabeth Gaskell, and Frances Trollope.

Health checkups on British army recruits for the Second Boer War, 1899–1902, confirmed a deeply unhealthy nation. Industrialization had created a public health disaster.

Where the Whigs Went Wrong

Thomas Macaulay's *History of England*, first published in 1848, sums up recent British history this way:

> For the history of our country during the last hundred and sixty years is eminently the history of physical, of moral, and of intellectual improvement. Those who compare the age on which their lot has fallen with a golden age which exists only in their imagination may talk of degeneracy and decay: but no man who is correctly informed as to the past will be disposed to take a morose or desponding view of the present.

This rosy view reflects what is known more broadly as the Whig interpretation of history and is related to the more modern economic assumption of a self-acting productivity bandwagon. Both perspectives are based on the idea that progress eventually brings good things to most people.

Andrew Ure's account of the spread of factories in Britain articulated the optimism of the 1830s—and anticipated the rhetoric of tech visionaries today. Even as he was describing skilled artisans losing their jobs, Ure wrote confidently that "such is the factory system, replete with prodigies in mechanics and political economy, which promises, in its future growth, to become the great minister of civilization to the terraqueous globe, enabling this country, as its heart, to diffuse along with its commerce, the life-blood of science and religion to myriads of people."

Alas, the world is more complicated than these accounts suggest, precisely because social and economic improvements are far from automatic, even as institutions change and new technologies are introduced.

Whig optimism was understandable, for it reflected the views of the ascendant social classes in England, including the gentry and new mercantile and later industrial interests. It was also superficially plausible, for industrialization did bring new people and ideas to the forefront. Yet this form of social change did not inexorably lift up most of the population, as we have seen in this chapter.

Industrialists such as Arkwright pried open existing hierarchies in the early 1800s not because they wanted to bring down social barriers or create true equality of opportunity, certainly not for the "meaner sort" of people. Rather, the rising middling-sort entrepreneurs wanted to pursue their own opportunities, move up, and become part of the upper crust of society. The vision they developed reflected and legitimized this drive. Efficiency was key, as the prevailing argument went, and it was in the national interest. New technological, economic, and political leaders were the vanguards of progress, and everybody would benefit from this progress, even if they did not fully understand it.

The views of Jeremy Bentham, like those of Saint-Simon, Enfantin, and Lesseps in France, are emblematic of this vision. In addition to a firm belief in technology and progress, Benthamites had two central ideas. The first was the government had no business interfering in contracts between consenting adults. If people agreed to work long hours under unhealthy conditions, that was their business. There was legitimate public concern for the lives of children, but adults were on their own.

The second was that the value of any policy could be assessed by adding up how much was gained or lost by the individuals involved. Hence, if the reform of working conditions for children would result in gains for them, this could and should be weighed against what the losses would be for their employers. In other words, even if the gains for the children from a new policy were substantial—for example, because of improved health or schooling—this policy should not be adopted if the losses for the employer, primarily in terms of profits, were larger.

To those with political voice, including the upwardly mobile middling sort of people, this seemed modern and efficient, and it justified their belief that the inevitable march of progress should not be halted, even if it created casualties along the way.

Through the early decades of the nineteenth century, such was the path of progress, warts and all. Anybody questioning it or standing in its way was regarded as a fool or worse.

Progress and Its Engines

Fifty years later, things looked very different.

In the second half of the nineteenth century, wages started growing steadily. From 1840 to 1900, output per worker rose by 90 percent while real wages increased by 123 percent. This included substantial income growth and improvements in diet and living conditions for unskilled laborers. For the first time in the modern era, productivity and wages rose at roughly the same rate.

Working conditions also improved. The average working day was down to nine hours for many workers (54 hours per week for builders and engineers, 56.5 hours per week in textiles, and 72 hours per week on the railways were standard), and almost no one worked on Sunday. Corporal punishment in the workplace had become rare, and as we noted, the Master and Servant Acts were finally repealed in 1875. Child labor laws greatly reduced the factory work of children, and there was a movement to make free elementary education available to most children.

Public health improved dramatically as well, although it took another half century to bring the London fog fully under control. Sanitary conditions in large cities improved, and there was broader progress on preventing epidemics. Life expectancy at birth began to creep up, from around forty years at mid-century to nearly forty-five years at the start of the 1900s. None of these improvements were confined to Britain. We see similar progress across most of Europe and other industrializing countries. Is this, then, a vindication of the Whig interpretation of history?

Not at all. There was nothing automatic about any of the improvements that ushered in a broader sharing of productivity gains and the cleanup of cities. They resulted from a contested process of political and economic reforms.

The productivity bandwagon needs two preconditions to operate: improvements in worker marginal productivity and sufficient bargaining power for labor. Both elements were largely absent for the first century of the British industrial revolution but started falling into place after the 1840s.

The first phase of the Industrial Revolution that so alarmed Lord Byron was one in which the main technological innovations were all about automation, most notably replacing spinners and weavers with new textile machinery. As we explained in Chapter 1, advances in automation do not preclude shared prosperity, but there is a problem if automation predominates—in the sense that workers are displaced from their existing work at the same time as there are insufficient new tasks in other productive positions.

This is what happened starting toward the end of the eighteenth century, with textile workers displaced from their jobs and having a hard time finding alternative employment at a wage close to what they earned before. This was a long and painful phase, as Lord Byron recognized and most of the working classes felt. However, by the second half of the nineteenth century, the direction of technology had shifted.

Arguably, the defining technology of the second half of the nineteenth century was railways. When Stephenson's *Rocket* won the Rainhill Trials in 1829, there were about thirty thousand people engaged in long-distance stagecoach transportation, with a thousand turnpike companies maintaining twenty thousand miles of road. Within a few decades, several hundred thousand people were working on building and running railways.

Steam-powered trains reduced transportation costs and destroyed some jobs—for example, in the horse-drawn coach business. But railways did much more than just automate work. To start with, advances in railways generated many new tasks in the transport industry, and the jobs demanded a range of skills, from construction to ticket sales, maintenance, engineering, and management. We saw in Chapter 5 that many of these jobs offered improved working conditions and premium wages when railway companies shared some of their high profits with their employees.

As we discussed in Chapter 1, technological advances can stimulate demand for workers in other sectors, and this effect is more powerful if they significantly increase productivity or create linkages to other sectors. Railways did this as passengers and freight started moving more cheaply and farther. Long-range coach travel declined to nearly nothing, but railways raised the demand for short-distance horse-drawn travel as the people and goods they transported long distances had to be moved back and forth in the cities.

More important were linkages from railways to other industries, meaning positive effects on other sectors supplying inputs to the transport industry or those that heavily used transport services and could expand as a result of improvements caused by railways. The growth of railways increased the demand for a range of inputs, especially higher-quality iron products used in stronger metal rails and more-powerful locomotives. Lowering the cost of moving coal also made it possible to expand the metal-smelting industry, improving the quality of iron.

Cheaper transportation for finished products helped the metal smelting industry, which, following the patenting of the Bessemer process in 1856, could manufacture massive amounts of steel. Another round of benefits followed as more steel and cheaper coal helped expand other industries, including textiles and a range of new products, such as processed food, furniture, and early household appliances. Railways gave a boost to wholesale and retail trades as well.

In sum, nineteenth-century British railways represent an archetype of a systemic transformative technology that increased productivity both in transport and across several other sectors, and also generated new opportunities for labor.

It was not just innovations in railways. Other emerging industries also contributed to higher worker marginal productivity. The new suite of manufacturing technologies generated demand for both skilled and unskilled workers. Metals, especially with advances in iron and steel, were at the forefront of this process. In the words of the president of the Institute of Civil Engineers in 1848,

> The rapid introduction of cast-iron, together with the invention of new machines and new processes, called for more workmen than the millwright's class could supply, and men trained more in the working of iron were brought into the field. A new class of workmen was formed, and manufacturing establishments arose, to which were attached iron and brass foundries, with tools and machines for constructing machinery of every description.

These new industries received a further boost from new communication tools, such as the telegraph in the 1840s and the telephone in the 1870s. They created many jobs in communications and manufacturing. They also generated novel synergies with the transport sector as they improved the efficiency of railways and logistics. Although the telegraph did replace other forms of long-distance communication, such as mail and special courier, the number of displaced workers did not compare to the new jobs in the communications industry.

Similarly, telephones replaced telegraphs, initially within cities and then for long-distance messages. But just like the telegraph and the railways, they were not pure automation technologies. Building and operating telephone systems was labor intensive and critically relied on a range of new tasks and occupations, such as switchboard operation, maintenance, and various new engineering tasks. Soon, telephone exchanges were employing large numbers of women both in public switchboards and in all organizations. Initially, all phone calls were connected by an operator. The first automatic dialing system in the United Kingdom did not open until 1912. The last manual switchboard in London continued to operate until 1960.

In fact, the development of the telephone happened alongside an expansion in the business of sending telegrams, in part because competition brought down prices. In 1870, before telephones, there were seven million telegrams sent in the UK. By 1886, this was up to fifty million per year. The US telegraph network handled more than nine million messages in 1870 and more than fifty-five million messages in 1890.

Overall, the implications for labor from these technologies were more favorable than the situation with automation of textiles in the first phase of the Industrial Revolution because they created new tasks and activated productivity improvements across a number of sectors, expanding the demand for labor. However, these outcomes were very much dependent on choices about how these production methods were developed and used, as we will see.

Gifts from Across the Atlantic

Something else greatly helped Britain move toward shared prosperity: new innovations from the other side of the Atlantic. Even though Americans were latecomers to industrial growth compared to their British counterparts, US industry surged in the second half of the nineteenth century. The American path of technology was targeted at increasing efficiency and contributed to higher worker marginal productivity. As this technology spread in Britain and Europe, it further raised labor demand in these economies.

The United States was abundant in land and capital, but scarce in labor, especially skilled labor. The small number of artisans who emigrated to America enjoyed higher wages and bargaining power than they did back at home. This high cost of skilled labor meant that American inventions often prioritized not just automation but also finding ways to boost the productivity of lower-skilled workers. In the words of Joseph Whitworth, a future president of the Institution of Mechanical Engineers, who visited American industry in 1851, "The labouring classes are comparatively few in number, but this is counterbalanced by, and indeed may be regarded as one of the chief causes of, the eagerness with which they call in the aid of machinery in almost every department of industry." And as E. Levasseur, a Frenchman visiting American steel mills, silk factories, and packinghouses,

put it in 1897, "The inventive genius of the American is perhaps a native gift, but it has been unquestionably stimulated by
the high rate of wages. For, the entrepreneur seeks to economize
human labor the more it costs him. On the other hand, when machinery gives greater productive force to the laborer it is possible
to pay him more." This was one outcome of Eli Whitney's focus
on interchangeable parts, which strove to build standard pieces
that could be combined in different ways, making it easier for
unskilled workers to produce guns. Whitney himself described
his objective as "to substitute correct and effective operations of
machinery for that skill of the artist which is acquired only by
long practice and experience; a species of skill which is not possessed in this country to any considerable extent."

Most European technology, including in Britain, relied on
skilled craftsmen to adjust parts depending on their use. The new
approach did not just reduce the need for skilled labor. Whitney
aimed at building a "systems approach," combining specialized
machinery and labor to increase efficiency. The gains were evident
to the British Parliamentary Committee inspecting American
arms factories using interchangeable parts: "The workman whose
business it is to 'assemble' or set up the arms, takes the different
parts promiscuously from a row of boxes, and uses nothing but
the turnscrew to put the musket together, excepting on the slott,
which contains the bandsprings, which have to be squared out at
one end with a small chisel." This was not a deskilling technology,
however. A former superintendent at Samuel Colt's armory noted
that interchangeable parts reduced labor requirements by "about
50 per cent" but required "first-class labour and the highest price
is paid for it." In fact, quality output could not be produced without the involvement of well-trained labor.

What became known, rather grandly, as the American System
of Manufacturing had a slow start. Whitney's first order of guns
was delivered to the federal government almost a decade late. All
the same, it expanded rapidly thereafter as arms production was
revolutionized in the first half of the nineteenth century. Next was
the turn of sewing machines. The company that manufacturer
Nathaniel Wheeler formed with inventor Allen B. Wilson started

out producing fewer than 800 machines in 1853 using traditional handcraft methods. By the 1870s, it had introduced interchangeable parts and new specialized machine tools, and its annual output exceeded 170,000 units. Soon the Singer sewing machine company went further, combining interchangeable parts, specialized machinery, and better designs, and was producing more than 500,000 units per year. Woodworking and then bicycles were the next industries to be transformed by the American System of Manufacturing.

In 1831 Cyrus McCormick invented a mechanical reaper. In 1848 he moved his production to Chicago, making more than 500 each year to sell to farmers on the prairie. Increasing farmer productivity boosted North American grain production, made food cheaper around the world, and pushed young people to move from rural areas to burgeoning cities.

According to the 1914 Census of Manufactures, there were already 409 machine-tool establishments in the United States. Many of the machines were superior to those produced anywhere else in the world. As early as in the 1850s, the report of the British Committee on the Machinery of the US noted:

> As regards the class of machinery usually employed by engineers and machine makers, they are upon the whole behind those of England, but in the adaptation of special apparatus to a single operation in almost all branches of industry, the Americans display an amount of ingenuity, combined with undaunted energy, which as a nation we would well do to imitate, if we mean to hold our present position in the great market of the world.

Soon aided by steamships and the telegraph, these machines were spreading in Britain, Canada, and Europe, raising wages for both skilled and unskilled workers, as they had done in the United States. In 1854 Samuel Colt opened an armory by the Thames in London. Singer established a factory in Scotland in 1869, capable of making four thousand machines per week, and another one in Montreal, Canada, following soon thereafter.

Indeed, the potential of new machinery to increase efficiency was long recognized in British metal and machine-tool industries. Following Watt's improvements in the steam engine and using the cotton machinery invented by Arkwright, a British expert noted,

> The only obstacle to the attainment of so desirable an end [increasing production of cotton and other goods] consisted in our almost entire dependence upon manual dexterity for the formation and production of such machines as were required, the necessity of more trustworthy and productive agents rendered some change in the system imperative. In short, a sudden demand for machinery of unwonted accuracy arose, while the stock of workmen then existing were neither adequate in respect to number or ability to meet the wants of the time.

These machinery and production methods were adopted and increased productivity in British industry, while also expanding the set of tasks and opportunities for workers.

Technological change is never enough by itself to raise wages, however. Workers also need to get more bargaining power vis-à-vis employers, which they did in the second half of the nineteenth century. As industry expanded, firms competed for market share and for workers. Workers began to obtain higher wages through collective bargaining. This was the culmination of a long process that had started at the beginning of the century and reached fruition only in 1871, when trade unions became fully legal. This institutional transformation strengthened and in turn was supported by a broader push for political representation.

The Age of Countervailing Powers

The first phase of the British industrial revolution was shaped by a vision that guided technology and determined how the benefits of new industrial machinery would be shared—or not. A different path for technology and distributing the gains from higher productivity necessarily implied a different vision.

A first step in this process was the realization that, in the name of progress, much of the population was being impoverished. A second was for people to organize and exercise countervailing powers against those who had control over the direction of technology and enriched themselves in the process.

In medieval society, such organization was difficult, not just because of persuasion by the society of orders but also because the coordination and exchange of ideas were hampered by the structure of agricultural economy. Industries and densely settled cities changed that. As the statement from the British writer and radical John Thelwall in the Prologue illustrates, factories helped workers organize because, again in Thelwall's words, "Now, though every workshop cannot have a Socrates within the pale of its own society, nor even every manufacturing town a man of such wisdom, virtue, and *opportunities* to instruct them, yet a sort of Socratic spirit will necessarily grow up, wherever large bodies of men assemble" (italics in original). Out of this concentration of workers in factories and cities came several movements agitating for better working conditions and political rights. Perhaps the most important was Chartism.

The People's Charter, drafted in 1838, focused on political rights. At that time, only about 18 percent of the adult male population in Britain had the right to vote, up from less than 10 percent before the 1832 Reform Act. The driving force of Chartism was the creation of a more radical Magna Carta, focused on the rights of ordinary people.

The six demands of the People's Charter were the right to vote for all men over the age of twenty-one, no property ownership requirement to become a member of Parliament, annual parliamentary elections, division of the country into three hundred equal electoral districts, payment of members of Parliament, and secret ballots. Chartists understood that these demands were critical for creating a fairer society. Leading Chartist J. R. Stephens argued in 1839 that "the question of universal suffrage is a knife and fork question, a bread and cheese question. . . . By universal suffrage I mean to say that every working man in the land has a right to a good coat on his back, a good hat on his head, a good roof for the shelter of his household, and a good dinner upon his table."

Chartist demands seem entirely reasonable today, and they gathered strong backing at the time, receiving more than three million signatures in support. But the Chartists ran into stiff opposition from the people who controlled the political system. All the Chartists' petitions were rejected by Parliament, which refused to consider any legislation to improve representation. After several Chartist leaders were arrested and imprisoned, the movement lost steam and disintegrated in the late 1840s.

Yet the demand for political representation among the working classes did not disappear with the demise of the Chartists. Their baton was picked up by the National Reform Union and the Reform League in the 1860s. In 1866 riots broke out in Hyde Park as people organized for political reform. In response, the Second Reform Act of 1867 extended the right to vote to male heads of households over age twenty-one and male lodgers paying at least ten pounds per year in rent, doubling the electorate. The 1872 Reform Act introduced the secret ballot. And in 1884, legislation extended the franchise further, enabling about two-thirds of men to vote.

The Chartists also broke new ground in terms of organizing workers, and the rise of the trade union movement proved to have staying power. Although workers organized and went on strike, forming labor unions for collective bargaining was in principle illegal in the first half of the nineteenth century. Changing this became one of the main political aims of the reform movement that started with the Chartists.

Their pressure was critical for the ultimate formation of the Royal Commission on Trade Unions in 1867, which led to the full legalization of union activities under the Trade Union Act of 1871. The Labour Representation Committee, formed under the auspices of new unionism, became the basis of the Labour Party, providing political voice and a more institutionalized foundation for the working people's ability to stand up to employers and to demand legislation.

This organization and the success of Chartism had much to do with the prevalence of industry and the fact that most people now worked and lived closely packed together in urban areas. By 1850, nearly 40 percent of all British people lived in cities; by 1900, urban residents constituted almost 70 percent of the total population. As Thelwall had anticipated, organizing workers proved much easier in big cities than had ever been the case in agricultural societies.

There were also major changes in how government operated. Pressure for democratization was an important element here. Fear of full-fledged democracy was a major motivation in the minds of even the most conservative politicians, encouraging incremental reforms through legislation. In advance of introducing the First Reform Act of 1832, which increased the electorate from 400,000 to more than 650,000 and reorganized constituencies to be more representative, Whig prime minister Earl Grey declared: "I do not support—I have never supported universal suffrage and annual Parliaments, nor any other of those very extensive changes which have been, I regret to say, too much promulgated in this country, and promulgated by gentlemen from whom better things might have been expected."

The same was true of the later reforms as well, especially when they were spearheaded by conservative politicians. For example, Benjamin Disraeli broke with Robert Peel's Tory government over repealing the Corn Law in 1848. He propelled himself to prominence and eventually became prime minister by aligning with landowners who wanted to keep grain prices high through continued tariffs on imports. At the same time, he courted broader support with political reforms, jingoism, and "one-nation conservatism." Disraeli was also the architect of the Second Reform Act of 1867, which doubled the electorate, and he did not oppose factory-reform legislation. Rural landowners who supported him did not want a revolution to start in manufacturing cities.

Together with political reform came sweeping changes in the civil service. Previously, many government jobs were treated as sinecures. When it came to policy, most officials took a harsh view on what the poor really needed—as seen in the design and implementation of the new Poor Law.

But starting in mid-century, some officials began to gain a degree of autonomy and pursue what could reasonably be regarded as a broader social interest. Benthamite ideas of social efficiency had previously been used to justify policies that can only be described as mean. As better data were collected, it became clear that the market process would not necessarily lead to improvement in social conditions, which was exactly the lesson drawn from the Royal Commission on children's work.

Sanitation is a perfect example of this shift. As we have seen, by the 1840s, burgeoning British manufacturing cities had become cesspools, and most people's living quarters were rife with deadly bacteria and other pathogens. With waste from backyard privies that were seldom emptied, the stench became unbearable, barely imaginable by anyone alive today. Sewers existed in some places, but these were designed mostly to handle rainwater and prevent flooding. For a long time, there were no attempts to improve public infrastructure. In fact, in many jurisdictions it was illegal to connect flushing toilets to sewers.

Edwin Chadwick changed all of that. He was a follower of Jeremy Bentham, but over time he started paying more attention to the plight of ordinary people. He undertook an extensive investigation into urban sanitary conditions, with a particular focus on new manufacturing cities. His report on this topic, which appeared in 1842, created a sensation and brought the issue to the top of the political agenda.

Chadwick also clarified that with different choices of technology and building of better sewer systems, waste could be cleaned and the spread of diseases much reduced. The idea was to bring water into the home on a continuous basis and use that water to flush human waste out through pipes to places where it could be processed safely. For this, the shape and construction of sewers needed to change. Previously, sewers in Britain were made from brick and were primarily designed to be places where sediment would collect. Periodically, municipal workers would dig down and remove enough brick to empty the sewer by hand. In contrast, with different choices on how to engineer the sewer system, wastewater could run continually through egg-shaped terra-cotta pipes, scouring and cleaning as it flowed. Although there was quite a bit of opposition to his innovations, Chadwick prevailed, and the organization of cities was revolutionized, resulting in tremendous improvements in public health.

Remarkably, the political consensus shifted during this process as well. Even Conservatives who stood for traditional values and the importance of protecting private property were persuaded by the need for better sanitation. Speaking in Manchester in April 1872, Disraeli spoke strongly in favor of "sanitary improvement" and improving public health more broadly:

> A land may be covered with historic trophies, with museums of science and galleries of art, with Universities and with libraries; the people may be civilised and ingenious; the country may be even famous in the annals and action of the world, but, gentlemen, if the population every ten years decreases, and the stature of the race every ten years diminishes, the history of that country will soon be the history of the past.

Policy became responsive to public pressure, and policy makers had to think about their social responsibilities. Nobody wanted epidemics or premature death, either from disease or dangerous working conditions—not when voters could kick politicians out of office and trade unions kept the pressure up.

Poverty for the Rest

We can follow the frontier of technological advances of the nineteenth century across Britain and the United States. But it would be wrong to think that innovations had their most consequential effects in these economies. Nor is it reasonable to presume that new technology's impact was similar across places. Countries made different choices on how to use the available technological know-how, with very distinct implications.

In fact, even technologies that created the beginnings of shared prosperity in Britain could, and did, plunge hundreds of millions of people around the world into deeper misery. This can be seen most clearly for people caught up in the rapidly expanding global web of raw materials and manufactured goods.

In 1700 India had some of the most advanced ceramics, metalworking, and printed textile products in the world, all produced by highly skilled artisans who were well paid by the standards of the time. The much coveted "Damascus steel" was from India, and its calico and muslin were greatly prized in England. In response, the English woolen goods industry lobbied successfully for import restrictions in order to keep out the high-quality Indian textiles.

Despite being established to pursue the spice trade, in its early years the East India Company's commercial success was based on bringing finished cotton textiles and clothing into Britain. The company also organized production of cotton clothing in India because that is where skilled workers and the raw materials were. In the first hundred or so years of British control over parts of India, exports of finished cotton goods to Europe rose.

Then came the innovation of harnessing waterpower for operating machines that could spin initially silk (to be used with cotton) and subsequently cotton itself. Britain had fast-moving water and plenty of capital willing to invest. The cost of transporting raw cotton to Liverpool was low relative to the price of the final product.

The East India Company had prevented the export of cotton goods back to India. But this part of its monopoly on trade ended in 1813, resulting in a massive inflow of textiles, particularly from Lancashire, into the Indian market. This was the beginning of the deindustrialization of the Indian economy. By the second half of the 1800s, domestic spinners supplied no more than 25 percent of the country's market, and probably less. Village artisans were driven out of business by cheap imports and had to fall back on growing food or other crops. India deurbanized from 1800 to 1850, with the share of population living in urban areas declining from around 10 percent to under 9 percent.

There was much more to come. Members of the British elite were convinced that they should remake Indian society, purportedly to civilize it, but in reality for their own ends. Lord Dalhousie, governor-general of India in the early 1850s, was adamant that India needed to adopt Western institutions, administration, and technology. Railways, Lord Dalhousie argued, "will afford to India the best security which can now be devised for the continued extension of these great measures of public improvement and for the consequent increase of the prosperity and wealth in the territories committed to its charge."

But instead of economic modernization, railways brought British economic interests, and they intensified control over the Indian population. In his memorandum of April 20, 1853, which shaped policy in the subcontinent for nearly a century, Dalhousie made the case for rail in three parts: to improve access to raw cotton for Britain; to sell "European" manufactured goods in more remote parts of India; and to attract British capital in railway undertakings, hoping that this would subsequently lead to engagement in other industrial activities.

The first train line was built in 1852–1853, using the latest available techniques. Modern engines were imported from England. Dalhousie was right about the value of increased access to raw cotton. Between 1848 and 1856, India further deindustrialized, and its export of raw cotton doubled, making the country primarily an exporter of agricultural products. India also became a significant exporter of items such as sugar, silk, saltpeter, and indigo, and greatly increased its exports of opium. From the mid-1800s until the 1880s, opium was India's largest export, sold mostly by the British to China.

Indian railways did increase internal trade, allowing for a reduction in price differences across faraway places. There was also some boost to agricultural incomes. Bullocks were not an effective substitute means of transportation, and the inland waterway system was not competitive. But there was no significant impact on the iron and steel industry, and most of the rolling stock for Indian railways was purchased from Britain. In 1921 India was still not able to build locomotives.

Even worse, the railways became an instrument of oppression, both by commission and omission. The commission was explicit: rail was used to move troops around the country in response to local trouble. A good railway network can reduce the cost of repression, and this was a key part of how a few thousand British officials could rule over a population of more than three hundred million.

The omission part was more horrific. When famine struck in various parts of the country, it would have been possible to bring food in by rail. However, at key moments in the 1870s and again in Bengal in the 1940s, under Winston Churchill's wartime administration, the British authorities declined to do so, and millions of Indians died.

Plenty of excuses were, and still are, given, but the fact remains that the British never invested enough in irrigation, inland waterways, and clean water, and they never focused the power of railways on feeding people at times when they had no other sources of food or could not afford what was provided by the market. The British attitude was well summarized by Churchill's response in 1929, when he was asked to meet leaders from India's independence movement to become better informed about changes in the country: "I am quite satisfied with my views of India. I don't want them disturbed by any bloody Indian."

Eventually, rail links became an effective element of famine-prevention policy. But not until after the British left India.

Technology has huge potential to raise productivity and can improve the lives of billions of people. But, as we have seen, the path of technology is often biased and tends to deliver benefits mainly to those who are socially powerful. Those without political participation or voice are often left behind.

Confronting Technology's Bias

The Whig view of history is comforting but misleading. There is nothing automatic about the "progress" part of technological progress. In Chapter 4 we discussed how many of the important new agricultural techniques of the last ten thousand years did basically nothing to alleviate misery and sometimes actually intensified poverty. The first century of industrialization proved almost as bleak, with a few people becoming very rich while most found their living standards pressed down hard, and diseases and pollution infested cities.

The second half of the nineteenth century was different, but not because there was an inexorable arc bending toward progress. What set this period apart was a change in the nature of technology and the rise of countervailing power, forcing the people in charge to get serious about sharing the benefits of higher productivity.

In contrast to the drive toward automation in the first phase of the Industrial Revolution, new technologies of the second phase started creating some new opportunities for both skilled and unskilled workers. Railways generated a slew of new tasks and stimulated the rest of the economy through linkages to other sectors. Even more importantly, the American path of technology focused on increasing efficiency, especially by expanding the set of tasks that could be performed by factory workers and new machinery, largely because of the shortage of skilled workers in the country. As these innovations spread in the United States and Europe, they created new opportunities for labor and raised worker marginal productivity throughout the industrializing world.

Equally, institutional changes moved in the direction of bolstering worker power so that this higher productivity would be shared between capital and labor. Industrial growth brought people together in workplaces in cities and allowed the organizing and developing of shared ideas. This changed politics both in the workplace and in the nation.

In Britain, Chartism and the rise of trade unions expanded political representation and transformed the scope of government action. In the United States, union organization combined with farmer protests did the same. Throughout Europe, the rise of factories meant that it was easier to organize workers.

More democracy helped greatly with the sharing of productivity gains as it facilitated collective bargaining for better working conditions and higher wages. With new industries, products, and tasks increasing worker productivity and rents being shared between employers and workers, wages increased.

Political representation also meant demands for less-polluted cities, and public health issues began to be taken more seriously.

None of this was automatic, and it often occurred only after a protracted struggle. Moreover, conditions improved only for those who had sufficient political voice. Women did not have the right to vote in most places during the nineteenth century; consequently, economic opportunities and broader rights for them were much slower to arrive.

Even more jarringly, conditions in most European colonies, rather than improving, significantly deteriorated. Some, such as India, were forcefully deindustrialized when British textiles flowed into the country. Others, India and parts of Africa included, were turned into raw-material suppliers to meet the ferocious appetite of growing industrial production in Europe. And yet others, like the US South, saw the intensification of the worst type of coercion toward labor in the form of slavery, as well as vicious discrimination against native populations and immigrants, all in the name of progress.

7

The Contested Path

I am young, I am twenty years old; yet I know nothing of life but despair, death, fear, and fatuous superficiality cast over an abyss of sorrow. I see how peoples are set against one another, and in silence, unknowingly, foolishly, obediently, innocently slay one another.

—ERICH MARIA REMARQUE,
All Quiet on the Western Front, 1929

There is unanimous agreement among the members on these fundamental points:

1. Automation and technological progress are essential to the general welfare, the economic strength, and the defense of the Nation.

2. This progress can and must be achieved without the sacrifice of human values.

3. Achievement of technological progress without sacrifice of human values requires a combination of private and governmental action, consonant with the principles of a free society.

—PRESIDENT'S ADVISORY COMMITTEE
ON LABOR-MANAGEMENT POLICY, 1962

After the reforms and redirection of technological change in the second half of the nineteenth century, a degree of hope seemed warranted. For the first time in thousands of years, there was a confluence of rapid technological progress and the institutional preconditions for the benefits to be shared beyond a narrow elite.

Yet fast-forward to 1919, and the foundations of this shared prosperity lay in tatters. For people coming of age in Europe during the early 1900s, the world was one of growing economic disparities and unprecedented carnage caused by the Great War, which killed about twenty million people. The tragic deaths of millions of young men and women were the result of brutally effective military technology, which ranged from new guns to more powerful bombs, tanks, aircraft, and poison gas.

It did not escape most people that this was a very dark side of technology. War had been a common occurrence for thousands of years, but the weapons of destruction evolved only slowly after the Middle Ages. When Napoleon was defeated at Waterloo in 1815, he and his opponents fought primarily with short-range muskets and smooth-bore cannons, largely unchanged for centuries. The killing tools of the twentieth century were much more advanced.

Misery did not end with the war. An unprecedented flu pandemic ravaged the world beginning in 1918, infecting more than five hundred million people and causing fifty million deaths. Although the postwar decade witnessed a resurgence of growth, especially in the United States, in 1929 the Great Depression plunged much of the world into the sharpest economic contraction ever experienced in the era of industrialization.

Recessions and economic meltdowns were not unknown. The United States had suffered banking panics and recessions in 1837, 1857, 1873, 1893, and 1907. But none could compare with the Great Depression in terms of disruption and ruined lives. Nor did earlier crises generate anything comparable to the levels of unemployment that the US and much of Europe experienced after 1930. It was not necessary to be exceptionally prescient in the 1930s to see that the world was sleepwalking into another massive technology-fueled carnage.

The Austrian novelist Stefan Zweig captured the desperation that many of his generation felt when he wrote this in his memoir, *The World of Yesterday*, before he and his wife took their lives in 1942:

> Even in the abyss of despair in which today, half-blinded, we grope about with distorted and broken souls, I look up again and again to those old star-patterns that shone over my childhood, and comfort myself with the inherited confidence that this collapse will appear, in days to come, as a mere interval in the eternal rhythm of the onward and onward.

One might have taken issue with Zweig's cautious optimism that this was an interval on the way to some sort of progress, onward and onward. In the 1930s few could express the optimism of the Whig version of history.

But events, at least in the medium run, proved Zweig right. After World War II, much of the Western world, and some countries in Asia, built new institutions supporting shared prosperity and enjoyed rapid growth that benefited almost all segments of their societies. The decades following 1945 came to be referred in France as the "*les trente glorieuses*," the thirty glorious years, and the feeling was widespread throughout the Western world.

This growth had two critical building blocks, similar to those that had begun to emerge in the United Kingdom during the second half of the nineteenth century: first, a direction of travel for new technology that generated not just cost savings through automation but also plenty of new tasks, products, and opportunities; and second, an institutional structure that bolstered countervailing powers from workers and government regulation.

These building blocks were put in place during the 1910s and 1920s, suggesting that we should view the first seven decades of the twentieth century as part of the same epoch, albeit with significant reversals along the way. Studying these two building blocks, and the vision that developed alongside, provides not only clues about how we can rebuild shared prosperity today; it also demonstrates how contingent and difficult this outcome really was. There was opposition at many critical points from powerful forces, driven by narrow visions and selfish interests. The opposition did not initially prevail, although it laid the foundations for the subsequent dramatic unraveling of shared prosperity, which we discuss in Chapter 8.

Electrifying Growth

The United States had a total GDP of about $98 billion in 1870, just after the Civil War. By 1913, it had reached $517 billion in constant prices. The US was not just the largest economy in the world, but together with Germany, France, and Britain, it was a scientific leader. Newly developed technology permeated the American economy and transformed people's lives.

But there was plenty to worry about—inequality, poor living conditions, as well as displacement and impoverishment of workers, in the same way that the British people had suffered in the decades following 1750. In fact, the danger may have been greater for the US, which in the middle of the nineteenth century was still largely agricultural. In 1860, 53 percent of the workforce was working in farming. Rapid agricultural mechanization could throw millions of people out of work.

And some new agricultural machines had this effect. The McCormick reaper, invented in the 1830s and continuously improved thereafter, reduced the need for labor at harvest time. Reapers, twine binders, improved threshing machines, mowers, and then combine harvesters completely changed US agriculture in the decades following 1860. These machines reduced the need for workers per acre at various parts of the crop-growing cycle. Labor requirements for corn production in 1855 using hand methods were more than 182 worker-hours per acre. Mechanization reduced the need for labor to less than 28 worker-hours per acre by 1894. The reduction was similar in cotton (from 168 to 79 worker-hours per acre) and in potatoes (from 109 to 38 worker-hours per acre) over 1841–1895 and 1866–1895 respectively. Potential gains were even more staggering for wheat, from more than 62 worker-hours per acre in 1829–1830 to just over 3 worker-hours per acre in 1895–1896.

The consequences of mechanization for workers were sweeping. The share of labor in agricultural value added was around 32.9 percent in 1850. By 1909–1910, it had fallen to 16.7 percent. The fraction of the US population employed in agriculture declined equally rapidly, reaching about 31 percent in 1910.

If industry had also moved in the direction of automating work and shedding labor, the implications for the US labor force would have been dire. Instead, something very different happened. As US industry innovated rapidly, the demand for labor increased significantly. The share of workers employed in US manufacturing increased from 14.5 percent in 1850 to 22 percent in 1910.

This moment was not just about more people finding jobs in manufacturing; the share of labor in national income increased, a telltale sign that technology was moving in a more worker-friendly direction. During the same time period, the share of labor in value added in manufacturing and services increased from about 46 percent to 53 percent (the rest went to owners of the machinery and financiers).

How did the US avoid the Luddite phase of British industrialization, with workers displaced and impoverished by machines and stagnant or declining wages?

Part of the answer is about the technological path that the US developed as machines came into wider use. As we saw in Chapter 6, the American path of technology strove to raise productivity to make better use of labor that was relatively in short supply. The interchangeable parts system was first and foremost an effort to simplify the production process so that workers lacking in artisanal skills could produce high-quality products. Efforts to improve productivity in similar fashion continued throughout the second half of the nineteenth century. One indication of this innovativeness was an explosion in patents. The United States had 2,193 patent applications in 1850. By 1911, this number had increased to 67,370.

More important than the number of patents was the direction this innovative energy took, building on two blocks: mass production and a systems approach, both founded on Whitney's lead. Mass production meant the use of machinery to produce a large amount of standardized, reliable output at lower cost. The systems approach focused on integrating engineering, design, manual labor, and machinery, and organizing different parts of the production process in the most efficient way.

The productivity bandwagon depends on new tasks and opportunities for workers and an institutional framework that enables them to share some of the productivity gains. We saw in Chapter 1 that it is also more likely to work when technological advances generate sizable improvements and these in turn stimulate greater demand for labor in other sectors—for example, via backward and forward linkages (linkages to input suppliers and to customer industries, respectively). The systems approach and mass production were particularly important in this respect because they aimed at large cost reductions and increased production significantly, generating demand for inputs from other sectors and a potential boost to their output.

This direction of technology increased worker marginal productivity and living standards, as E. Levasseur, the French visitor we quoted in Chapter 6, appreciated:

> The manufacturers judge that the movement [of adopting industrial machines] has been advantageous to workmen, as sellers of labor, because the level of salaries has been raised, as consumers of products, because they purchase more with the same sum, and as laborers, because their task has become less onerous, the machine doing nearly everything which requires great strength; the workman, instead of bringing his muscles into play, has become an inspector, using his intelligence.

Although these tendencies were visible in the 1870s, two interrelated changes intensified them and transformed American industry: electricity and the greater use of information, engineering, and planning in the production process.

There were advances in the science of electricity starting in the late eighteenth century, but the big breakthroughs that reshaped the world began in the 1880s. Thomas Edison not only advanced the scientific understanding of light but also spearheaded its mass-scale adoption. Filament bulbs increased the amount of available light for reading during nighttime darkness by a factor of about twenty (compared with candles).

Electricity is particularly important because it is a general-purpose technology. This new, versatile power source enabled the creation of many new devices. It also spurred fundamentally different organizations. And choices available for developing and using electrical technologies produced widely different distributional effects.

The new communication devices made possible by electricity, especially the telegraph, telephone, and radio, had a momentous impact on US industry, as well as on the domestic consumer. With better communications, logistics and planning improved, and these proved critical for the success of the systems approach.

Arguably, the most consequential application of electricity in the production process was the transformation of how factories operated. Andrew Ure described the essence of the early British factory in 1835 this way: "The term *Factory*, in technology, designates the combined operation of many orders of work-people, adult and young, in tending with assiduous skill a system of productive machines, continuously impelled by a central power" (italics in original).

The use of "central power" in this fashion was a breakthrough, Ure argued, because it increased efficiency and coordinated work. But relying on a single central source of power, whether from wind, water, or steam, was also a bottleneck. It limited the division of labor, forced machinery to be clustered around the central power source, did not permit some machinery to use more power when it needed to do so, and led to frequent work stoppages that affected the entire production process. There was essentially no way of sequencing machines in the order in which tasks would need to be completed because the locations of machines were dictated by their power needs. For example, machines driven by overhead shafts had to be placed very close to the central source: they lost power when they were farther away. This meant that conveyor belts could not be introduced and semifinished products had to be moved back and forth between machines located in different parts of the factory.

This all changed with the municipal distribution of electric power into households and workplaces starting in 1882. Electricity spread rapidly. In 1889, about 1 percent of factory power came from electricity. By 1919, this share exceeded 50 percent.

With the advent of electricity, factories became much more productive. Better lighting enabled workers to better see their environs and operate machinery with greater precision. Electricity additionally enabled better ventilation and easier maintenance. As an architect noted in 1895, "Incandescent electric light is the acme of all methods of lighting; requiring no care; always ready; never impairing the air in the room; no heat; no odors; perfections of neatness, and steady as clocks." Electricity also promised new applications including electrical time clocks, control devices, and new furnaces that could be integrated with other machinery to improve the precision of mechanical work.

Even more important was the reorganization of the factory thanks to flexible sequencing of machinery. Each equipment could now have its own, dedicated local power source. Westinghouse Electric & Manufacturing Company was at the forefront of many of these developments. In 1903, as a Westinghouse engineering executive emphasized,

> But the greatest advantage of the electric drive lies in its greater flexibility and the freedom which it affords for better planning of the whole establishment and the arrangement of tools. Large tools equipped with independent motors may be located in any position most convenient for work without regard to such limitations as obtain when the power must be taken from line shafting, and, as has already been pointed out, there is the immense advantage of ability to use large portable tools. The absence of overhead shafting also gives a free space for the operation of cranes, so that they may be used to the best advantage. A shop without overhead networks of shafting and belts is also much lighter and more

inviting, and it has been demonstrated by experience that the output of labor is materially increased when working in well lighted and well ventilated shops.

These expectations were not wrong. The flexible location and modular structure of the factory enabled the number of specialized machines to increase rapidly. One of the very first factories to make use of them was Columbia Mills in South Carolina. Columbia Mills was designed around electricity in the early 1890s and immediately started demonstrating the advantages of electric motors and better lighting. In such mills and in early Westinghouse-built factories, we see how a dedicated power source for different types of machinery allowed a simpler factory layout, less transport of goods within the factory, and much easier control of power going to a particular machine.

Electric power also meant less need for repairs and a more modular structure, where minor repairs could be conducted without stopping the overall production process. This type of factory reorganization, electrical machinery, and associated conveyor belts were adopted in many industries and resulted in striking productivity gains. Foundries that introduced these methods were estimated to produce as much as ten times more iron using less space.

The sizable productivity improvements from electric power were vital for the expansion of the economy and higher demand for workers coming from sectors beyond manufacturing. In addition, they had major benefits for workers because of the way in which electricity was used to reorganize factories.

New Tasks from New Engineers

In theory, any new energy source can automate some existing tasks so that the need for workers does not increase much or at all. More advanced machinery and greater mechanical power certainly implied some automation. Nevertheless, demand for employees increased significantly in American industry around the turn of the twentieth century. In fact, the numbers on the labor share of national income indicate that labor became more central to the production process in the early twentieth century, raising its share of national income. Why?

Another fundamental change in the organization of production is a large part of the answer. Concurrent with the rise of electric power in manufacturing came an elevated role for engineers and white-collar workers, who restructured factories and the production process, with better consequences for productivity and labor.

US factories in the 1850s looked like their British counterparts. An entrepreneur who had invested the capital and arranged to install the machinery managed the workforce. Some early manufacturers, like Richard Arkwright, excelled at introducing new production techniques. But in general, there was little by way of production planning, information collection, efficiency analysis, and continuous improvement. Accounting and inventory control were haphazard. There was insufficient attention to design and almost none to marketing. The organizational aspects of industry started changing in the last decades of the nineteenth century, heralding the onset of the age of engineer-managers.

In 1860 white-collar workers, including managers and engineers, made up less than 3 percent of all employees in manufacturing in the United States. By 1910, this had increased to almost 13 percent. At the same time, the total manufacturing workforce expanded from less than one million to more than 7.5 million workers. The share of white-collar workers continued to grow after World War I, reaching almost 21 percent of the workforce by 1940.

These white-collar workers remade the factory in a more efficient form, and in the process raised labor demand, not just for themselves but also for their blue-collar brethren now performing new tasks. Managers collected information, sought to increase efficiency, started improving designs, and continuously readjusted production methods by introducing novel functions and tasks. The combination of engineers' role in production, information collected by white-collar workers, and electricity was critical in the installation of specialized electrical devices and the new tasks that accompanied them, such as welding, punching, and specialized machine operation.

In this way, the reorganization of manufacturing enabled by white-collar workers created relatively high-paying jobs for blue-collar workers. As the scale of production expanded, there was even more demand for white-collar employees.

Another dimension of job growth came from the linkages that new factories created for the retail and wholesale sectors. As they churned out more output and mass production became widespread, new jobs for engineers, managers, sales clerks, and administrators appeared in these sectors as well.

It is also notable that many of these white-collar tasks required more skills than the majority of the jobs available in the nineteenth century. For example, clerical workers needed decent numeracy and literacy skills to measure production, inventories, and financial accounts, and to communicate their findings accurately. This is where another trend in the US economy, the rapid expansion of the number of workers with a high school degree, came to the rescue. In 1910, less than 10 percent of people aged eighteen had high school diplomas. By 1940, this number had risen to 50 percent. This was the result of large investments in schooling in the second half of the nineteenth century, with local "common schools" offering elementary education throughout the country. By the 1880s, approximately 90 percent of White children between the ages of eight and twelve were in school in the Northeast and the Midwest. (Schooling for Black children lagged far behind.) Statistical analysis confirms the vital role that new tasks and industries played in the growth in labor demand. One study documents that new industries with a more diverse set of occupations were at the forefront of both overall employment growth and the expansion of white-collar occupations in US manufacturing during this period. Another study estimates that between 1909 and 1949, US productivity growth was associated with increases in employment, and this pattern was manifest first in new industries relying on electrical machinery and electronics.

It is worth reiterating two critical aspects of the direction of technology during this era. First, companies continued to automate parts of the production process. In fact, not just in agriculture, but throughout the economy, new machinery substituted for labor in some tasks. The key difference from the first phase of the British industrial revolution was that the reduction in labor requirements driven by automation was offset, sometimes more than one for one, with other aspects of technology that created opportunities for workers, especially for those with some basic schooling, to be employed in manufacturing or services.

Second, although some of the benefits for workers from the expansion of various sectors were natural given the productivity improvements and linkages from new factories, others were a result of choices made by companies and the new cadre of engineer-managers. The direction of progress during this era was not an inexorable consequence of the nature of the leading scientific breakthroughs of the age. In fact, electricity, as a general-purpose technology, allowed different applications and distinct development paths.

Managers and engineers could have chosen to double down on automation as a method to cut costs in existing industries. Instead, they built on the American path of technology and pushed to build new systems and machinery, increasing efficiency and in the process augmenting the capabilities of both skilled and unskilled labor. These technological choices were foundational to the increase in demand for workers in industry, which more than made up for declining labor intensity in agriculture and in some manufacturing tasks.

In the Driving Seat

There is no better example to illustrate how electricity, engineering, the systems approach, and new tasks came together than the automobile industry, especially the Ford Motor Company.

US automobile manufacturing started in 1896. The Ford Motor Company, led by its iconic owner and manager, Henry Ford, was established in 1903. Its early vehicles, known as models A, B, C, F, K, R, and S, were produced using techniques common in the industry, combining elements of the interchangeable parts system with artisanal skills. These were medium-priced automobiles serving a niche market.

Henry Ford's ambition from early on was to produce many more cars and sell them at a lower price in order to reach a mass market. Although Model N was a first step in this direction, it did not break the mold. It was produced in the company's Piquette Avenue plant in Detroit, which employed the same architecture and structure as factories powered from a central source and did not contain the full suite of electrical machinery.

The sea change came with the famous Model T, which Ford launched in 1908 as a "car for the masses." What enabled it was a perfect blend of advances that had taken place in other industries, adapted to car manufacturing. The Ford factory moved to a new Highland Park facility, which opened a machine-tool plant that was organized on a single floor and had a range of new electrical machinery. The plant combined the novel factory organization with full-scale adoption of interchangeable parts and, later, conveyor belts to achieve mass production. Around that time the company boasted: "We are making 40,000 cylinders, 10,000 engines, 40,000 wheels, 20,000 axles, 10,000 bodies, 10,000 of every part that goes into the car . . . all *exactly alike*" (italics in original).

Mass production allowed further expansion. The output of the company soon exceeded two hundred thousand automobiles per year, which was mind-boggling to contemporaries.

The spirit of Ford's approach to production was captured by a reporter from the *Detroit Journal* who visited the new Highland Park factory, just outside Detroit, where the Model T was being mass-produced. He summarized its essence as "System, system, system!" An in-depth study by Fred Colvin, published in the *American Machinist*, arrived at the same conclusion:

> So thoroughly is the sequence of operations followed that we not only find drilling machines sandwiched in between heavy millers and even punch presses, but also carbonizing furnaces and babbitting equipment in the midst of the machines. This

reduces the handling of work to the minimum; for, when a piece has reached the carbonizing stage, it has also arrived at the furnace which carbonizes it, and, in the case of work to be finished by grinding, the grinders are within easy reach when it comes from the carbonizing treatment.

Henry Ford himself was quite clear on this issue:

> The provision of a whole new system of electric generation emancipated industry from the leather belt and the line shaft, for it eventually became possible to provide each tool with its own electric motor. This may seem only a detail of minor importance. In fact, modern industry could not be carried on with the belt and line shaft for a number of reasons. The motor enabled machinery to be arranged according to the sequence of the work, and that alone has probably doubled the efficiency of industry, for it has cut out a tremendous amount of useless handling and holding. The belt and shaft were also very wasteful of power—so wasteful, indeed, that no factory could be really large, for even the longest line shaft was small according to modern requirements.

Ford emphasized that preelectricity methods could not keep up with what was now required: "Also high-speed tools were impossible under the old conditions—neither the pulleys nor the belts could stand modern speeds. Without high-speed tools and the finer steels which they brought about, there could be nothing of what we call modern industry."

This organization of production, combined with advances in electrical machinery, enabled a much cheaper product that was reliable and could be operated without special knowledge of engines and the mechanical parts of the car. The Model T was initially priced around $850 (equivalent to about $25,000 today), compared to the cost of other cars, which were typically around $1,500.

Ford's subsequent assessment captured the spirit of the breakthroughs that the car industry pioneered:

> Mass production is not merely quantity production, for this may be had with none of the requisites of mass production. Nor is it merely machine production, which also may exist without any resemblance to mass production. Mass production is the focussing upon a manufacturing project of the principles of power, accuracy, economy, system, continuity, and speed. The interpretation of these principles, through studies of operation and machine development and their co-ordination, is the conspicuous task of management.

The implications for labor were similar to those that followed from the introduction of the factory system in other industries, but much amplified for several reasons. Mass production of cars meant a very large increase for inputs and a sizable boost to many other sectors that depended on transportation for consumers and goods. Technologically, the automobile industry was among the most advanced sectors of the economy and thus made more extensive use of engineering, designs, planning, and other information-intensive activities. It was, as a result, at the forefront of the creation of new white-collar tasks.

Ford was also a leader in introducing new blue-collar tasks, for the nature of assembly, painting, welding, and machine operation transformed with the reorganization of the factory. This change was not without costs for workers, who often found it challenging to cope with the requirements of Ford factories.

Worker adjustment problems had several consequences, the most important one being very high rates of absenteeism and turnover. High turnover was particularly challenging to Henry Ford and his engineers because it made the assembly line and production planning much more difficult. For example, turnover rates in the Highland Park plant reached a staggering 380 percent per annum in 1913. It was impossible to keep workers in place. The reason for this discontent was well summarized by a letter from the wife of a worker to Henry Ford: "The chain system you have is a *slave driver*! *My God*!, Mr. Ford. My husband has come home & thrown himself down & won't eat his supper—so done out!" Such reactions induced Ford and his engineers to start increasing pay for workers, first to $2.34 per day and then reaching the famous $5 per day, which was remarkably high in an economy where most of the workforce toiled for much less. As pay increased, turnover and absenteeism declined, and Ford also believed that worker productivity increased.

An important source of productivity gains was training. Ford factories required specialized skills, but these skills were not difficult to acquire. The flexible factory had created a modular task structure, with most machinery requiring only a well-defined set of steps and troubleshooting knowledge to operate. As Colvin emphasized, "The keynote of the whole work is simplicity." This meant that workers could be easily trained to acquire the necessary skills, and Ford, like many other companies around this time, started offering extensive training in order to increase worker productivity.

The link between advanced machinery and new skills training as a foundation for creating novel opportunities and increased demand for workers would turn out to be critical in the postwar era as well. In 1967 a manager at Ford described the company's hiring strategy this way: "If we had a vacancy, we would look outside in the plant waiting room to see if there were any warm bodies standing there. If someone was there and they looked physically okay and they weren't an obvious alcoholic, they were hired." What this meant in practice was unprecedented opportunities for workers who did not have high levels of schooling and specialized know-how when they came into the job market. Unskilled workers could be hired, trained, and used productively with advanced machinery, expanding demand for all workers. The implications were far-reaching: they created a powerful force toward more widely shared prosperity—some of the most highly paid jobs were open to unskilled workers.

There was another reason why Ford was receptive to higher wages. As Magnus Alexander, an electrical engineer who had helped design the production systems at Westinghouse and General Electric, put it, "Productivity *creates* purchasing power." And purchasing power was vital for mass production.

These developments were not confined to the Ford Motor Company and became part and parcel of American industry. General Motors was soon besting Ford at its game, investing more in machinery and developing a more flexible production structure. Mass production meant mass market, but mass market did not necessarily mean everybody buying the same car in the same color. GM understood this ahead of Ford, and while Ford persisted in offering the Model T to everybody, regardless of their tastes and needs, GM started using its flexible production structure to offer more versatile models.

An Incomplete New Vision

The vision of the middling-sort entrepreneurs that powered the early phase of the British industrial revolution was one based on increasing efficiency in order to reduce costs so that they could generate more profits. How this affected the "meaner sort" of people they employed was of little concern to these ambitious entrepreneurs. Profits were central for the industrialists of the American system as well, and the early phases of the country's industrialization were associated with a sizable increase in inequality, as industrialists such as Andrew Carnegie in steel and John D. Rockefeller in oil adopted new techniques, dominated their sectors, and made huge fortunes.

Many of these industrial tycoons were hostile to workers who wanted to organize. For instance, Andrew Carnegie used violence against union members during the Homestead Strike of 1892.

Nevertheless, several industrialists recognized that a more cooperative relationship with their workforces and their communities would be beneficial for the company in the age of electric power. Henry Ford was a pioneer in this as well. Together with the $5 day, the company introduced a pension program, other amenities, and a range of benefits for families, signaling an intent to share some of the substantial profits it was earning from new technologies and mass production of cars.

Ford was not motivated by altruism. He adopted these measures because he believed that higher wages would reduce turnover, limit strikes, prevent costly stoppages of the assembly line, and increase productivity. Many leading companies followed suit, introducing their own version of high-wage policies and amenity programs. Alexander summarized the essence of this new approach, arguing that "whereas *laissez faire* and intensive individualism marked the economic life of the first half of the history of the United States, the emphasis is now shifting towards a voluntary assumption of social obligations, implied in the direction of economic activities and of national and international and cooperative effort in the common interest."

Another person crystallizing this vision was the American economist John R. Commons, who advocated for a type of "welfare capitalism," where productivity increases would benefit workers based on a bond of loyalty and reciprocity between employers and employees. According to Commons, focusing on reducing costs at the expense of workers was a losing proposition.

Nevertheless, this type of welfare capitalism was bound to remain no more than an aspiration unless there were institutional changes enabling workers to organize and exercise countervailing powers. This started happening after the Great Depression, and initially far away from the United States.

Nordic Choices

The Great Depression opened with sharp falls in US stock prices in 1929, wiping out half of the stock market value within a few months. It brought first the US and then the world economy to a standstill. By 1933, US GDP had fallen by 30 percent, and unemployment was up to 20 percent. There were widespread bank failures, destroying people's life savings.

The shock from the stock-market meltdown and the resulting economic chaos was palpable. Rumors spread that stockbrokers jumped out of windows of high-rises as the market collapsed. It would turn out later that this was not quite the case. When the chief medical examiner of New York City investigated the data, he found that suicides were down, year on year, in October and November 1929. Even if financiers throwing themselves onto the concrete pavement was an exaggeration, the macroeconomic quagmire the country found itself in was not.

Though originating in the United States, economic troubles rapidly spread around the world. By 1930, most of Europe was in a sharp economic contraction. How different countries responded to economic hardships differed, with distinct political and social consequences. Germany was already beset by political polarization, with several right-wing parties trying to limit the ability of social democrats to govern the country. Lawmakers came up with no comprehensive responses, and some of their policy reactions further deepened the crisis. Soon German industrial production was in freefall, dropping to just over half of its value in 1929, and unemployment reached over 30 percent.

The economic hardships and incompetence, and in the eyes of many, indifferent policy reactions, paved the way to an almost complete loss of legitimacy of established parties and the rise of the National Socialist (Nazi) Party. The Nazis were no more than a fringe political movement, receiving only 2.6 percent of the national vote share in the 1928 election before the Depression. Their vote share shot up in the first election after the Depression and reached 37.3 percent in July 1932. In November 1932, the Nazis lost ground but still won 33.1 percent of the vote, and in January 1933 Adolf Hitler became chancellor.

Similar dynamics played out in France, which also experienced a debilitating economic meltdown, incoherent and ineffective policy responses, and extremist parties gaining power, even if the democratically elected government held on.

The reaction in the small and economically still backward country of Sweden was very different. In the late 1920s the Swedish economy was predominantly agricultural, with approximately half of the population working in farming. The country had introduced full universal suffrage only in 1921, and industrial workers had limited political power. However, the party that represented them, the Swedish Social Democratic Workers' Party (SAP), had one major advantage. Going back to the late nineteenth century, the party's leadership had recognized that it had to reform Sweden's institutions. To achieve this objective, the party had to come to power democratically, which meant moving away from Marxism, and its leadership endeavored to form a coalition with rural workers and the middle classes. As one of the party's most influential leaders, Hjalmar Branting, argued in 1886, "In a backward land like Sweden we cannot close our eyes to the fact that the middle class increasingly plays a very important role. The working class needs the help it can get from this direction, just like the middle class for its part needs the workers behind it, in order to be able to hold out against [our] common enemies."

After the beginning of the Great Depression, the SAP started campaigning for a robust policy response that had both a macroeconomic leg (greater government spending, higher wages in industry to prop up demand, and expansionary monetary policy by leaving the gold standard) and an institutional leg (providing foundations for consistent sharing of profits between labor and capital, redistribution via taxation, and social insurance programs).

To achieve this, the party started seeking coalition partners. It looked like a hopeless task, at least at first. The center-right had no intention of working with the SAP, and worker and agrarian parties were often at loggerheads, not just in Sweden but throughout much of Western Europe during this period. The SAP, which was organically linked to trade unions, was determined to maintain high wages in industry and expand manufacturing employment. Trade unions viewed higher food prices as undermining these plans, for they would raise the cost of the much-needed government programs and erode the real take-home pay of workers. Agrarian interests prioritized higher food prices and did not want the government's resources to be directed toward industrial programs.

The SAP's leadership understood the critical importance of a coalition that would give the party a robust majority in Parliament. This was in part a response to the dire economic conditions, for poverty and unemployment had started rising rapidly in

Sweden in 1930. But it was also because the party's leadership was observing how inaction was pushing other European countries into the arms of extremists.

In the run-up to the 1932 national elections, the party's leader, Per Albin Hansson, consistently presented the party as the "people's home," embracing all working people and the middle classes. As the party's program stated, "The party does not aim to support and help [one] working class at the expense of the others. It does not differentiate in its work for the future between the industrial working class and the agricultural class or between workers of the hands and workers of the brain." The appeal worked, and the party increased its vote share from 37 percent in 1928 to almost 42 percent in 1932. It also convinced the Agrarian Party to join Hansson in a coalition. This was based on a deal that is now referred to as a "cow trade," in which the SAP received support from agrarian interests to increase spending, including for the industrial sector, in return for greater protection for the agricultural sector and higher government-set prices.

As important as the macroeconomic responses was the new institutional structure that the SAP was building. The solution it came up with for institutionalizing rent sharing was to bring the government, trade unions, and businesses together to reach mutually beneficial bargains, which would secure equitable distribution of productivity gains between capital and labor.

The business community was at first opposed to this corporatist model, viewing the labor movement in the same way that their German and American counterparts did—something to be avoided to keep costs low and maintain control in workplaces. But that started changing after the 1936 elections, which witnessed further gains for the SAP. The business community saw the writing on the wall; it would not be able to bring down the SAP by sheer opposition.

In a famous meeting in the resort town of Saltsjöbaden in 1938, a deal was struck with a significant fraction of the business community, agreeing on the basic ingredients of the Scandinavian social democratic system. The most important elements were industry-level wage setting, ensuring that profits and output increases were shared with workers, and significant expansion of redistributive and social insurance programs, along with government regulations. This was not a deal to expropriate the business community, however. There was general agreement that private businesses had to remain productive, and this would be achieved by technological investments.

Two elements of this deal are particularly noteworthy. First, corporations would have to pay high wages and negotiate employment and working conditions with unions, precluding mass layoffs to reduce labor costs. They would then have incentives to increase the marginal productivity of workers, locking in a natural bias toward worker-friendly technologies.

Second, industry-level bargains created incentives for corporations to raise productivity without fear that these increases would lead to further wage hikes. To put it simply, if a company managed to boost its productivity above that of its competitors, under industry-level wage setting it would continue to pay more or less the same wages, and thus the full increase in productivity would translate into greater profits. This realization provided a powerful motive for businesses to innovate and invest in new machinery. When this happened throughout an industry, it led to higher wages, creating benefits for both labor and capital.

Remarkably, therefore, the corporatist model that SAP and the trade unions put together in Sweden achieved some of the aspirations of the vision of welfare capitalism that people such as J. R. Commons were articulating in the United States. The difference was that welfare capitalism coming purely as a voluntary gift from corporations was highly contingent and often ran into resistance from managers intent on increasing profits and reducing wages. When embedded in an institutional framework bolstering the countervailing powers of workers and including the regulatory capacity of the state, it was on much firmer ground.

Trade unions played a central role in the building of the state's regulatory capacity as well. They implemented and monitored the expanded welfare programs, enabling communication between workers and the management when new technologies were being introduced or some establishments were being downsized.

At the beginning of the twentieth century, Sweden was an extremely unequal place. The share of the richest 1 percent of the country in national income was over 30 percent, which made it more unequal than most European countries. In the decades after the basic institutional structure of this new coalition was established, employment and productivity grew rapidly, but inequality declined. By the 1960s, Sweden had become one of the most equal countries in the world, with the share of the top 1 percent of the population hovering around 10 percent of national income.

New Deal Aspirations

Like the Social Democratic Workers' Party in Sweden, US president Franklin Delano Roosevelt (FDR) was elected with the promise of confronting the Great Depression. FDR's vision had a lot in common with the Swedes. Macroeconomic response in the form of higher spending, support for agricultural prices, public works, and other policies to prop up demand were critical. In 1933 FDR's administration was the first in national history to introduce a minimum wage, which was viewed not just as a poverty-reducing legislation but as a means of macroeconomic stabilization, for it would create additional purchasing power in the hands of workers. Equally pivotal was institutional overhaul, which centered on creating countervailing powers against businesses, both in the form of government regulation and a stronger labor movement.

In this institutional overhaul, New Dealers were building on the reforms that the Progressive movement had implemented (which we will discuss in more detail in Chapter 11). But their plans went further.

Economist Rexford Tugwell, a member of FDR's "Brain Trust," captured the essence of New Dealers' regulatory approach: "A strong government with an executive amply empowered by legislative delegation is the one way out of our dilemma, and forward to the realization of our vast social and economic possibilities." Based on this philosophy, the administration introduced what the *New York Times* named "forty alphabet New Deal agencies," ranging from AAA (Agricultural Adjustment Administration) to

USES (United States Employment Service), and proceeded to implement several policies similar to those adopted by the Swedish Social Democratic Workers' Party, including wage and price controls, protection for workers under "codes of fair practices," and actions against child labor.

Measures aimed at strengthening the labor movement were arguably even more important. They were based on the belief that, despite Progressive Era reforms, businesses were still not sharing productivity and profit gains with their workers, and low wages were creating both inequality and macroeconomic problems. Inequality was high and growing. By 1913, the richest 1 percent of households were already capturing about 20 percent of national income, and this number kept on rising, exceeding 22 percent by the end of the 1920s.

A major policy initiative of the FDR administration was the Wagner Act of 1935, which recognized the right of workers to collectively organize (without intimidation and threat of firing from employers) and introduced various arbitration procedures to resolve disputes. Even before the Depression, some intellectuals and businesspeople were acknowledging that without collective bargaining, productivity gains would not be shared fairly, even if companies such as Ford raised wages to reduce turnover.

In 1928 the pioneering American engineer Morris Llewellyn Cooke spoke to the Taylor Society, a group dedicated to "scientific management":

> The interests of society, including those of the workers, suggest some measure of collective bargaining in industry to the end that the weaker side may be represented in negotiations as to hours, wages, status and working conditions. Collective bargaining implies the organization of the workers on a basis extensive enough—say nation-wide—as to make this bargaining power effective.

Cooke, who later became a senior government official under presidents Roosevelt and Truman, argued that given the prevalence of large modern corporations, it was critical that workers become organized and "to look upon some organization of the workers, such as labor unions, as a deep social need."

Carle Conway, chairman of the board of Continental Can and a "hero of capitalist endeavor," according to the Harvard Business School (unofficial scorekeeper), was surprisingly pro-union:

> Certainly anyone who has been in business during [the past 30 years] would have to be naive to think that management by and large desired collective bargaining or certain of the other reforms which labor has finally won. . . . But isn't it also likely that better understanding of the basic fundamentals involved in the struggle over the last thirty years between labor and management can work toward harmonizing the two viewpoints into a common objective and so make collective bargaining and many of the other reforms operate in the interests of both labor and management?

However, New Dealers' aspirations, in contrast to those of the Swedish Social Democratic Workers' Party, would not be fully realized. One nexus of resistance came from southern Democrats, who were worried that New Deal policies would challenge Jim Crow segregation and hence worked to make the legislation less comprehensive than in Sweden.

Aspects of the New Deal plan aimed at greater spending and expansion of collective bargaining faced stiff resistance as well and were often blocked by the US Supreme Court. Nevertheless, FDR's policies did manage to stop the slide of the macroeconomy and gave a huge boost to the labor movement. Both these elements would play an important role in the postwar era.

It was vital that these major institutional overhauls, both in Sweden and in the United States, happened within the context of a democratic system. FDR himself attempted to centralize power in his hands and even tried to circumvent resistance to his policies from the Supreme Court by increasing the number of justices. But his own party blocked the court-packing attempts.

The Allies won World War II on the back of the United States throwing its entire economy into war production. Factories that had made washing machines now produced munitions. Landing craft were manufactured by the thousands. The US had started the war with six aircraft carriers. By early 1945, it was producing one highly effective, even if smaller, carrier every month.

The US military struggled to build robust logistics to support its troops overseas. In September 1942, as General Eisenhower's forces prepared to invade North Africa, Ike complained to Washington that appropriate supplies had not arrived in England. The War Department replied caustically: "It appears that we have shipped all items at least twice and most items three times." For some years there was chaotic oversupply in transatlantic shipments, albeit not in a way that prevented the US from winning. As one general quipped, "The American Army does not solve its problems, it overwhelms them."

All this production required workers, and workers had to keep at it hard. After victory in 1945, what would be their reward for these extraordinary efforts?

Glorious Years

Although the foundations of shared prosperity were laid in the first four decades of the twentieth century, most Americans would not have recognized them clearly. The first half of the century witnessed the two most brutal, destructive, and murderous wars of human history and a massive depression that instilled fear and uncertainty in the people who survived it. These fears were deep and long-lasting. Recent research documents that people who lived through the Great Depression were often permanently scarred and remained unwilling to take economic risk for the rest of their lives. There were periods of robust economic growth in the first half of the century, but these were, as often as not, associated with much of the benefits being captured by the wealthy, so inequality remained high and sometimes even increased.

Against this background, the decades after 1940 were striking. US aggregate output (gross domestic product, or GDP) per capita grew at an average rate of more than 3.1 percent between 1940 and

1973. This growth was fueled by productivity improvements, both during and after the war. In addition to GDP per capita, total factor productivity (TFP) growth is an informative measure of economic growth, in part because it takes out the contribution of increases in the capital stock (machinery and buildings). The TFP growth rate is therefore a better measure of technological progress, for it picks up how much GDP growth comes from technological changes and efficiency improvements. US TFP growth (in the nonagricultural, nongovernment sector) between 1891 and 1939 averaged less than 1 percent per year. Between 1940 and 1973, it rose to an average of almost 2.2 percent per year. This was not driven just by the boom during and in the immediate aftermath of the war. The average rate of annual TFP growth between 1950 and 1973 was still above 1.7 percent.

This unprecedented rate of expansion of the productive capacity of the economy was based on technological breakthroughs that had started in the 1920s and 1930s, but it was also vital that they were swiftly adopted and effectively organized.

Mass-production methods were already well established in the automobile industry, and they spread throughout American industry after the war. Car manufacturing itself continued to expand rapidly. In the 1930s, the United States produced an average of about three million automobiles every year. By the 1960s, production had increased to almost eight million. It is not an exaggeration to say that America made the automobile but then the automobile remade America.

Backward and forward linkages to other industries were critical in improving the productive capacity of the economy. Mass production of cars generated growing demands for inputs from almost every sector of the economy. Even more importantly, as more highways and roads were built and more of the population had access to cars and other forms of modern transport, the geography of cities was transformed, with rapid growth of suburbs. Better transport also enabled service and entertainment options via shopping centers, larger stores, and bigger movie theaters.

As remarkable as the speed of overall growth and productivity improvements was the inclusive nature of prosperity.

The first half of the twentieth century had a hard time making growth broadly shared. Bursts of growth came with much inequality. The pattern of the postwar decades stands in stark contrast to that.

For one, inequality fell rapidly during and after World War II. The share of the top 1 percent of the income distribution was down to less than 13 percent by 1960, from its high of 22 percent in the 1920s. Other aspects of inequality during the postwar years declined as well, in part because of tighter regulations and price controls. Two researchers who studied this episode were so struck by the declines in inequality during this era that they dubbed it the "Great Compression."

Even more remarkable was the pattern of subsequent growth. Average real wages rose as fast and sometimes faster than productivity, recording an overall growth rate of almost 3 percent between 1949 and 1973. And, this growth was broadly shared. For example, the real wage growth of low- and high-education men was similarly close to 3 percent per year during this period.

What was the secret sauce of shared prosperity in the decades following World War II? The answer lies in the two elements we emphasized earlier in this chapter: a direction of technology that created new tasks and jobs for workers of all skill levels and an institutional framework enabling workers to share productivity increases with employers and managers.

The direction of technology built on what began in the first half of the century. In fact, most of the technologies foundational to the era of shared prosperity were invented decades before and were implemented in the 1950s and 1960s. This is quite clear in the case of the internal-combustion engine, which underwent further improvements, but the basic technology remained largely unchanged.

Robust postwar US growth did not immediately guarantee that these technologies would benefit workers. The sharing of prosperity was contested from the day World War II came to an end. Ensuring that economic growth benefited a broad cross section of society took hard work, as we explain next.

Clash over Automation and Wages

Concerns about technological unemployment voiced by John Maynard Keynes, discussed in Chapter 1, were perhaps even more relevant in the decades following World War II. Machine tools continued to improve, and striking advances in numerically controlled machinery built on and perfected the ideas that dated back to Jacquard's loom. Designed by Joseph-Marie Jacquard in 1804, this loom was one of the most important weaving automation devices of the nineteenth century, performing tasks that even skilled weavers found challenging. Its breakthrough was conceptualizing and designing a machine that wove fabric according to the designs entered via a set of punch cards.

The numerically controlled machinery of the 1950s and 1960s pushed this idea one step further, linking a variety of machines first to punch cards and then to computers. Now drills, lathes, mills, and other machinery could be instructed to implement production tasks previously performed by workers.

Fortune magazine captured the enthusiasm about programmable machine-tool automation (also known as numerical control) in 1946, with an issue on "The Automatic Factory," announcing that "the threat and promise of laborless machines is closer than ever." A feature article in the same issue, "Machines Without Men," opened with these lines: "Imagine, if you will, a factory as clean, spacious, and continuously operating as a hydroelectric plant. The production floor is barren of men." The factory of the future would be operated by engineers and technicians, and without (many) blue-collar workers. The promise resonated with numerous American managers, who were only too happy to have new ways to reduce labor costs.

Numerical control also received substantial investment from the navy and the air force, which viewed advances in automation to be of strategic importance. More important than the government's direct investment in automation technologies was its leadership and incentives for the development of digital technologies. The war effort multiplied what the Department of Defense was willing to spend on science and technology, and a significant portion went to computers and advancing the digital infrastructure.

Policy makers took note and came to view the challenge of job creation in the midst of rapid automation as a defining one for the age. As President Kennedy responded in 1962 when asked about automation, "I regard it as the major domestic challenge, really, of the '60s, to maintain full employment at a time when automation, of course, is replacing men."

Indeed, throughout this period, advances in automation technologies continued, even beyond numerically controlled machinery and outside of manufacturing. For example, telephone switchboards were operated manually in the 1920s, often by young women. AT&T was the largest US employer of women under the age of twenty. Over the next three decades, automatic switchboards were introduced around the country. Most manual operators were displaced, and by 1960, there were almost none left. In local markets where automatic switchboards were introduced, there were fewer jobs for young women.

Yet fears of dwindling job opportunities did not materialize; labor fared quite well, and demand for workers of all different skills continued to increase throughout the 1950s, 1960s, and early 1970s. Most of the women displaced from the Bell Company switchboards, for example, found opportunities in the expanding service sectors and business offices in the decades that followed.

In essence, technologies of the era created as many opportunities for workers as the ones they displaced. This was for the same reasons we have seen in the context of mass production in the automobile industry. Improvements in communications, transport, and manufacturing technologies gave a boost to other sectors. But even more importantly, these advances also generated new jobs in the sectors in which they were introduced. Neither numerical control nor other automatic machinery completely removed the human operator, in part because the machines were not fully automatic and created a range of additional tasks as they mechanized production.

Recent research that studies the evolution of US occupations since 1940 finds that new job titles and tasks were plentiful in many blue-collar occupations, including glaziers, mechanics, truck and tractor operators, cement and concrete finishers, and craft workers in the 1950s. In the 1960s, plenty of novel tasks for sawyers, mechanics, graders and sorters, metal molders, truck and tractor operators, and oilers and greasers were among those added. Manufacturing continued to create new jobs for technicians, engineers, and clerical workers as well.

In other industries, expanding tasks went beyond technical ones. Retail and wholesale industries were growing rapidly, offering a variety of jobs in customer service, marketing, and back-office support functions. Throughout the US economy, administrative, clerical, and professional occupations grew faster than essentially all others during this era. Most of the tasks that workers in these occupations performed did not exist in the 1940s. As in manufacturing, when these jobs required specialized knowledge, most companies followed the practice of the first half of the century and continued to hire workers without formal qualifications. Trained to perform the necessary tasks, workers benefited from the higher wages paid for these jobs.

Similar to the prewar period, many expanding tasks demanded more numeracy and literacy skills, but also social skills to communicate in complex organizations and to solve problems that arose in interactions with customers and in the operation of advanced machinery. This meant that new tasks would be fully forthcoming only when workers had the necessary general skills to be trained to deal with them. Fortunately, as in the earlier era, US education expanded rapidly, and the necessary skills for these new roles became readily available. Many blue-collar workers now had a high school education, and engineering, technical, design, and clerical positions could be filled with people who had some post-secondary schooling.

Yet it would be incorrect to think that postwar technology was preordained to go in a direction that created new tasks to compensate for the ones that were being rapidly automated away. The contest over the direction of technology heated up as an integral part of the struggles between labor and management, and advances in worker-friendly technologies cannot be separated from the institutional setup that induced companies to move in this direction, especially because of the countervailing powers of the labor movement.

The Wagner Act and trade unions' critical role in the war effort strengthened labor, and there was every expectation that unions would be a mainstay of the institutional fabric of postwar America. Harold Ickes, FDR's secretary of the interior, confirmed this expectation when telling a trade union convention as the war was drawing to a close, "You are on your way and you must let no one stop you or even slow up your march."

The labor movement listened and showed that it meant business right after the war. The United Auto Workers (UAW) demanded large wage hikes from General Motors in their first postwar contract negotiation. When GM did not accept, a major strike ensued. The automobile sector was not alone. The same year, 1946, witnessed a broader wave of strikes, which the Bureau of Labor Statistics called "the most concentrated period of labor-management strife in the country's history." For example, an electrical workers' strike paralyzed another behemoth of American manufacturing, General Electric.

The labor movement was not uniformly against automation, precisely because there was an understanding that automation was inevitable and that, with the right choices, reducing costs would be beneficial for all stakeholders. What was being demanded was the use of technological advances to create new tasks for workers and allow them to share some of the cost reductions and productivity increases. As the UAW, for example, declared in 1955, "We offer our cooperation . . . in a common search for policies and programs . . . that will insure that greater technological progress will result in greater human progress."

In 1960, GM installed a numerically controlled drill in its Fisher Body Division in Detroit and classified the job of the machine's operator at the same rate as that for the operation of the manual turret drill. The union disagreed, arguing that this was a new task with additional responsibilities and requiring additional skills. But the questions were deeper. The union wanted to set a precedent, establishing that existing skilled or semiskilled workers had a vested right to the new tasks, and this was the most troubling interpretation for management because it would mean losing control of the production process and organizational choices. The two parties could not reach an agreement, and the case went to arbitration. In 1961 the arbitrator ruled in the union's favor, concluding, "This is not a case where a management decision has eliminated a function or otherwise changed methods, processes, or means of manufacture."

The ruling's implications were sweeping. GM was required to provide additional training and pay higher wages to the operators of the numerically controlled machinery. The general lesson was that the operator "has to acquire additional skills to handle the numerical control systems," and "the increased effort required of the workers put on automated machines entitles them to higher rates of pay." In fact, for unions the central issue was worker training. They insisted on training provisions to ensure that workers could be brought up to the necessary skill level to operate the new machinery and benefit from it.

The role of labor unions in how automation technologies were adopted and how workers fared in the process can also be seen from another iconic technology of the era: containers. The introduction of large metal containers in long-distance shipping in the 1950s revolutionized the transport industry by massively reducing freight costs across the globe. It simplified and eliminated many of the manual tasks that longshoremen used to perform, such as packing, unpacking, and repacking pallets. It also enabled the introduction of other heavy equipment for lifting and transport. In many cases, such as in the New York port, containers significantly reduced the number of longshoremen jobs.

On the West Coast, however, things transpired quite differently. By the time the container arrived, there was already trouble in the Pacific ports. A congressional investigation in 1955 had revealed endemic inefficiencies caused by work practices, often under the auspices of the International Longshore and Warehouse Union (ILWU). Harry Bridges, a veteran and independent labor organizer heading the local ILWU branch, understood that reform of work rules was necessary for the union and longshoremen jobs to survive, arguing, "Those guys who think we can go on holding back mechanization are still back in the thirties, fighting the fight we won way back then." This led to the ILWU policy of encouraging the introduction of new technology but in a way that was beneficial to the workers, especially their members. In 1956 the negotiating committee of the union recommended: "We believe that it is possible to encourage mechanization in the industry and at the same time establish and reaffirm our work jurisdiction, along with practical minimal manning scales, so that the ILWU will have all of the work from the railroad tracks outside the piers into the holds of the ships."

In essence, this was a similar approach to that of the UAW in its negotiations with GM: allow automation but make sure that there are also new jobs for workers. What made this approach work was Bridges's credibility with the rank and file and his efforts to communicate with the management on technology choices. Although not all union members were at first as open to new technology, Bridges and the local union leadership ultimately convinced them. In the words of a labor journalist covering events in the late 1950s, "Every longshoreman started talking about what can be done under mechanization and still maintain jobs and income, benefits, pensions, and so forth."

Containers automated work, but they also increased productivity and expanded the amount of cargo going through Pacific ports. Ships could be loaded faster and with much larger quantities of goods. As traffic grew, so did demand for longshoremen, and the union started asking for faster adoption of cranes and other machines. As Bridges told management in 1963, "The days of sweating on these jobs should be gone and that is our objective."

Autos and shipping were not exceptional. There was steady automation throughout the economy during the postwar decades, but in many cases, new opportunities for labor were created at the same time. Recent research estimates that, by itself, such automation would have reduced the labor share of national income by 0.5 percentage points every year in the 1950s, 1960s, and 1970s. Notably, however, the displacement effects of automation technologies were almost perfectly counterbalanced by other technological advances that created new tasks and opportunities for workers. As a result, across each of the most major sectors of the economy—manufacturing, services, construction, and transport—labor's share remained steady. This balanced pattern ensured that productivity increases translated into average wage growth, as well as into growth in the earnings of workers from different skill groups.

New tasks during this era played a critical role both in driving productivity growth and in spreading the gains across the skill distribution. In industries with new tasks we see higher productivity growth as well as higher demand for lower-skilled workers, who thus also benefited from technological progress.

American choices about technology and rent sharing during these decades were defining in many ways. But to Europeans, any problems in North America were trivial compared to their more existential struggles.

Abolition of Want

Germany's population suffered heavily from the war. Many cities, including Hamburg, Cologne, Düsseldorf, Dresden, and even Berlin, had been leveled by Allied bombing. More than 10 percent of the German population had perished, and possibly twenty million Germans were homeless. Several million German speakers were forced to move westward.

France, Belgium, the Netherlands, and Denmark, which had been occupied and savagely treated by the Nazis, lay in ruins as well. Much of the road networks in these countries had been destroyed. As in Germany, most of the resources had been directed toward armaments, and shortages were endemic.

Britain, even if spared the ravages of occupation, was also suffering from the aftermath of the war. The nation had fallen behind in terms of adopting modern appliances. Few households had refrigerators and ovens, already standard in North America, and only half of the houses had indoor plumbing with hot water.

Out of these ashes of war came something quite unexpected. The next three decades witnessed breakneck-paced economic growth in much of Europe, from Scandinavia to Germany, France, and Britain. GDP per capita in real terms increased at an average rate of around 5.5 percent in Germany between 1950 and 1973. The same number was just over 5 percent for France, 3.7 percent for Sweden, and 2.9 percent for the UK. In all these cases, growth was remarkably broadly shared. The share of top 1 percent households in national income, which in the late 1910s hovered above 20 percent in Germany, France, and the UK, fell to less than 10 percent in the 1970s in all three countries.

The foundations of this shared prosperity were no different than what happened in the United States. The first leg was provided by technologies that were broadly labor-friendly, creating new tasks at the same time as they were automating work. Here, Europe followed the United States, which had pulled even further ahead of the Continent in terms of industrial technology. Advances that were implemented in America spread to Europe, and industrial technology and mass-production methods were adopted rapidly. There were all sorts of incentives for European companies to embrace these technologies, and the postwar reconstruction program under the auspices of the Marshall Plan provided an important framework for technology transfer. So did many European governments' generous support for research and development.

In this way, a direction of technology that sought to make best use of both skilled and unskilled workers spread from the United States to Europe. Many more countries thus started investing both in manufacturing and services for their growing mass markets.

In most of Europe, as in the US, this path of economic development was bolstered by increasing educational investments and worker-training programs, which ensured that there were workers with the skills to fill the new positions. As high-earning workers became the middle class, they boosted the demand for the new products and services that their industries were starting to mass-produce.

However, there was no uniformity in technological choices across countries. Each organized its economy in unique ways, and these choices naturally affected how new industrial knowledge was used and further developed. Whereas in Nordic countries technological investments were made in the context of the corporatist model, German industry developed a distinctive system of apprenticeship training, which structured both labor-management relations and technology choices (as we will discuss in greater detail in Chapter 8).

Equally critical was the second leg of shared prosperity: the power of the labor movement and the overall institutional foundation that emerged in Europe after the war.

The US started strengthening the labor movement and building a regulatory state with some timidity in the 1930s. The same pattern of small steps, interspersed with several reversals, characterized the evolution of US institutions in the postwar era as well. Other pillars of the modern social safety net and regulations were introduced slowly, culminating in President Lyndon Johnson's Great Society program during the 1960s.

Shaken by two world wars, many European nations had a greater appetite for building new institutions, and perhaps they were even readier to learn from the Scandinavian example.

In Britain a government commission led by William Beveridge published a landmark report in 1942. It declared that "a revolutionary moment in the world's history is a time for revolutions, not for patching." The report identified five giant problems for British society as want, disease, ignorance, squalor, and idleness, and started by stating, "Abolition of want requires, first, improvement of State insurance, that is to say provision against interruption and loss of purchasing power." The report offered a blueprint for a state-run insurance program protecting people "from cradle to grave," with redistributive taxation, social security, unemployment insurance, worker compensation, disability insurance, child benefits, and nationalized health care.

These proposals were an immediate sensation. The British public embraced them in the middle of the war. When the report's news reached the troops, they reportedly cheered and were energized. Immediately after the war, the Labour Party, campaigning on a promise to implement the report in full, swept to power.

Similar state insurance arrangements were adopted in most European countries. Japan implemented its own version.

Social Progress and Its Limits

In the long sweep of history, the decades that followed the end of World War II are unique. There has never been, as far as anyone knows, another epoch of such rapid and shared prosperity.

Ancient Greeks and Romans experienced hundreds of years of growth before the modern era, but this growth was much slower, in the range of about 0.1–0.2 percent a year. It was also based on savage exploitation of excluded groups, most importantly an army of slaves and large numbers of noncitizens working as forced laborers in both Greece and Rome. Patrician or aristocratic classes were the main beneficiary of this growth, although a broader cadre of citizens experienced some prosperity as well.

Growth during the Middle Ages was slow and unequal, as we saw in Chapter 4. The rate of growth picked up after the onset of the British industrial revolution, starting around 1750, but this was lower than the growth rates experienced in the 1950s and 1960s, which averaged over 2.5 percent per annum in much of the Western world.

Other aspects of postwar growth were equally distinctive. Secondary and postsecondary education used to be a privilege of the very wealthy and the upper-middle classes. This changed after the war, and by the 1970s, secondary education, and even higher education, became much more democratic in almost all of the West.

The health of the population improved tremendously as well. Conditions were not as bad in the UK and elsewhere as they had been in the early nineteenth century. Nevertheless, infectious diseases were common in the first half of the twentieth century, and their burden fell much more heavily on the poor. This changed in the decades following World War II. Life expectancy at birth in Britain increased from fifty years in 1900 to seventy-two in 1970. In the United States, the increase was similar, from forty-seven in 1900 to seventy-one in 1970, and in France, from forty-seven to seventy-two. In all cases, improvements in health care and health conditions for the working classes, thanks to investments in public health and hospitals and clinics, drove the change.

We should not get carried away with this upbeat assessment. Even as an unparalleled episode of shared prosperity transpired in the Western world, three groups were excluded from both political power and some of the economic benefits: women; minorities, especially Black Americans in the US; and immigrants.

Many women were still locked in patriarchal power relations within families and their communities. This had begun changing after enfranchisement earlier in the century and then accelerated with the entry of women into the labor force during and after World War II and with broader changes in social attitudes. As a result, in the postwar decades women's economic conditions improved, and the gender pay gap narrowed. Nevertheless, discrimination in the family, schools, and workplaces continued. Greater gender parity in managerial positions and in terms of pay, as well as greater social liberation, has been slow in coming.

Minorities fared worse. Although the economic conditions of Black Americans had started improving and the wage gap between them and White Americans narrowed in the 1950s and 1960s, the US remained a racist society, especially in the South. Black workers were often excluded from good jobs, sometimes by unions. Lynchings continued well into the 1960s, and many politicians from both parties ran on overtly or covertly racist platforms throughout much of this period.

Some immigrants were also excluded from the core coalition. Guest workers from Turkey and southern Europe, brought into Germany because of the country's labor shortage after the war, remained second-class citizens throughout this era. The US turned to Mexican immigrants to work its agricultural fields, and they often toiled under harsh conditions for very low pay and without benefits. Immigrants were no longer welcome when economic conditions or the political tides turned. For example, the *bracero* program, which at its peak in the late 1950s brought more than 400,000 Mexicans per year to work on US farms, was discontinued in 1964, when Congress became concerned about immigrants taking American jobs.

The biggest excluded groups from the shared prosperity of these decades were not inside but outside Europe and North America.

A few non-Western countries, such as Japan and South Korea, grew fast and achieved some amount of shared prosperity. Notably, this was based on adopting and sometimes improving on the large-scale industrial production systems developed in the United States. It was also supported by domestic arrangements that encouraged an equitable split of the fruits of growth. In Japan, long-term employment relationships and accompanying high-wage policies were critical for the sharing of the gains from growth. In South Korea, shared prosperity owed much to the threat from North Korea and the labor movement's strength, especially after the country democratized in 1988.

But the East Asian experience was the exception, not the rule. The populations of remaining European colonies had little voice and little chance at shared prosperity. Independence, which arrived for most colonies between 1945 and 1973, did not mean the end of misery, violence, and repression. Many of the former European colonies soon discovered that colonial institutions fell into the hands of authoritarian rulers, who used the system they inherited to enrich themselves and their cronies, and squeeze everyone else. Europe stood back from this, sometimes providing support to kleptocrats in order to access natural resources. America's Central Intelligence Agency stepped in to help coups against democratically elected politicians—for example, in Iran, Congo, and Guatemala—and was always ready to support US-friendly rulers, whether they were corrupt or even murderous. Most of the non-Western world remained far behind in terms of economic development.

Meanwhile, another, equally fateful limit to progress was brewing at home. The economic model underlying shared prosperity was being increasingly challenged in the United States, and the balance of power gradually shifted away from labor and government regulation after the direction of technology moved toward greater automation. Shared prosperity started unraveling soon thereafter, as we will see in Chapter 8.

8

Digital Damage

The good news about computers is that they do what you tell them to do. The bad news about computers is that they do what you tell them to do.

—attributed to TED NELSON

One might say that the process by which progressive introduction of new computerized, automated, and robotized equipment can be expected to reduce the role of labor is similar to the process by which the introduction of tractors and other machinery first reduced and then completely eliminated horses and other draft animals in agriculture.

—WASSILY LEONTIEF, "Technological Advance, Economic Growth, and the Distribution of Income," 1983

The beginnings of the computer revolution can be found on the ninth floor of MIT's Tech Square building. In 1959–1960, a group of often-unkempt young men coded there in assembly language into the early hours of the morning. They were driven by a vision, sometimes referred to as the "hacker ethic," which foreshadowed what came to energize Silicon Valley entrepreneurs.

Key to this ethic was decentralization and freedom. Hackers felt great disdain for the major computer company of that era, IBM (International Business Machines). In their view, IBM wanted to control and bureaucratize information, whereas they believed that access to computers should be completely free and unlimited. Anticipating a mantra that would later become much misused by tech entrepreneurs, hackers argued that "all information should be free." Hackers mistrusted authority, so much so that there was an almost anarchist element to their thinking.

What became the more famous branch of the hacker community, emerging in Northern California in the early 1970s, was similarly distrustful of large companies. One of its luminaries, Lee Felsenstein, was a political activist who viewed computers as a means of liberating people and liked to quote "Secrecy is the keystone of all tyranny" from the science-fiction novel *Revolt in 2100*. Felsenstein worked on hardware improvements to democratize computing and break the grip of IBM and other incumbents.

Another Northern California hacker, Ted Nelson, published what can be considered as a handbook of hacking, "Computer Lib," which starts with the motto "THE PUBLIC DOES NOT HAVE TO TAKE WHAT IS DISHED OUT" and continues:

> *THIS BOOK IS FOR PERSONAL FREEDOM.*
> *AND AGAINST RESTRICTION AND COERCION . . .*
> *A chant you can take to the streets:*
> *COMPUTER POWER TO THE PEOPLE!*
> *DOWN WITH CYBERCRUD!*

Cybercrud here is Nelson's term for the lies that powerful people tell about computers and information—about how their experts had to control them.

The hackers were not misfits at the margins of the computer revolution. They were instrumental in many advances in both software and hardware. They symbolized the values and attitudes that many computer scientists and entrepreneurs held, even when they did not share the hackers' work and sanitary habits.

The view that the future of computing and information lay with decentralization was not confined to the scruffy, male hackers of Tech Square at MIT and Berkeley. Another pioneer, Grace Hopper, was pushing for greater decentralization in computing at the Department of Defense in the 1970s. Hopper played an important role as a software innovator, devising early programming conventions that culminated in the new language COBOL. Hopper also viewed computing as a way of broadening access to information, and she influenced how computing was used in one of the largest organizations in the world, the US armed forces.

With the most promising technology of the era in the hands of visionaries like this, an astute contemporary could have reasonably predicted that the next several decades would further bolster countervailing powers against big business, create new productive tools for workers, and lay the foundations of even stronger shared prosperity.

In the event, something very different transpired, and digital technologies became the graveyard of shared prosperity. Wage growth slowed down, the labor share of national income declined sharply, and wage inequality surged starting around 1980. Although many factors, including globalization and the weakening of the labor movement, contributed to this transformation, the change in the direction of technology was most important. Digital technologies automated work and disadvantaged labor vis-à-vis capital and lower-skilled workers vis-à-vis those with college or postgraduate degrees.

This redirection cannot be understood without recognizing the broader social changes taking place in the United States. Businesses became better organized against labor and government regulations, but even more importantly, a new vision maintaining that maximizing profits and shareholder values was for the common good became an organizing principle for much of society. This vision, and the massive enrichment it offered, pushed the tech community in a direction very different from the one envisaged by the early hackers. The new vision was of a "digital utopia," based on the top-down design of software to automate and control labor. The resulting path of technology not only manufactured inequality but also failed to deliver on its promise of spectacular productivity growth, as we will see.

A Reversal

Any hopes that the decades after the initial phases of the computer revolution would bring more shared prosperity were dashed rather swiftly. Economic growth after the mid-1970s would look nothing like growth in the 1950s or 1960s. Some of the slowdown was a result of the oil crises of 1973 and then 1979, which triggered high levels of unemployment and inflation—stagflation—throughout the Western world. But the more fundamental transformation, in the structure of economic growth, was about to come.

US median real wages (hourly compensation) grew at above 2.5 percent per year between 1949 and 1973. Then from 1980 onward, median wages all but stopped growing—increasing only 0.45 percent per year, even though the average productivity of workers continued to rise (with an annual average growth rate of over 1.5 percent from 1980 to the present).

This growth slowdown was far from equally shared. Workers with postgraduate degrees still enjoyed rapid growth, but men with a high school diploma or less saw their wages fall by about 0.45 percent, on average, every year between 1980 and 2018.

It was not just a widening gap between workers with postgraduate degrees and those with low levels of education. Every dimension of inequality skyrocketed from 1980 onward. For example, the share of the richest 1 percent of US households in national income rose from around 10 percent in 1980 to 19 percent in 2019.

Wage and income inequality tells only part of the story. The United States used to pride itself for its "American dream," which meant people from modest backgrounds rising in terms of income and children doing better than their parents. From the 1980s onward, this dream came under growing pressure. For children born in 1940, 90 percent of them earned more than their parents did, in inflation-adjusted terms. But for children born in 1984, the percentage was only 50 percent. The US public is fully aware of the bleak prospects for most workers. A recent survey by the Pew Research Center found that 68 percent of Americans think that today's children will be financially worse off than their parents' generation.

Other dimensions of economic progress were reversed, too. In 1940, Black men and women earned less than half what White Americans earned. By 1979, hourly wages for Black men rose to 86 percent of the level for White men. After that time, the gap widened, with Black men now earning only 72 percent as much as White men. There is a similar reversal for Black women.

The distribution of income between capital and labor also changed significantly. Throughout most of the twentieth century, about 67–70 percent of national income went to workers, and the rest went to capital (in the form of payments for machinery and profits). From the 1980s onward, things started getting much better for capital and much worse for workers. By 2019, labor's share of national income had dropped to under 60 percent.

These broad trends are not confined to the United States, although for various reasons they have been less pronounced in other countries. Already by 1980, the US was more unequal than most other industrialized economies and subsequently had one of the sharpest rises in inequality. Several others were not far behind.

Labor's share of national income has been on a protracted downward trend in most industrialized economies. In Germany, for example, it fell from close to 70 percent in the early 1980s to around 60 percent in 2015. At the same time, the income distribution became more skewed in favor of the very richest people. From 1980 to 2020, the share of the top 1 percent increased from about 10 percent to 13 percent in Germany, and from 7 percent to almost 13 percent in the UK. During the same period, inequality increased even in Nordic countries: the share of the top 1 percent rose from about 7 percent to 11 percent in Sweden and from 7 percent to 13 percent in Denmark.

What Happened?

At some level, what happened is clear. There were two pillars of shared prosperity in the postwar period: alongside automation, new opportunities were created for all kinds of workers, and robust rent sharing (meaning the splitting of productivity and profit gains between capital and labor) kept wages buoyant. After about 1970, both pillars collapsed, most spectacularly in the United States.

Even at the best of times, the directions of technology and high wages are contested. Left to their own devices, many managers would try to reduce labor costs by limiting wage raises and also by prioritizing automation, which eliminates labor from some tasks and weakens the bargaining power of workers. These biases then influence the direction of innovation, pushing technology more toward automation. As we saw in Chapter 7, these tendencies were partly contained by collective bargaining during the decades that followed World War II, and unions further encouraged companies to introduce more skilled tasks and systematic training together with new machinery.

The emaciation of the labor movement over the last several decades has been a double whammy for shared prosperity. Wage growth slowed down partly because US labor unions became weaker and could not negotiate the same terms for their workers. Even more importantly, without strong unions, worker voice on the direction of technology disappeared.

Two other changes amplified the decline of labor and inequality. First, without countervailing powers from the labor movement, corporations and their managers developed a very different vision. Cutting labor costs became a priority, and sharing productivity gains with workers came to be viewed as akin to a failure of management. In addition to taking a harder line in wage negotiations, corporations shifted production toward nonunionized plants in the United States and increasingly abroad. Many firms introduced incentive pay, which rewarded managers and high performers, but at the expense of lower-skill workers. Outsourcing became fashionable as another cost-cutting strategy. Many low-skill functions, including cafeteria work, cleaning, and security, used to be performed by employees of large organizations such as General Motors or General Electric. These employees used to benefit from the overall wage increases that these companies' workforces enjoyed. In the cost-cutting vision of the post-1980s, however, this practice was seen as a waste, so managers outsourced these functions to low-wage outside providers, severing another channel of wage growth for workers.

Second, it was not only companies choosing more automation from a given menu of technologies. With the new direction of the digital industry, the menu itself shifted powerfully toward greater automation and away from worker-friendly technologies. With a whole slew of digital tools enabling new ways of substituting machines and algorithms for labor, and little countervailing powers to oppose this move, many corporations embraced automation enthusiastically and turned their back on creating new tasks and opportunities for workers, especially those without a college degree. Consequently, although productivity (output per worker) continued to increase in the US economy, worker marginal productivity (how much that an additional hour of labor boosts production) did not keep up.

It bears repeating that shared prosperity was not destroyed by automation per se, but by an unbalanced technology portfolio prioritizing automation and ignoring the creation of new tasks for workers. Automation was also rapid in the decades following World War II but was counterbalanced by other technological changes that raised the demand for labor. Recent research finds that from 1980 onward, automation accelerated; more significantly, there were fewer new tasks and technologies that created opportunities for people. This change accounts for much of the deterioration of workers' position in the economy. The labor share in manufacturing, where the acceleration of automation and the slowdown in the creation of new tasks has been most pronounced, declined from around 65 percent in the mid-1980s to about 46 percent in the late 2010s.

Automation has also been a major booster of inequality because it concentrates on tasks typically performed by low- and

middle-skill workers in factories and offices. Almost all the demographic groups that experienced real wage declines since 1980 are
those that once specialized in tasks that have since been automated. Estimates from recent research suggest that automation
accounts for as much as three-quarters of the overall increase in
inequality between different demographic groups in the United
States.

The automotive industry is indicative of these trends. US
car companies were some of the most dynamic employers in the
country in the first eight decades of the twentieth century, and
as we saw in Chapter 7, they were at the forefront of not just
automation but also the introduction of new tasks and jobs for
workers. Blue-collar work in the automotive industry was plentiful and well paid. Workers without college degrees and sometimes even without high school diplomas were hired and trained
to operate new, sophisticated machinery, and they received quite
attractive wages.

The nature and availability of work in the automobile industry changed fundamentally in recent decades, however. Many of
the production tasks in the body shop, such as painting, welding,
and precision work, as well as a range of assembly jobs, have been
automated using robots and specialized software. The wages of
blue-collar workers in the industry have not increased much since
1980. Achieving the American dream through the automotive
industry is much harder today than in the 1950s or 1960s.

One can see the implications of this change in technology and
organization of production in the hiring strategies of the industry.
Since the 1980s, the US automotive giants stopped hiring and
training low-education workers for complex production tasks and
started accepting just higher-skilled applicants with formal qualifications, and only after a battery of aptitude and personality tests
and interviews. This new human-resource strategy was enabled by
the fact that there were many more applicants than available jobs
and many of them had postsecondary education.

The effects of automation technologies on the American
dream are not confined to the automotive industry. Blue-collar
jobs on other factory floors and clerical jobs in offices, which

used to provide opportunities for upward mobility to people from disadvantaged backgrounds, have been the main target of automation by robots and software throughout the US economy. In the 1970s, 52 percent of US workers were employed in these "middle-class" occupations. By 2018, this number had fallen to 33 percent. Workers who once occupied these jobs were often pushed toward lower-paying positions, such as construction work, cleaning, or food preparation, and witnessed their real earnings plummet. As these jobs disappeared throughout the economy, so did many of the opportunities for workers with less than a postgraduate degree.

Although the abatement of rent sharing and the automation focus of new technologies have been the most important drivers of inequality and the decline of the labor share, other factors have also played a role. Offshoring has contributed to worsening conditions for labor: numerous jobs in car manufacturing and electronics have been shifted to lower-wage economies, such as China or Mexico. Even more important has been rising merchandise imports from China that have adversely affected many US manufacturing industries and the communities in which they were concentrated. The total number of jobs lost to Chinese competition between 1990 and 2007, just before the Great Recession, may be as high as three million. However, the effects of automation technologies and the eclipse of rent sharing on inequality have been even more extensive than the consequences of this "China shock."

Import competition from China impacted mostly low-value-added manufacturing sectors, such as textiles, apparel, and toys. Automation, on the other hand, has concentrated in higher-value-added and higher-wage manufacturing sectors, such as cars, electronics, metals, chemicals, and office work. It is the dwindling of this latter set of jobs that has played a more central role in the surge in inequality. As a result, although competition from China and other low-wage countries has reduced overall manufacturing employment and depressed wage growth, it has been the direction of technological change that has been the major driver of wage inequality.

These technology and trade trends have sometimes devastated local communities as well. Many areas in the industrial heartland of the United States, such as Flint and Lansing in Michigan, Defiance in Ohio, and Beaumont in Texas, used to specialize in heavy industry and offered employment opportunities to tens of thousands of blue-collar workers. After 1970, however, these places were pushed into decline as workers were displaced from their jobs by automation. Other metropolitan areas, such as Des Moines in Iowa and Raleigh-Durham and Hickory in North Carolina, that used to specialize in textiles, apparel, and furniture were equally adversely affected by competition from cheap Chinese imports. Whether from automation or import competition, job losses in manufacturing put downward pressure on worker incomes throughout the local economy and reduced demand for retail, wholesale, and other services, in some cases plunging an entire region into a deep, long-lasting recession.

The fallout from these regional effects has gone beyond economics and gives us a microcosm of the problems that the US economy has been facing more broadly. As manufacturing jobs disappeared, social problems multiplied. Marriage rates fell, out-of-wedlock childbirth increased, and mental health problems rose in the worst-affected communities. More broadly, job losses and worsening economic opportunities, especially for Americans without a college degree, appear to have been a major driver of the rise in what economists Anne Case and Angus Deaton call "deaths of despair"—premature deaths caused by drugs, alcohol, and suicide. Partly as a result of these deaths, US life expectancy at birth has declined for several years in a row, which is unparalleled in the recent history of Western nations.

In some popular discussions of rising inequality, globalization is pitted against technology as competing explanations. It is often implied that technology represents the inevitable forces leading to inequality, while there is some degree of choice about how much globalization and import competition from low-wage countries the United States (and other advanced economies) should have allowed.

This is a false dichotomy. Technology does not have a preordained direction, and nothing about it is inevitable. Technology has increased inequality largely because of the choices that companies and other powerful actors have made. Globalization is not separate from technology in any case. The huge boom in imports from countries thousands of miles away and the complex global supply chains involved in the offshoring of jobs to China or Mexico are enabled by advances in communication technologies. With better digital tools to track and coordinate activities in faraway facilities, companies reorganized production and sent offshore many of the assembly and production tasks they used to perform in-house. In the process, they also eliminated many middle-skill, blue-collar jobs, exacerbating inequality.

In fact, globalization and automation have been synergistic, both driven by the same urge to cut labor costs and sideline workers. They have both been facilitated by the lack of countervailing powers in workplaces and in the political process since 1980.

Automation, offshoring, and import competition from China have also impacted other advanced economies, but in more nuanced forms. Collective bargaining did not decline as much in most of Europe. In the Nordic countries, union coverage has remained high. Not coincidentally, even though their inequality levels also increased, they did not experience the declines in real wages that have been such a major part of US labor market trends. In Germany, as we will see, companies often shifted workers from blue-collar occupations into new tasks, charting a somewhat different, more labor-friendly direction of technology. In France, too, minimum wages and unions have limited the rise in inequality, albeit at the cost of greater joblessness.

These caveats notwithstanding, technology trends have been broadly similar across most Western countries and have had analogous implications. Most tellingly, jobs in blue-collar and clerical occupations have declined in almost all industrialized economies.

All of this then begs two obvious questions: How did businesses manage to become so powerful vis-à-vis labor and to cripple rent sharing? And why did technology turn antilabor? The answer to the first question, as we will see below, is related to a series of institutional transformations in the United States and other Western nations. The answer to the second also builds on these institutional changes but crucially involves the emergence of a new utopian (but in reality, largely dystopian) digital vision, which pushed technologies and practices in an increasingly antilabor direction. In the next several sections we start with the institutional developments and return to how the idealistic hacker ethic of the 1960s and 1970s morphed into an agenda for automation and worker disempowerment.

The Liberal Establishment and Its Discontents

We saw in Chapter 7 how a sort of balance between business and organized labor emerged in the United States after the 1930s. It was undergirded by robust wage growth across jobs ranging from the unskilled to the highly skilled, and by a broadly worker-friendly direction of technology. In consequence, the political and economic landscape of the United States looked very different by the 1970s than in the early decades of the twentieth century. Gone was the overwhelming political and economic clout of mega-businesses, such as the Carnegie Steel Company and John D. Rockefeller's Standard Oil.

Emblematic of these changes was the consumer protection activism led by Ralph Nader, whose book *Unsafe at Any Speed*, published in 1965, was a manifesto for keeping corporations accountable. In this instance, activism focused on automobile manufacturers, although Nader's target was all misbehaviors by business, especially big business.

Several iconic government regulations resulted from consumer activism. The National Traffic and Motor Vehicle Safety Act of 1966, which set the first safety standards for automobiles, was a direct response to the issues that Nader publicized. The Environmental Protection Agency was launched in 1970, with an explicit remit to prevent pollution and environmental damage by businesses. The Occupational Safety and Health Administration (OSHA) came into existence in December of the same year to protect the health and well-being of workers. Although some of these problems were previously monitored by the Bureau of Labor Standards, OSHA gained much greater authority over businesses. The Consumer Product Safety Act, enacted in 1972, was even more far-reaching, giving an independent agency authority to set standards, recall products, and bring lawsuits against companies to protect consumers against the risk of injury or death.

Title VII of the Civil Rights Act of 1964 had already banned employment discrimination on the basis of race, gender, color, religion, and national origin, but this act had little bite without an agency enforcing it. That changed with the launch of the Equal Employment Opportunity Act of 1972, tasked with going after individual employers for discrimination against Black Americans and other minorities.

The Food and Drug Administration (FDA), which had been around since the beginning of the century, significantly increased its powers because of the Kefauver-Harris amendment of 1962 and the US Public Health Service reorganizations of 1966–1973. The impetus for these changes came from a number of highly publicized scandals in Europe and the United States, convincing lawmakers that the agency needed to be more independent and approve only drugs that were safe and effective. The year 1974 also witnessed the beginning of the Department of Justice's action to break up AT&T, which had dominated the telephone sector in the US.

These changes reflected a new, more muscular regulatory approach. Many were implemented under a Republican president, Richard Nixon. Nixon's embrace of regulation was not a sharp break with the postwar Republican establishment. Dwight Eisenhower had already moved in the same direction, defining himself as a "modern Republican," meaning that he was going to maintain most of what was left of the New Deal.

It was not just regulation. The 1960s witnessed the success of the civil rights movement and greater mobilization among left-wing Americans supporting civil rights and further political reforms. Lyndon Johnson initiated the Great Society program and the War on Poverty, adapting some key tenets of a European-style social safety net to the US context.

Not everyone saw these changes as beneficial. Constraints on business conduct often benefited workers and consumers but were resented by business owners and executives. Segments of the business sector had been organizing against regulations and legislation strengthening labor unions since the beginning of the twentieth century. Their activity accelerated during the New Deal, when executives from some of the largest corporations, including DuPont, Eli Lilly, General Motors, General Mills, and Bristol-Myers, founded organizations such as the American Enterprise Association (which later became the American Enterprise Institute, or AEI) and the American Liberty League to formulate criticisms of and alternatives to New Deal policies.

After the war, many businesspeople continued to be animated by a belief that the country was being lost to the "liberals." In his 1965 book, *The Liberal Establishment: Who Runs America and How*, M. Stanton Evans wrote that "the chief point about the Liberal Establishment is that it is in control."

Early pro-business, right-wing organizations and think tanks received funding from executives and wealthy Americans philosophically opposed to the New Deal. As is often the case, philosophy was mixed with material interests. Tax-exempt philanthropic and charitable donations by large US corporations have tended to support causes aligned with their strategic interests (for example, energy companies philanthropically funding anti-climate-science think tanks).

The pernicious role of money in US politics has been much discussed. But the story is more nuanced than what is sometimes presumed. Corruption at the federal level is not unknown, and political stances sometimes change because of campaign contributions from wealthy donors. Most of the time, however, politicians and their staff need to be persuaded that a particular approach to public policy serves either the public interest or their constituency. Copious amounts of money alone cannot achieve this unless an alternative vision of how the market economy should be organized becomes accepted. During the 1950s and 1960s, elements of such a vision started to come together.

What Is Good for General Motors

In 1953 President Dwight Eisenhower nominated Charles Wilson, then the president of General Motors, as secretary of defense. During his confirmation hearing, Wilson had to defend his controversial decision to hold on to substantial shares of GM, and he coined the aphorism "What was good for our country was good for General Motors, and vice versa."

Wilson was arguing that he could not imagine a situation in which he would have to do something good for the country that would not be good for GM. But people misconstrue him as claiming that what was good for GM was good for the country, for understandable reasons. By the 1980s, the view that what was good for business, or even large corporations, was good for the country had become commonplace. This was an about-face from the prevailing attitudes of the 1930s, and the idea was now taking hold that shifting the rules to favor companies and to boost profits was the best possible way to help everyone.

This intellectual reversal was rooted in a lot of hard work by political entrepreneurs and organizations. An intellectual leader in this endeavor was the conservative magazine *National Review*, founded by William F. Buckley Jr. in 1955. Buckley intended his publication to counter the trends from the Left because "in its maturity, literate America rejected conservatism in favor of radical social experimentation." He continued: "Since ideas rule the world, the ideologues, having won over the intellectual class, simply walked in and started to run things."

The Business Roundtable, an influential business organization, agreed that "business has very serious problems with the intellectual community, the media and the youth. . . . The continuing hostility of these groups menaces all business." The roundtable's 1975 advertisement in *Reader's Digest* read, "The way we earn our 'daily bread' in this country is under attack as never before" and identified the threat as arguments such as the "free enterprise system makes us selfish and materialistic" and "free enterprise concentrates wealth and power in the hands of a few." The Chamber of Commerce, representing in theory all US businesses, joined the Business Roundtable and also started pushing against government regulations.

George H. W. Bush's 1978 speech to top executives at a conference in Boston, while Bush was seeking the Republican presidential nomination, captured this mood: "Less than fifty years ago, Calvin Coolidge could say that the business of America is business. Today, the business of America seems to be the regulation of business."

Efforts by various think tanks and leaders notwithstanding, still missing was a coherent paradigm that what was good for business was good for everybody. The productivity bandwagon was a key part of this new vision, but with its logic extended even further. Organizational changes or laws that are good for business must also be good for society at large because, with a similar reasoning, they will increase demand for workers and translate into shared prosperity. Take it one step further, and you get "trickle-down economics," a term identified today with President Ronald Reagan's economic policies in the 1980s, including the idea of cutting taxes on the very rich: when the rich face lower taxes, they will invest more, increasing productivity and benefiting everybody in society.

Applying this perspective to regulation leads to conclusions that are diametrically opposed to the ideas that energized Ralph Nader and other consumer activists. According to this free-market view, if the market economy is working well, regulation is at best unnecessary. If incumbent firms are marketing unsafe or low-quality products, consumers will be upset, and this creates an opportunity for other companies or new entrants to offer better alternatives, to which consumers will enthusiastically switch.

The same competitive process that underlies the productivity bandwagon can then act as a force to discipline product quality as well. Seen through these lenses, regulation may even be counterproductive, harming consumers and workers. If the market process was already incentivizing businesses to offer safe and high-quality products, additional regulations would only divert effort and reduce profitability, forcing businesses to increase prices or reduce labor demand.

These ideas about the idealized market process have been part of economic theory ever since Adam Smith's *The Wealth of Nations* introduced the notion of the invisible hand—a metaphor for the notion that the market provides good outcomes for everyone, if there is enough competition. There has always been debate on this point, with the other side taken by people like John Maynard Keynes, who points out that markets do not function in an idealized way. For example, as we have seen, the productivity bandwagon collapses when there is not enough competition in the labor market. The same is true without sufficient competition in the product market. Nor can we count on the market process delivering high quality when consumers have a hard time distinguishing unsafe products from better ones.

The pendulum has periodically swung between more market-friendly and market-skeptical perspectives in academia and in policy circles. The postwar decades were decidedly on the market-skeptical side, partly under the influence of Keynes's ideas and the policies and regulations introduced during the New Deal era. But there were many pockets of diehard pro-market economists—for example, at the University of Chicago and Stanford University's Hoover Institution.

These ideas started coalescing into a more coherent whole in the 1970s. There were many contributing factors at play here. Some intellectuals, such as Friedrich Hayek, offered widely read critiques of the postwar policy consensus. Hayek developed his theories in interwar Vienna, where free-market notions were popular and the disaster of central planning in the nearby Soviet Union only too visible. Hayek left Austria in the early 1930s and landed at the London School of Economics, where he further developed many of his ideas. In 1950 he moved to the University of Chicago, where his influence grew.

Particularly important was Hayek's view that markets, as a decentralized system, were much better at using the dispersed information in society. In contrast, whenever central planning or government regulation was used to allocate resources, there was a loss of information about what consumers truly wanted and about how productivity improvements could be implemented.

To be sure, regulation is never an easy process, and the postwar era was filled with unintended consequences and inefficiencies created by regulators. For example, the airline industry was tightly regulated by the Civil Aeronautics Board during much of this time. The board set schedules, routes, and airline fares, and decided which new airlines could enter new markets. As civil aviation technology improved and demand for air travel grew, these regulations became more arcane and contributed to massive inefficiencies in the industry. The Airline Deregulation Act of 1978 allowed airlines themselves to set fares. This made it easier for new airlines to enter the market, increasing competition and driving down prices in ways that were generally welcomed by consumers.

On the Side of Angels and Shareholders

The idea that unregulated markets work in the interest of the nation and the common good became the basis for a new approach to public policy. Missing from this emerging consensus was a clear set of recommendations for business leaders—how should they behave, and what would justify their actions? The answers came from two economists at the University of Chicago, George Stigler and Milton Friedman. Stigler's and Friedman's views about economics and politics overlapped with Hayek's, but in some ways went further. Both Stigler and Friedman were more opposed to regulations than Hayek was.

Friedman, who, like Hayek and Stigler, was awarded the Nobel Prize in economics, made important contributions to many areas, including macroeconomics, price theory, and monetary policy. Arguably, however, his most influential work did not appear in an academic journal but in a short piece published in September 1970 in the *New York Times Magazine*, immodestly titled "A Friedman Doctrine." Friedman argued that the "social responsibility" of business was misconstrued. Business should care only about making profits and generating high returns for their share-holders. Simply put, "The social responsibility of business is to increase its profits."

Friedman articulated an idea that was already in the air. The previous decades had witnessed stinging criticisms of government regulations and more voices in favor of the market mechanism. Nevertheless, the impact of the Friedman doctrine is hard to exaggerate. At one fell swoop, it crystallized a new vision in which big businesses that made money were heroes, not the villains that Ralph Nader and his allies painted them as. It also gave business executives a clear mandate: raise profits.

The doctrine also received support from a different angle. Another economist, Michael Jensen, argued that managers of publicly listed corporations were not sufficiently committed to their shareholders and were instead pursuing projects that glorified themselves or built wasteful empires. Jensen maintained that these managers needed to be controlled more tightly, but because that was difficult, the more natural path was to have their compensation tied to the value they created for shareholders. This meant giving managers big bonuses and stock options in order to focus them on boosting the company's stock price.

The Friedman doctrine, along with the Jensen amendment, brought us the "shareholder value revolution": corporations and managers should strive to maximize market value. Unregulated markets, combined with the productivity bandwagon, would then work for the common good.

The Business Roundtable agreed and suggested that citizens should be educated in "economics" because greater economic knowledge would make them more favorable to business and supportive of policies such as lower taxes that would boost economic growth and benefit everybody. In 1980 it stated: "The Business Roundtable believes that future changes in tax policy should aim at improving the investment or supply side of the economy in order to increase the quality and scope of our productive capacity."

Two additional implications of this doctrine may have been even more important. First, it justified all sorts of efforts at moneymaking, for boosting profits was in alignment with the common good. Some companies pushed this even further. The combination of the Friedman doctrine and the lavish stock options to top executives motivated several executives to venture into gray areas and then into the red. The journey of the energy giant Enron, a darling of the stock market, is emblematic. The Houston-based company was selected as "America's Most Innovative Company" six years in a row by *Fortune* magazine. But in 2001 it was revealed that Enron's financial success was in large part a result of systematic misreporting and fraud, which boosted the company's stock market performance (and made hundreds of millions of dollars for its executives). Although Enron was the culprit that is most keenly remembered today, many other corporations and executives were involved in similar shenanigans, and several more scandals were revealed in the early 2000s.

Second, the doctrine altered the balance between managers and workers. Sharing of productivity gains between companies and workers was a key pillar of broad-based prosperity after 1945. It was bolstered by labor's collective bargaining power to make corporations pay high wages, by social norms of sharing the benefits of growth, and even by ideas of "welfare capitalism," as we saw in Chapter 7. The Friedman doctrine pushed in a different direction: good CEOs did not have to pay high wages. Their social responsibility was solely to the shareholders. Many high-profile CEOs, such as General Electric's Jack Welch, heeded the advice and took a tough stance against wage raises.

Nowhere can the impact of the Friedman doctrine be seen more clearly than in business schools. The 1970s were the beginning of the professionalization of managers, and the share of managers trained in business schools increased rapidly during this period. In 1980, about 25 percent of CEOs in publicly listed firms had a business degree. By 2020, this number exceeded 43 percent. Many faculty at business schools embraced the Friedman doctrine and shared this vision with aspiring managers.

Recent research shows that managers who attended business schools started implementing the Friedman doctrine, especially when it came to wage setting. They stopped wage growth in their firms, compared to similar companies run by managers who did not attend business schools. Managers in the United States and Denmark without an MBA share with their workers about 20 percent of any increase in value added. For managers inculcated in business schools, this number is zero. Somewhat disappointingly for business schools and for economists from the Friedman-Jensen school, there is no evidence that business school–trained managers increase productivity, sales, exports, or investment. But they do increase shareholder value because they cut wages. They also pay themselves more handsomely than other managers.

Resistance to the New Deal, accompanied by the antiregulation, antilabor philosophical stances of some business executives and the Friedman doctrine, was not enough, however. In the early 1970s, wholesale deregulation and dismantling the labor movement were fringe ideas, even if more businesses were becoming vocal about the burdens of growing regulations. That changed with the oil-price shock of 1973 and the stagflation that followed, which were interpreted as a failure of the existing system and signs that the US economy was not working anymore. A course correction was needed, and the Friedman doctrine and its bolstering of the power of businesses against regulations and organized labor came to be seen as the answer.

Ideas that used to be advocated by think tanks outside of the mainstream started gaining adherents among lawmakers and businesses. Barry Goldwater, the Republican presidential candidate in the 1964 election, failed to get the support of the broader business community in part because his antiregulation ideas appeared extreme at that time. By 1979, Goldwater was boasting, "Now that almost every one of the principles I advocated in nineteen sixty-four have become the gospel of the whole spread of the spectrum of politics, there really isn't a heck of a lot left." Ronald Reagan reaffirmed this conclusion shortly after his election, when he told a crowd of conservative activists, "Had there not been a Barry Goldwater willing to make that lonely walk, we would not be talking of a celebration tonight."

Big Is Beautiful

Even if one bought into the view that the market mechanism works seamlessly, regulations are mostly unnecessary, and the business of business should be maximizing shareholder value, there was still a tricky issue from the viewpoint of large corporations.

Many businesses have considerable ability to set their price because they dominate parts of the market or have a loyal clientele. Think of the market power of Coca-Cola, for example, which controls 45 percent of the carbonated soft drinks market and can significantly shape the industry's prices. Monopoly means that the market mechanism starts breaking down. Things are even worse when these corporations can block entry by new competitors or when they are able to acquire competing businesses, as the robber barons of late nineteenth-century America understood all too well.

Adam Smith, the original proponent of the market-mechanism magic, was damning in his account of how even small groups of businessmen getting together could damage the common good. In a famous passage in *The Wealth of Nations*, he wrote, "People of the same trade seldom meet together, even for merriment and diversion, but the conversation ends in a conspiracy against the public, or in some contrivance to raise prices." Building on Smith's ideas, many free-market advocates have remained skeptical of large corporations, and some of them raise alarms when mergers and acquisitions increase the power of big players.

Thwarting the workings of the market is not the only reason for being suspicious of big businesses. A well-known proposition in economics is the Arrow replacement effect, named after the Nobel Prize–winning economist Kenneth Arrow and later popularized by the business scholar Clayton Christensen as the "innovator's dilemma." It states that large corporations are timid innovators because they are afraid of eroding their own profits from existing offerings. If a new product will eat into the revenues a corporation enjoys from what it is already doing, why go there? In contrast, a new entrant could be very keen on doing something quite different because it cares only about those new profits. The available evidence bears out this conjecture. Among innovative firms, younger and smaller ones invest almost twice as much in research as a fraction of their sales and subsequently tend to grow much faster than older and larger businesses.

Even more important is the impact of large corporations on political and social power. US Supreme Court justice Louis Brandeis nailed this when he stated, "We may have democracy, or we may have wealth concentrated in the hands of a few, but we can't have both." He was opposed to large corporations not just because they increased market concentration and created conditions of monopoly, undercutting the market mechanism. He maintained that as they became very large, they exercised disproportionate political power, and the wealth they created for their owners further degraded the political process. Brandeis did not focus as much on social power—for example, whose ideas and vision we listen to—but his reasoning extends to that domain as well. When a few companies and their executives achieve higher status and greater power, it becomes harder to counter their vision.

By the 1960s, however, several economists were already articulating ideas that were more skeptical of the utility of antitrust measures, aimed at limiting the power of big business. Particularly important in this was George Stigler, who saw antitrust action as part of the overall meddling of governments, just like regulations. Stigler's ideas influenced legal scholars with some knowledge of economics, most notably Robert Bork.

Bork's influence and persona extended far beyond academia. He was Richard Nixon's solicitor general and then became acting attorney general after his predecessor and his deputy resigned rather than accepting the pressure from the president to fire Archibald Cox, the independent prosecutor going after the Watergate scandal. Bork did not have the same qualms and relieved Cox of his duties as soon as he took up the post.

Bork's greater influence was through his scholarship, however. He took Stigler's and related ideas and articulated a new approach to antitrust and regulation of monopoly. At the center was the idea that large corporations dominating their market were not necessarily a problem that required government intervention.

The key question was whether they harmed consumers by raising prices, and the onus was on government authorities to prove that they were doing so. Otherwise, these companies could be presumed to benefit consumers through greater efficiency, and public policy should stand aside. So big companies like Google and Amazon may look like and walk like monopolies, but according to this doctrine, no government action was needed until they could be proven to have increased prices.

The Manne Economics Institute for Federal Judges, founded in 1976 with corporate funding, instructed scores of judges in economics during intensive training camps, but the economics they taught was a very specific version based on Friedman's, Stigler's, and Bork's ideas. Judges who attended these training sessions became influenced by their teaching and began using more of the language of economics in their opinions. Strikingly, they also started issuing more conservative decisions and ruling consistently against regulatory agencies and antitrust action. The Federalist Society, founded in 1982 with similarly generous support from antiregulation executives, had a similar aim—grooming pro-business, antiregulation law students, judges, and Supreme Court justices. It has been phenomenally successful; six of the current Supreme Court justices are among its alumni.

The consequences of the new approach to big business were sweeping. Today the United States has some of the largest and most dominant corporations ever: Google, Facebook, Apple, Amazon, and Microsoft are jointly worth about one-fifth of US GDP. The value of the largest five corporations at the beginning of the twentieth century—when the public and reformers were up in arms about the problem of monopoly—was not more than one-tenth of GDP. This is not just about the tech sector. From 1980 to today, concentration (market power of the largest firms) increased in more than three-quarters of US industries.

The new antitrust approach has been critical in this. The Department of Justice has blocked only a handful of mergers and acquisitions over the last four decades. This hands-off approach allowed Facebook to buy WhatsApp and Instagram, Amazon to acquire Whole Foods, Time Warner and America Online to join together, and Exxon to merge with Mobil, reversing part of the breakup of Standard Oil. In the meantime, Google and Microsoft have purchased scores of start-ups and small companies that could have turned into their rivals.

The implications of the rapid growth of big businesses are wide-ranging. Many economists argue that they are now enjoying greater market power, which they are exercising both to thwart innovation from rivals and to enrich their top executives and shareholders. Gargantuan monopolies are often bad news for consumers because they distort prices and innovation. They also spell trouble for the productivity bandwagon because they reduce competition for workers. They powerfully multiply inequality at the top by enriching their already-wealthy shareholders. Large corporations have sometimes boosted the earnings of their employees by sharing their profits with them. But another part of the institutional changes of the last several decades meant that this was not likely to happen: the eclipse of worker power.

A Lost Cause

The effects of the Friedman doctrine on wage setting may have been as important as its direct impact. If managers maximizing shareholder value were on the side of the angels, then anything standing on their path was a distraction or—worse—an impediment to the common good. Hence, the Friedman doctrine gave an additional impetus to managers to campaign against the labor movement.

Despite American unions' important role in the shared prosperity of the decades that followed World War II, their relationship with management was always strained. When unions win elections for representation in a plant, we see a striking increase in the likelihood that the plant will close. This is partly because of multiplant corporations shifting their production to nonunionized establishments. Executives delay unionization votes and adopt various tactics to convince workers to reject unions; if this fails, the jobs are moved elsewhere.

The conflict inherent in this relationship has both idiosyncratic and institutional roots. Some unions developed close ties with organized crime because of their presence in activities that were controlled by the Mafia. Leaders such as Jimmy Hoffa, president of the International Brotherhood of Teamsters, came to signify this dark side and likely contributed to the decline in public support for labor organizations. Hoffa served time in prison for bribery and various other crimes, and was probably murdered by the Mafia.

More important than the flaws of the union leaders has been the way in which American unions were organized. We saw in Chapter 7 that collective agreements in Sweden and other Nordic countries were organized in the context of the corporatist model, which attempted to cultivate greater communication and cooperation between management and workers. They also set wages at the industry level. The German system combines industry-level wage bargaining together with work councils at the firm level, which represent the worker voice on corporate boards. In the United States, on the other hand, the 1947 Taft-Hartley Act weakened some of the pro-union provisions of the Wagner Act and legislated that collective bargaining had to take place at the business-unit level. It also banned secondary industrial action, such as boycotts in sympathy with strikers. Consequently, American unions organize and negotiate wages in their immediate workplaces, with no industry coordination. This arrangement breeds more conflictual relations between business and labor. When managers think that a hard line against unions can reduce wages and create a cost advantage relative to competitors, they are less likely to accept union demands.

Starting around 1980, the balance of power shifted further away from the labor movement. Particularly important was Ronald Reagan's tough stance against the Professional Air Traffic Controllers Organization in 1981. When the organization's negotiations with the Federal Aviation Administration stalled, it called a strike, even though industrial action by government employees was illegal. President Reagan was swift in firing striking workers, calling them a "peril to national safety." Where Reagan led, private businesses followed, and several large employers hired new workers when confronted with industrial actions rather than giving in to union demands.

Even before Reagan and the corporate pushback, the United States was past peak unionization. Nevertheless, in the early 1980s there were still about eighteen million union workers, and 20 percent of wage and salary workers belonged to unions. Since then, there has been a steady decline, partly because of the tougher antiunion stance of businesses and politicians, and partly because employment in the more heavily unionized manufacturing sector has dwindled. In 2021 only 10 percent of workers were union members. Additionally, by the 1980s, most of the cost-of-living escalator clauses that ensured automatic wage increases without full-scale agreements had been negotiated out of union contracts, further weakening the hands of labor and the prospect for sharing of productivity gains with workers.

This antilabor shift is not unique to the United States. Margaret Thatcher, who was elected British prime minister in 1979, prioritized deregulation, enacted myriad pro-business laws, and vigorously fought unions, so British unions have also lost much of their earlier strength.

A Grim Reengineering

Rising industrial concentration and the waning of rent sharing were a first salvo against the shared prosperity model of the 1950s and 1960s, but by themselves would not have created the tremendous turnaround we witnessed. For that, the direction of technology would also have to move in an antilabor direction. This is where digital technologies enter the story, in a big way.

The Friedman doctrine encouraged corporations to increase profits by whatever means necessary, and by the 1980s, this idea was embraced by the corporate sector. Executive compensation, in the form of stock options, strongly supported this shift. The culture at the top of corporations began to change. In the 1980s a big story for corporate America was rivalry from efficient Japanese manufacturers, first in consumer electronics and then in the auto industry. The people who ran US firms felt a pressing need to respond.

As a result of the broadly balanced investments in automation and new tasks in the 1950s and 1960s, worker marginal productivity had increased, and the labor share of income in manufacturing remained broadly constant, hovering close to 70 percent between 1950 and the early 1980s. But by the 1980s, many American managers came to see labor as a cost, not as a resource, and to withstand foreign competition, these costs needed to be cut. This meant reducing the amount of labor used in production through automation. Recall that automation increases output per worker but, by sidelining labor, it limits and may even reduce worker marginal productivity. When this happens on a large enough scale, there is less demand for workers and lower wage growth.

To cut labor costs, US businesses needed a new vision and new technologies, which came, respectively, from business schools and the nascent tech sector. The main ideas on cost cutting are well summarized in a 1993 book by Michael Hammer and James Champy, *Reengineering the Corporation: A Manifesto for Business Revolution*. The book argues that US corporations had become highly inefficient, especially because there were too many middle managers and white-collar workers. The US corporation should therefore be reengineered to compete more vigorously, and new software could provide the tools.

To be fair, Hammer and Champy emphasized that reengineering was not just automation, but they also took the view that more effective use of software would eliminate many unskilled tasks: "Much of the old, routine work is eliminated or automated. If the old model was simple tasks for simple people, the new one is complex jobs for smart people, which raises the bar for entry into the workforce. Few simple, routine, unskilled jobs are to be found in a reengineered environment." In practice, the smart people for the complex jobs were almost always workers with college or postgraduate degrees. Well-paying jobs for noncollege workers became scant in reengineered environments.

The high priests of the emerging vision came from the newly burgeoning management-consulting field. Management consulting barely existed in the 1950s, and its growth coincides with efforts to remake corporations through "better" use of digital technology. Together with business schools, leading management-consulting companies such as McKinsey and Arthur Andersen also pushed cost cutting. As these ideas were increasingly preached by articulate management experts, it became harder for workers to resist.

Just like the Friedman doctrine, *Reengineering the Corporation* crystallized ideas and practices that were already being implemented. By the time the book came out, several large US corporations had used software tools to downsize their workforces or expand operations without having to hire new employees. By 1971, IBM was prominently advertising its "word-processing machines" as a tool for managers to increase their productivity and automate various office jobs.

In 1981 IBM launched its standardized personal computer, with a range of additional capabilities, and soon new software programs for automation of clerical work, including administrative and back-office functions, were being developed. As far back as 1980, Michael Hammer anticipated more extensive "office automation":

> Office automation is simply an extension of the kinds of things that data processing has been doing for years, updated to take advantage of new hardware and software possibilities. Distributed processing to replace mail, source data capture to reduce retyping, and end-user oriented systems are the ways in which "office automation" will be brought beyond the traditional applications and to the aid of all segments of the office.

A vice president of Xerox Company around the same time was predicting, "We may, in fact, witness the full blossoming of the postindustrial revolution when routine intellectual work becomes as automated as heavy mechanical work did during the 19th century." Other commentators were more worried about these developments but still expected "the automation of all phases of information manipulation from gathering to dissemination."

Interviews from the 1980s with workers both on shop floors and in offices indicated their anxiety in the face of new digital technologies. As one worker put it, "We don't know what will be happening to us in the future. Modern technology is taking over. What will be our place?"

It was the arrival of these early digital technologies that made Wassily Leontief, another Nobel Prize–winning economist, worry in 1983 that human labor would go the way of the horses and become mostly unnecessary for modern production.

These expectations were not completely off the mark. A case study of the introduction of new computer software in a large bank finds that the new technologies adapted in the 1980s and early 1990s led to a significant reduction in the number of workers employed in check processing. Back-office tasks were automated equally rapidly across various industries during the same time.

As these technologies spread, many relatively high-wage occupations started declining. In 1970 about 33 percent of American women were in clerical jobs, which paid decent salaries. Over the next six decades, this number declined steadily and is now down to 19 percent. Recent research documents that these automation trends have been a powerful contributor to the wage stagnation and declines for low- and middle-skill office workers.

But where did the software to support downsizing come from? Not from the early hackers, who were steadfastly opposed to corporate control over computers. Designing software to fire workers would have been anathema to them. Lee Felsenstein anticipated this type of demand and railed against it: "The industrial approach is grim and doesn't work: the design motto is 'Design by Geniuses for Use by Idiots,' and the watchword for dealing with the untrained and unwashed public is KEEP THEIR HANDS OFF!" Instead, he insisted on the importance of "the user's ability to learn about and gain some control over the tool." In the words of one of his associates, Bob Marsh, "We wanted to make the microcomputer accessible to human beings."

William (Bill) Henry Gates III had a different idea. Gates enrolled at Harvard to study pre-law and then mathematics, but left school in 1975 to found Microsoft with Paul Allen. Allen and Gates built on the pathbreaking work of many other hackers to produce a rudimentary compiler, using BASIC, for the Altair, which they then turned into an operating system for IBM. Gates had his eye on monetization from the beginning. In a 1976 open letter, he accused hackers of stealing software programmed by Allen and himself: "As the majority of hobbyists must be aware, most of you steal your software."

Gates was determined to find a way of making a lot of money from software. Selling to large, established companies was the obvious way forward. Where Microsoft and Bill Gates led, much of the rest of the industry followed. By the early 1990s, a major part of the computer industry, including emerging household names such as Lotus, SAP, and Oracle, supplied office software to big corporations and were spearheading the next phase of office automation.

Although automation based on office software was likely more important for employment, the overall trends can also be seen in the effects of another iconic technology of the era: industrial robots.

Robots are the quintessential automation tool, targeting the performance of repetitive manual tasks, including the moving of objects, assembly, painting, and welding. Autonomous machines performing human-like tasks have captured popular imaginations since Greek mythology. The idea came into clearer focus with *R.U.R.*, an imaginative 1920 play by the Czech writer Karel Čapek that introduced the word *robot*. In this science-fiction tale, robots run factories and work for humans, but it does not take them long to turn against their masters. Fears of robots doing all sorts of bad things have been part of the public conversation ever since. Science fiction aside, one thing is for certain: robots do automate work.

The United States was a laggard in robotics in the 1980s, in part because it was not under the same demographic pressures that affected countries such as Germany and Japan. In the 1990s, robots started spreading rapidly in US manufacturing. Just like automation software in offices, robots did what their designers intended them to do—they reduced the labor intensity of production. For example, the automotive industry was completely revolutionized by robots and, as a result, now employs many fewer workers in traditional blue-collar tasks.

Robots increase productivity. However, in US manufacturing, rather than launching the productivity bandwagon, they have reduced employment and wages. As with the automation of white-collar jobs with office software, the elimination of blue-collar jobs by robotics technology was swift. Some of the best jobs available to workers without a college degree in the 1950s and 1960s were in welding, painting, material handling, and assembly, and these jobs have steadily disappeared. In 1960 almost 50 percent of American men were in blue-collar occupations. This number has subsequently fallen to about 33 percent.

Once Again, a Matter of Choice

Could the turn toward automation, starting around 1980, be the inevitable result of technology's progress? Perhaps advances in computers were by their nature more amenable to automation. Although it is difficult to dismiss this possibility entirely, there is plenty of evidence that the direction of technology and the emphasis on cost cutting were choices.

Digital technologies, even more so than electricity, discussed in Chapter 7, are general purpose, enabling a wide range of applications. Different choices on their direction will likely translate into gains and losses for different segments of the population. In fact, many of the early hackers thought that computers could empower workers and enrich their work rather than automate it. We will see in Chapter 9 that they were not wrong: several important digital tools powerfully complemented human labor. Unfortunately, however, most efforts in the blossoming computer industry went toward automation.

Moreover, although they had access to the same software tools and robotics technology, other countries made very different choices than their American counterparts. For example, German manufacturing firms still had to negotiate with unions and explain their decisions to worker representatives on their corporate boards. They were also understandably wary of laying off workers who had gone through years of apprenticeship in the company and developed a range of relevant skills. They thus made technological and organizational adjustments to increase the marginal productivity of the workers they had already trained, blunting automation's impact.

Consequently, even though industrial automation has been faster in Germany, with the number of robots per industrial workers more than twice that in the United States, companies made efforts to retrain blue-collar workers and reallocate them to new tasks, often in technical, supervisory, or white-collar occupations. This creative use of worker talent is also visible in how German companies use new software in manufacturing. At the center of programs such as Industry 4.0 or Digital Factory, which became popular in German manufacturing in the 1990s and 2000s, was the use of computer-assisted design and computer-aided quality control that enabled well-trained workers to contribute to design and inspection—for instance, by working on virtual prototypes or by using software tools to detect problems. These efforts ensured that worker marginal productivity increased, even as the German industry rapidly introduced new robots and software tools. Tellingly, following robot adoption, the reallocation of blue-collar workers to new, technical tasks is more pronounced in German workplaces, where labor unions are stronger.

Germany started the postwar era with labor shortages because of the significant fraction of its male population that had perished in the war. Labor scarcity continued as birthrates declined in Germany faster than in the rest of Europe, creating an acute need for working-age people in the country by the 1980s. In the same way that a shortage of skilled labor encouraged more worker-friendly uses of technology in the nineteenth-century United States, it induced German firms to find ways of making best use of their employees' capabilities by investing in skills during apprenticeship programs that now run three or four years.

It also encouraged the retraining of workers for more technical tasks as automation technologies were adopted.

As a result of these priorities and adjustments, the number of workers in the auto industry in Germany rose between 2000 and 2018. These gains have been accompanied by an increase in the fraction of white-collar and technical occupations, such as engineering, design, and repair, in the industry from 30 percent to 40 percent. In the meantime, US automakers, whose output followed a similar trajectory to their German counterparts, reduced employment by about 25 percent and did not undertake similar occupational upgrading.

This was not just a German story. Japanese firms, also facing a declining labor force, have been even faster in adopting robots. But they too combined automation with the creation of new tasks. With the emphasis on flexible production and quality, Japanese companies did not automate all of the jobs on the factory floor, instead creating a range of complex and well-paid tasks for their employees. They also invested as much in software for flexible planning, supply-chain management, and design tasks as software tools used for automation. Overall, during the same time period, Japanese automakers did not reduce their workforces in the same way that their American peers did.

In Finland, Norway, and Sweden, where collective bargaining remained important and a large share of the industrial workforce is still covered by collective agreements, corporations have continued to share productivity gains with workers, and automation has often been combined with other technological adaptations more favorable to labor.

In the 1950s and 1960s, US labor unions could also object to excessive automation technologies or demand other changes to protect workers, as in Germany. But by the 1990s, the US labor movement was enfeebled. With the prevailing vision emphasizing cost cutting and the superiority of fully automated processes, American labor came to be seen as something to be eliminated from the production process, rather than as people with skills who could become more valuable with training and appropriate technological investments. These automation and labor-cutting choices then became self-reinforcing, for automation also reduced the number of unionized blue-collar workers, delivering another blow to the labor movement.

Government policy also contributed to these developments. The US tax system has always favored capital relative to labor, imposing lower effective taxes on capital earnings than labor income. Starting in the 1990s, the asymmetry of capital taxation and labor-income taxation intensified, especially for equipment and software capital. Successive administrations reduced corporate income and federal income taxes on the richest Americans, pushing down the tax rate on capital (because returns on capital investments in corporate profits go disproportionately to those people). Starting in 2000, capital tax cuts went into overdrive with increasingly generous depreciation allowances on equipment and software capital. Although these were at first supposed to be temporary, they were often extended and then made more generous.

Overall, whereas the average tax rate on labor income, based on payroll and federal income taxes, remained over 25 percent for the last thirty years, the effective tax rates on equipment and software capital (including all capital gains and income taxes) fell from around 15 percent to less than 5 percent in 2018. These tax incentives meant that businesses had even a greater appetite for automation equipment, and their demand fueled further development of automation technologies in a self-reinforcing cycle.

The evolution of federal research and science policy may have been another contributing factor. Starting from before World War II, government funding of science and private-sector research was generous, especially in areas that were national defense priorities. This was a powerful inducement to new critical areas, such as antibiotics, semiconductors, satellites, aerospace, sensors, and the internet.

Over the last five decades, both government strategic technology leadership and funding declined. Federal spending on research and development fell from around 2 percent of GDP in the mid-1960s to about 0.6 percent today. The government also became more likely to support the research priorities set by leading corporations. This new landscape then allowed large corporations, especially in the digital area, to determine the direction of technology. Their incentives and mind-set pushed toward more and more automation.

US technologies and business strategies spread more broadly, even if countries differed in how they adopted and configured automation technologies, as we have seen. The Friedman doctrine and ideas related to the use of digital tools in order to cut costs influenced business practices in the United Kingdom and the rest of Europe. For example, the effects of managers trained in business schools are remarkably similar in Denmark and the United States. Management consulting expanded throughout the Western world, and new digital technologies and robots were adopted rapidly. Automation and globalization reduced the fraction of the labor force working in blue-collar and clerical occupations in essentially all industrialized nations. Thus, despite variation across countries, the direction of progress in the US has had a significant global impact.

Digital Utopia

The direction of technology that prioritized automation cannot be understood unless we recognize the new digital vision that emerged in the 1980s. This vision combined the drive to cut labor costs, rooted in the Friedman doctrine, with elements of the hacker ethic, but abandoned the philosophy of early hackers such as Lee Felsenstein that was antielitist and suspicious of corporate power. Felsenstein admonished IBM and other big corporations because they were trying to misuse technology with their ideology of "design by geniuses for use by idiots." The new vision instead embraced the top-down design of digital technologies aimed at eliminating people from the production process.

There was a euphoria, reminiscent of the way that Ferdinand de Lesseps used to talk about building the Suez and Panama Canals, about what technology could achieve, provided that it was shepherded by talented programmers and engineers. Bill Gates summed up this techno-optimism when he proclaimed,

"Show me a problem, and I'll look for technology to fix it." That technology may be socially biased—in their favor and against most people—does not seem to have occurred to Gates and his confederates.

The transformation from the hacker ethic to corporate digital utopia was largely about following the money and social power. By the 1980s, software engineers could either have their ideals or gain tremendous riches by signing up with companies that were becoming larger and more powerful. Many chose the latter.

Meanwhile, antiauthoritarianism morphed into a fascination with "disruption," meaning that disrupting existing practices and livelihoods was welcome or even encouraged. The precise words differed, but the underlying thinking was reminiscent of British entrepreneurs in the early 1800s, who felt fully justified in ignoring any collateral damage they created along their path, especially on workers. Later, Mark Zuckerberg would make "Move fast and break things" a mantra for Facebook.

An elitist approach came to dominate almost the entire industry. Software and programming were things in which very talented people excelled, and the less able were of limited use. Journalist Gregory Ferenstein interviewed dozens of tech start-up founders and leaders who expressed these opinions. One founder stated that "very few are contributing enormous amounts to the greater good, be it by starting important companies or leading important causes." It was also generally accepted that those few seen as contributing to the public good by launching new businesses should be handsomely rewarded. As the Silicon Valley entrepreneur Paul Graham, one of *Businessweek*'s "twenty-five most influential people on the web," put it, "I've become an expert on how to increase economic inequality, and I've spent the past decade working hard to do it. . . . You can't prevent great variations in wealth without preventing people from getting rich, and you can't do that without preventing them from starting startups."

Even more consequential was the elitism of this vision when it came to the nature of work. Most people were not smart enough to even excel at the jobs that were assigned to them, so the use of software designed by technology leaders to reduce corporations' dependence on these fallible people was fully justified. Hence, automating work became an integral part of this vision, and perhaps its most powerful implication.

Not in the Productivity Statistics

The productivity bandwagon is foundational to this vision of a digital utopia. If many workers are made worse off by technological improvements, it becomes much harder to claim that productivity gains are in the common good.

The bandwagon is less likely to operate when employers have too much power relative to workers, when technology is moving in an antilabor direction, and when productivity gains do not translate into employment growth in other sectors. But there is an even more fundamental problem. During the last several decades, there has been less productivity growth to share, even though we are bombarded with new products and apps every day.

Generations that lived in the 1960s and 1970s used the same (rotary dial) telephone and the same TV set for decades, until they broke down and buying new equipment became inevitable. Today, most middle-class families upgrade their mobile phones, TVs, or other electronics every year or two: new models are faster, glossier, and more capable because of their myriad new features. For example, Apple releases a new iPhone almost every year.

Indeed, the rate of overall innovation appears to have skyrocketed. In 1980, there were 62,000 domestic patents filed with the United States Patent and Trademark Office. By 2018, this number had increased to 285,000, a nearly fivefold rise. Over the same period, the population of the United States rose by less than 50 percent.

Moreover, much of the growth in both patenting and research spending is driven by new patents in electronics, communication, and software, the fields that were supposed to propel us forward. But look a little closer, and the fruits of the digital revolution are much harder to see. In 1987, Nobel Prize winner Robert Solow wrote: "You can see the computer age everywhere but in the productivity statistics," pointing out the small gains from investments in digital technologies.

Those more optimistic about computers told Solow that he had to be patient; productivity growth would soon be upon us. More than thirty-five years have passed, and we are still waiting. In fact, the US and most other Western economies have had some of the most unimpressive decades in terms of productivity growth since the beginning of the Industrial Revolution.

Focusing on the same measure of productivity we discussed in Chapter 7, total factor productivity (TFP), US average growth since 1980 has been less than 0.7 percent, compared to TFP growth of approximately 2.2 percent between the 1940s and 1970s. This is a remarkable difference: it means that if TFP growth had remained as high as it had been in the 1950s and 1960s, every year since 1980 the US economy would have had a 1.5 percent higher GDP growth rate. The productivity slowdown is not just a problem of the era following the global financial crisis of 2008. US productivity growth between the booming years of 2000 and 2007 was less than 1 percent.

This evidence notwithstanding, technology leaders maintain that we are lucky to be alive in this age of technology and innovation. The journalist Neil Irwin summarizes this optimistic view succinctly in the *New York Times*: "We're in the golden age of innovation, an era in which digital technology is transforming the underpinnings of human existence."

Slow productivity growth is then simply a problem of not fully recognizing all the benefits we are getting from new innovations. For example, Google's chief economist, Hal Varian, argues that slow productivity growth is rooted in mismeasurement: we are not accurately incorporating consumer benefits from products such as smartphones that simultaneously act like cameras, computers, global positioning devices, and music players. Nor are we appreciating the true productivity gains from better search engines and abundant information on the web. The chief economist of Goldman Sachs, Jan Hatzius, agrees: "We think it is more likely that the statisticians are having a harder and harder time accurately measuring productivity growth, especially in the technology sector." He reckons that the true productivity growth of the US economy since 2000 could be several times greater than statistical agencies' estimates.

In principle, consumer and productivity benefits from new technologies should be in the TFP numbers we reported, which are based on GDP growth adjusted for changes in prices, quality, and product variety. Thus, products that significantly increase consumer welfare should translate into much higher TFP growth. In practice, of course, such adjustments are imperfect, and mismeasurement can arise. Nevertheless, these problems are unlikely to explain away the productivity slowdown.

The same problem of undercounting quality improvements and broader social benefits from new products has been around ever since national income statistics were first devised. It is far from clear that digital technologies have worsened this problem. Indoor plumbing, antibiotics, and the highway system generated a panoply of new services and indirect effects that were only imperfectly measured in national statistics. Moreover, measurement problems cannot account for the current productivity slowdown; industries with greater investment in digital technologies show neither differential productivity deceleration nor any evidence of faster quality improvements than those that are less digital.

A few economists, such as Tyler Cowen and Robert Gordon, believe that this disappointing productivity performance reflects dwindling opportunities for revolutionary breakthroughs. In contrast to techno-optimists, they claim, the great innovations are behind us, and improvements from now on will be incremental, leading only to slow productivity growth.

There is no consensus among economists about what exactly is going on, but there is little support for the notion that the world is running out of ideas. In fact, as we saw in Chapter 1, there have been tremendous advances in the tools of scientific and technical inquiry and in communication and information acquisition. Rather than being afflicted by a shortage of ideas, quite a bit of evidence suggests that the US and Western economies are squandering the available opportunities and scientific know-how. There is a lot of research and innovation. Yet these economies are not getting the expected returns on these activities.

The simple fact is that the US research and innovation portfolio has become highly imbalanced. Although more resources keep pouring into computers and electronics, almost all other manufacturing sectors are lagging. Recent research shows that new innovations appear to benefit more-productive larger firms, whereas the second- and third-tier firms are falling behind across the industrialized world, most likely because their investments in digital technologies are not paying off.

More fundamentally, productivity gains from automation may always be somewhat limited, especially compared to the introduction of new products and tasks that transform the production process, such as those in the early Ford factories. Automation is about substituting cheaper machines or algorithms for human labor, and reducing production costs by 10 or even 20 percent in a few tasks will have relatively small consequences for TFP or the efficiency of the production process. In contrast, introducing new technologies, such as electrification, novel designs, or new production tasks, has been at the root of transformative TFP gains throughout much of the twentieth century.

As innovation has turned its back on boosting worker marginal productivity and creating new tasks for humans over the last forty years, it has also left many "low-hanging fruits." One place we can get a sense of these forgone productivity opportunities is in the automobile industry. Although the introduction of robots and specialized software has increased output per worker in the industry, there is evidence that investing more in humans would have boosted productivity by more. This is what Japanese car companies, such as Toyota, discovered starting in the 1980s. When they automated more and more tasks, they saw that productivity was not increasing by much because, without the workers in the loop, they were losing flexibility and the ability to adapt to changes in demand and production conditions. In response, the company took a step back and reinstated workers' central role in crucial production tasks.

Toyota has demonstrated the same possibilities in the United States, too. GM's Fremont, California, plant suffered from low productivity, unreliable quality, and labor conflict, and it shut down in 1982. In 1983, Toyota and GM launched a joint venture to produce cars for both companies and reopened the Fremont facility, retaining its previous union leadership and workforce. But Toyota applied its own management principles, including the approach of combining advanced machinery with worker training, flexibility, and initiative. Soon Fremont reached productivity and quality levels comparable to those of Toyota's Japanese plants, much higher than those of US automakers.

The Tesla electric car company, led by Elon Musk, learned the same lessons more recently. Driven by Musk's digital utopia, Tesla originally planned to automate almost every part of car production. It did not work. As costs multiplied and delays prevented Tesla from meeting demand, Musk himself admitted, "Yes, excessive automation at Tesla was a mistake. To be precise, my mistake. Humans are underrated."

This should not have been a big surprise. Karel Čapek, who christened robots, also recognized their limitations and inability to do the finer things that humans do: "Only years of practice will teach you the mysteries and bold certainty of a real gardener, who treads at random, and yet tramples on nothing."

Unexploited low-hanging fruit is even more consequential in the realm of innovation than in the way factories are organized. In the quest for greater automation, managers have ignored technological investments that could boost worker productivity by providing better information and platforms for collaboration and creating new tasks, as we discuss in Chapter 9. With a more balanced portfolio of innovations, rather than the excessive automation focus fueled by the digital utopia, the economy could have achieved faster productivity growth.

Toward Dystopia

The most important driver of the increase in inequality and the loss of ground for most American workers is the new social bias of technology. We have seen throughout that we should not bank on technology inexorably benefiting everybody. The productivity bandwagon works only under specific circumstances. It does not operate when there is insufficient competition between employers, little or no worker power, and ceaseless automation.

In the decades that followed World War II, automation was rapid but went together with equally innovative technologies that increased worker marginal productivity and the demand for labor. It was the combination of these two forces, as well as an environment that encouraged competition between corporations and collective bargaining, that made the productivity bandwagon effective.

Things look very different from 1980 onward. During this era, we see faster automation but only a few technologies counterbalancing the antilabor bias of automation. Wage growth also slowed down as the labor movement became increasingly impaired. In fact, lack of resistance from the labor movement was likely an important cause of the greater emphasis on automation. Many managers, even during periods of relatively shared prosperity, have a bias toward automation, for this enables them to reduce labor costs and diminish the bargaining power of workers. Once countervailing powers from the labor movement and government regulation weakened, rent sharing subsided, and a natural bias toward automation set in. Now the productivity bandwagon had far fewer people on board.

Worse, without countervailing powers, digital technologies became engulfed in a new digital utopia, elevating the use of software and machinery to empower companies and sideline labor. Digital solutions imposed from above by technology leaders came to be regarded almost by definition to be in the public interest. Yet what most workers got was much more dystopian: they lost their jobs and their livelihoods.

There were other ways of developing and using digital technologies. Early hackers, guided by a different vision, pushed the technology frontier toward greater decentralization and out of the hands of large corporations. Several notable successes were based on this alternative approach, even if it remained mostly marginal to the main developments of the tech industry, as we will soon see.

Hence, the bias of technology was very much a choice—and a socially constructed one. Then things started getting much worse, economically, politically, and socially, as tech visionaries found a new tool to remake society: artificial intelligence.

9

Artificial Struggle

Nothing has been written on this topic which can be considered as decisive—and accordingly we find everywhere men of mechanical genius, of great general acuteness and discriminative understanding, who make no scruple in pronouncing the Automaton a *pure machine*, unconnected with human agency in its movements, and consequently, beyond all comparison, the most astonishing of the inventions of mankind.

    —EDGAR ALLAN POE, "Maelzel's Chess Player,"
            1836 (italics in original)

The world of the future will be an ever more demanding struggle against limitations of our intelligence, not a comfortable hammock in which we can lie down to be waited upon by our robot slaves.

    —NORBERT WIENER, *God and Golem, Inc.*, 1964

In its special report on the future of work in April 2021, the *Economist* magazine took to task those worrying about inequality and dwindling job opportunities for workers: "Since the dawn of capitalism people have lamented the world of work, always believing that the past was better than the present and that the workers of the day were uniquely badly treated."

Fears about AI-driven automation are particularly overblown, and "popular perceptions about the world of work are largely misleading." The report proceeded to provide a clear restatement of the productivity bandwagon: "In fact, by lowering costs of production, automation can create more demand for goods and services, boosting jobs that are hard to automate. The economy may need fewer checkout attendants at supermarkets, but more massage therapists."

The report's overall assessment: "A bright future for the world of work."

The management consulting company McKinsey expressed a similar conclusion in early 2022 as part of its strategic partnership with the annual World Economic Forum in Davos:

> For many members of the world's workforces, change can sometimes be seen as a threat, particularly when it comes to technology. This is often coupled with fears that automation will replace people. But a look beyond the headlines shows that the reverse is proving to be true, with Fourth Industrial Revolution (4IR) technologies driving productivity and growth across manufacturing and production at brownfield and greenfield sites. These technologies are creating more and different jobs that are transforming manufacturing and helping to build fulfilling, rewarding, and sustainable careers.

The *Economist* and McKinsey were articulating views of many tech entrepreneurs and experts that concerns about AI and automation are exaggerated. The Pew Research Center surveyed academics and technology leaders, and reported statements from more than a hundred of them, with the overwhelming majority stating that although there were downsides, AI would bring widespread economic and societal benefits.

According to the prevailing perspective, there may be some disruption along the way—for example, in terms of jobs lost—but such transition costs are unavoidable. In the words of one of the experts quoted by the Pew Research Center, "In the coming 12 years AI will enable all sorts of professions to do their work

more efficiently, especially those involving 'saving life': individualized medicine, policing, even warfare (where attacks will focus
on disabling infrastructure and less in killing enemy combatants
and civilians)." The same person also conceded, "Of course, there
will be some downsides: greater unemployment in certain 'rote'
jobs (e.g., transportation drivers, food service, robots and automation, etc.)."

But we should not worry too much about these downsides,
for we have the same tech entrepreneurs to ease the burden with
their philanthropy. As Bill Gates articulated at the 2008 World
Economic Forum, these successful people have an opportunity
to do good while doing well for their businesses, by helping the
less fortunate with new products and technologies. He declared
that "the challenge is to design a system where market incentives,
including profits and recognition, drive the change," with the
goal of "improving lives for those who don't fully benefit from
market forces." He dubbed this system "creative capitalism" and
set the philanthropic goal for everybody "to take on a project of
creative capitalism in the coming year" as a way of alleviating the
world's problems.

We will argue in this chapter that this vision of almost inexorable benefits from new technology, including intelligent machines,
led by talented entrepreneurs is an illusion—the AI illusion. Like
Lesseps's conviction that canals would benefit both investors and
global commerce, it is a vision rooted in ideas, but it receives a
further boost because it enriches and empowers elites corralling
technology toward automation and surveillance.

Even the framing of digital capabilities in terms of intelligent
machines is an unhelpful aspect of this vision. Digital technologies are general purpose and can be developed in many different
ways. In steering their direction, we should focus on how useful
they are to human objectives—what we will call "machine usefulness." Encouraging the use of machines and algorithms to complement human capabilities and empower people has, in the past,
led to breakthrough innovations with high machine usefulness.
In contrast, infatuation with machine intelligence encourages
mass-scale data collection, the disempowerment of workers and

citizens, and a scramble to automate work, even when this is no more than so-so automation—meaning that it has only small productivity benefits. Not coincidentally, automation and large-scale data collection enrich those who control digital technologies.

From the Field of AI Dreams

People are right to be excited about advances in digital technologies. New machine capabilities can massively expand the things we do and can transform many aspects of our lives for the better. And there have also been tremendous advances. For example, the Generative Pre-trained Transformer 3 (GPT-3), released in 2020 by OpenAI, and ChatGPT released in 2022 by the same company, are natural-language processing systems with remarkable capabilities. Already trained and optimized on massive amounts of text data from the internet, these programs can generate almost human-like articles, including poetry; communicate in typical human language; and, most impressively, turn natural-language instructions into computer code.

Software programs have a simple logic. A program, or algorithm, is a recipe that instructs a machine to take a prespecified set of inputs and perform a set of step-by-step computations. For example, Jacquard's loom took several punched cards as its input and activated an elegantly designed mechanical process, which moved a beam and wove cloth to produce the designs specified in the cards. Different cards created distinct designs, some of them strikingly complex.

Modern computers are referred to as "digital" because the inputs are represented in discrete form, taking one of a finite set of values (most commonly as zeros and ones). But they share with Jacquard's loom the general principle that they implement exactly the sequence of computations or actions that are specified by a programmer.

What about artificial intelligence? Unfortunately, there is no commonly agreed upon definition. Some experts define artificial intelligence as machines or algorithms demonstrating "intelligent behavior" or "high-level capabilities," although what these are is often open to debate. Others provide definitions motivated by programs such as GPT-3, equating intelligent machines with those that have goals, observe their environment, obtain other inputs, and attempt to achieve their objectives. For example, GPT-3 receives distinct goals in different applications and tries to accomplish them as successfully as possible.

Whatever the exact definition of modern machine intelligence, it is clear that new digital algorithms are being applied widely to every domain of our lives. Rather than attempting to arbitrate between different definitions of machine intelligence, we will use "modern AI" to capture the currently prevailing approach in this domain.

Applying digital technologies to the production process—for example, with numerically controlled machinery—long predates modern AI. The major computing breakthroughs of the past seventy years came from finding ways of performing tasks using software in areas such as document preparation, database management, accounting, and inventory control. Software can also create new production capabilities. In computer-assisted design, it improves the precision and ease with which workers perform design tasks. It makes the work of cashiers and other consumer-facing employees potentially more productive. As we emphasized in Chapter 8, it also enables automation.

To be automated by traditional software, a task needs to be "routine," meaning that it must involve predictable steps that are implemented in a defined sequence. Routine tasks are performed repetitively, embedded in a predictable environment. For example, typing is routine. So are knitting and other simple production tasks that involve a significant amount of repetitive activity. Software has been combined with machinery that interacts with the physical world to automate various routine tasks, exactly as Jacquard intended, and modern numerically controlled equipment, such as printers or computer-assisted lathes, regularly accomplish this. Software is also an integral part of robotics technology used extensively for industrial automation.

But only a small fraction of human tasks is truly routine. Most of the things our species do involve some amount of problem solving. We deal with new situations or challenges by coming up with solutions that draw analogies on the basis of past experience and knowledge. We employ flexibility when the relevant environment changes constantly. We rely heavily on social interaction, such as communication and explanation or simply the camaraderie that many coworkers and customers enjoy in the process of economic transactions. Collectively, we are a pretty creative species.

Customer service, for example, requires a combination of social and problem-solving skills. There are tens of thousands of problems a customer may encounter, some that are rare or entirely idiosyncratic. It is relatively easy to help a customer who has missed a flight and would like to take the next available plane. But what if the traveler has ended up in the wrong airport or now needs to fly to a new destination?

Modern AI approaches have been used to extend automation into a broader range of routine tasks, such as bank-teller services. Pre-AI automation—for example, using automated teller machines (ATMs)—was extensive by the 1990s, with a focus on simple tasks, such as dispensing cash. Depositing checks was only partially automated. ATMs accepted deposits, and magnetic-ink-character-recognition technology was used to sort checks according to their bank code and bank account number. But humans were still necessary for other routine tasks, such as recognizing handwriting, organizing accounts, and monitoring overdrafts. Based on more recent advances in AI-based handwriting-recognition and decision-making tools, checks can now be processed without human involvement.

More significantly, the ambition of AI is to expand automation to nonroutine tasks, including customer service, tax preparation, and even financial advice. Many of the tasks involved in these services are predictable and can be automated straightforwardly. For example, information from wage and tax statements (such as the W-2 form in the United States) can be scanned and automatically entered in the relevant fields to compute tax obligations, or the relevant information about deposits and balances can be provided to a bank customer. Recently, AI has ventured into more complex tasks as well. Sophisticated tax-preparation software can query users about expenses or items that look suspicious, and customers can be presented with voice-activated menus to categorize their problem (even if this often works imperfectly, ends up shifting some of the work to users, and causes longer delays as customers wait for a human to provide the necessary help).

In robotic process automation (RPA), for example, software implements tasks after watching human actions in the application's graphical user interface. RPA bots are now deployed in banking, lending decisions, e-commerce, and various software-support functions. Prominent examples include automated voice-recognition systems and chatbots that learn from remote IT-support practices. Many experts believe this kind of automation will spread to myriad tasks currently performed by white-collar workers. *New York Times* journalist Kevin Roose summarizes RPAs' potential as follows: "Recent advances in A.I. and machine learning have created algorithms capable of outperforming doctors, lawyers and bankers at certain parts of their jobs. And as bots learn to do higher-value tasks, they are climbing the corporate ladder."

Supposedly, we will all be the beneficiaries of these spectacular new capabilities. The current CEOs of Amazon, Facebook, Google, and Microsoft have all claimed that AI will beneficially transform technology in the next decades. As Kai-Fu Lee, former president of Google China, puts it, "And like most technologies, AI will eventually produce more positive than negative impacts on our society."

The evidence does not fully support these lofty promises, however. Although talk of intelligent machines has been around for two decades, these technologies started spreading only after 2015. The takeoff is visible in the amount that firms spend on AI-related activities and in the number of job postings for workers with specialized AI skills (including machine learning, machine vision, deep learning, image recognition, natural-language processing, neural networks, support vector machines, and latent semantic analysis).

Tracking this indelible footprint, we can see that AI investments and the hiring of AI specialists concentrate in organizations that rely on tasks that can be performed by these technologies, such as actuarial and accounting functions, procurement and purchasing analysis, and various other clerical jobs that involve pattern recognition, computation, and basic speech recognition. However, the same organizations also lower their overall hiring substantially—for example, reducing their postings for all sorts of other positions.

Indeed, the evidence indicates that AI so far has been predominantly focused on automation. Moreover, claims that AI and RPAs are expanding into nonroutine, higher-skilled tasks notwithstanding, most of the burden of AI automation to date has fallen on less-educated workers, already disadvantaged by earlier forms of digital automation. Nor is there any evidence that lower-skilled workers are benefiting from AI applications, although obviously the people who run these firms see some gain for themselves and their shareholders.

Reassuringly, AI does not appear to be advancing so much that it will create mass joblessness. Like the industrial robots we discussed in Chapter 8, current technology thus far can perform only a small set of tasks, and its impact on employment is limited. Nevertheless, it is heading in a direction that is biased against workers and is destroying some jobs. Its most major likely impact is to further lower wages for many people, not create a completely workless future. The problem is that although AI fails in most of what it promises, it still manages to reduce the demand for workers.

The Imitation Fallacy

Why, then, all this emphasis on machine intelligence? What we should care about is whether machines and algorithms are useful to us. For example, according to most definitions, the global positioning system (GPS) may not be intelligent because it is based on the implementation of a straightforward search algorithm (the A* search algorithm, first devised in 1968). Yet GPS devices do provide a tremendously useful service to humans. Almost no expert would classify pocket calculators as intelligent, but they perform tasks that most humans would find impossible (such as quickly multiplying two seven-digit numbers).

Instead of fixating on machine intelligence, we should ask how useful machines are to people, which is how we define machine usefulness (MU). Focusing on MU would guide us toward a more socially beneficial trajectory, especially for workers and citizens. Before developing this case, however, we should understand where the current focus on machine intelligence comes from, which takes us to a vision articulated by the British mathematician Alan Turing.

Turing was fascinated by machine capabilities throughout his career. In 1936 he made a fundamental contribution to the question of what it means for something to be "computable." Kurt Gödel and Alonzo Church had recently tackled the question of how to define the set of computable functions, meaning the set of functions whose values can be calculated by an algorithm. Turing developed the most powerful way of thinking about this question.

He imagined an abstract computer, now called a Turing machine, that can carry out computations according to the inputs specified on a possibly infinite tape—for example, instructions to implement basic mathematical operations. He then defined a function to be computable if such a machine could compute its values. A machine is said to be a universal Turing machine if it can compute any number that can be calculated by any Turing machine. Notably, if the human mind is in essence a very sophisticated computer and the tasks that it performs are within the class of computable functions, then a universal Turing machine could replicate all human capabilities. Before World War II, however, Turing did not venture into the question of whether machines could really think and how far they could go in performing human tasks.

During the war, Turing joined the top-secret Bletchley Park research facility, where mathematicians and other experts worked to understand encrypted German radio messages. He devised a clever algorithm—and designed a machine—to speed up the breaking of enemy ciphers. This then helped British intelligence to quickly decipher encrypted communications that the Germans had presumed to be unbreakable.

After Bletchley, Turing took the next step in his prewar work on computation. In 1947 he declared to a meeting of the London Mathematical Society that machines could be intelligent. Undeterred by the hostile reactions of participants, Turing continued to work on the problem. In 1951 he wrote: "'You cannot make a machine think for you.' This is a commonplace that is usually accepted without question. It will be the purpose of this paper to question it."

His seminal 1950 paper, "Computing Machinery and Intelligence," defines one notion of what it means for a machine to be intelligent. Turing imagined an "imitation game" (now called a Turing test) in which an evaluator engages in a conversation with two entities, one human and one machine. By asking a series of questions communicated via a computer keyboard and screen, the evaluator attempts to tell which one is which. A machine is intelligent if it can evade detection.

No machine is currently intelligent according to this definition, but one could turn it into a less categorical ranking of machine intelligence. The better a machine can imitate humans, the more intelligent it is. To make this operational, one can define the notion of "human parity" at a task, which would be achieved if a machine can perform that task at least as well as humans. Then, the more tasks a machine can reach human parity in, the more intelligent it is.

Turing's own thoughts on this subject were subtler. He understood that passing this test might not mean true thinking capacity: "I do not wish to give the impression that I think there is no mystery about consciousness. There is, for instance, something of a paradox connected with any attempt to localise it." Despite this reservation, the modern field of AI followed in Turing's footsteps and focused on artificial intelligence, defined as machines acting autonomously, reaching human parity, and subsequently outperforming humans.

Boom and Mostly Bust

Fascination with machine intelligence often leads to exaggeration. The eighteenth-century French innovator Jacques de Vaucanson would have had a well-deserved place in the history of technology for his many innovations, including the design of the first automatic loom and an all-metal-cutting slide lathe, which was pathbreaking for the early machine-tool industry. Yet today he is remembered as a fraudster for his "digesting duck," which flapped its wings, ate, drank, and defecated. It was all an illusion, with food and water going into one of the many compartments, which then released already digested food as excrement.

Soon after de Vaucanson's duck came the Hungarian inventor Wolfgang von Kempelen's Mechanical Turk, an automated chess-playing machine, whose name originated from the life-size model sitting on top, dressed in an Ottoman robe and turban. The Turk defeated many notable chess players, including Napoleon Bonaparte and Benjamin Franklin; solved the well-known chess puzzle, where a knight must move around touching each square of the board once and only once; and even responded to questions using a letter board. Its success, unfortunately, was thanks to an expert chess player concealed inside the structure.

Claims that machines would soon replicate human intelligence generated great hype in the 1950s as well. The defining event, the first step in the current AI approach and the origin of the term *artificial intelligence*, was a 1956 conference at Dartmouth College, funded by the Rockefeller Foundation. Brilliant young scientists working on related topics convened at Dartmouth during the summer. Herbert Simon, a psychologist and an economist who was later awarded the Nobel Prize, captured the optimism when he wrote that "machines will be capable, within twenty years, of doing any work a man can do."

In 1970 Marvin Minsky, co-organizer of the Dartmouth conference, was still confident when speaking to *Life* magazine:

> In from three to eight years we will have a machine with the general intelligence of an average human being. I mean a

machine that will be able to read Shakespeare, grease a car, play office politics, tell a joke, have a fight. At that point the machine will begin to educate itself with fantastic speed. In a few months it will be at genius level and a few months after that its powers will be incalculable.

These hopes of human-level intelligence, sometimes also called "artificial general intelligence" (AGI), were soon dashed. Tellingly, nothing of great value came from the Dartmouth conference. As the spectacular promises made by AI researchers were all unmet, funding for the field dried up, and what came to be called the first "AI winter" set in.

There was renewed enthusiasm in the early 1980s based on advances in computing technology and some limited success of expert systems, which promised to provide expert-like advice and recommendations. A few successful applications were developed in the context of identifying infectious diseases and some unknown molecules. Soon, claims about artificial intelligence reaching human-level expertise were circulating again, and funding resumed. By the end of the 1980s, a second AI winter was upon the field because the promises were again unfulfilled.

The third wave of euphoria started in the early 2000s, focusing on what is sometimes called "narrow AI," where the objective is to develop mastery in specific tasks, such as identifying an object in pictures, translating text from a different language, or playing a game such as chess or Go. Reaching or surpassing human parity remained the overarching objective.

This time, rather than mathematical and logic-based approaches intended to replicate human cognition, researchers turned various human tasks into prediction or classification problems. For example, recognizing an image can be conceived as predicting which one of a long list of categories the image belongs to. AI programs can then rely on statistical techniques applied to massive data sets to make increasingly accurate classifications. Social media messages that pass among billions of people are an exemplar of this type of data.

Take the problem of recognizing whether there is a cat in a picture. The old approach would have required a machine to model the complete decision-making process used by humans to spot cats. The modern approach bypasses the step of modeling or even understanding how humans make decisions. Instead, it relies on a large data set of humans making correct recognition decisions based on images. It then fits a statistical model to large data sets of image features to predict when humans say that there is a cat in the frame. It subsequently applies the estimated statistical model to new pictures to predict whether there is a cat there or not.

Progress was made possible by faster computer processor speed, as well as new graphics processing units (GPUs), originally used to generate high-resolution graphics in video games, which proved to be a powerful tool for data crunching. There have also been major advances in data storage, reducing the cost of storing and accessing massive data sets, and improvements in the ability to perform large amounts of computation distributed across many devices, aided by rapid advances in microprocessors and cloud computing.

Equally important has been progress in machine learning, especially "deep learning," by using multilayer statistical models, such as neural networks. In traditional statistical analysis a researcher typically starts with a theory specifying a causal relationship. A hypothesis linking the valuation of the US stock market to interest rates is a simple example of such a causal relationship, and it naturally lends itself to statistical analysis for investigating whether it fits the data and for forecasting future movements. Theory comes from human reasoning and knowledge, often based on synthesis of past insights and some creative thinking, and specifies the set of possible relationships among several variables. Combining this theory with a relevant data set, researchers fit a line or a curve to a cloud of points in their data set and make inferences and forecasts on the basis of these estimates. Depending on the success of this first approach, there will be additional human input in the form of a revised theory or a complete change of focus.

In contrast, in modern AI applications the inquiry does not start with clear, causal hypotheses. For example, researchers do not specify which characteristics in the digital version of an image are relevant for recognizing it. Multilayer models, applied to vast amounts of data, attempt to compensate for this lack of prior hypotheses. Each different layer may primarily deal with a different level of abstraction; one layer may represent the edges of the picture and identify its broad outlines, whereas another one may focus on other aspects, such as whether an eye or a paw is present in there. These sophisticated tools notwithstanding, without human-machine collaboration, it is difficult to draw the right inferences from data, and this deficiency motivates the need for ever greater amounts of data and computational power to find patterns.

Typical machine-learning algorithms start by fitting a flexible model to a sample data set and then making predictions that are applied to a larger data set. In image recognition, for example, a machine-learning algorithm can be trained on a sample of tagged images that may indicate whether the image contains a cat. This first step leads to a model that can make predictions on a much larger data set, and the performance of these predictions fuels the next round of algorithmic improvements.

This new AI approach has already had three important implications. First, it has intertwined AI with the use of massive quantities of data. In the words of an AI scientist, Alberto Romero, who became disillusioned with the industry and left it in 2021, "If you work in AI you are most likely collecting data, cleaning data, labeling data, splitting data, training with data, evaluating with data. Data, data, data. All for a model to say: *It's a cat.*" This focus on vast quantities of data is a fundamental consequence of the Turing-inspired emphasis on autonomy.

Second, this approach has made modern AI appear highly scalable and transferable, and of course, in domains much more interesting and important than recognizing cats. Once the problem of recognizing cats in a picture is "solved," we can move on to doing the same for more complex image-recognition tasks or to seemingly unrelated problems, such as determining the meaning of sentences in a foreign language. The potential, therefore, is for truly pervasive use of AI in the economy and in our lives—for good but often also for bad.

In the extreme, the aim becomes the development of completely autonomous, general intelligence, which can do *everything* that humans can do. In the words of DeepMind cofounder and CEO Demis Hassabis, the objective is "solving intelligence, and then using that to solve everything else." But is this the best way to develop digital technologies? This question typically remains unasked.

Third and more problematically, this approach has pushed the field even further in the direction of automation. If machines can be autonomous and intelligent, then it is natural for them to take over more tasks from workers. Companies can break down existing jobs into narrower tasks, use AI programs and abundant data to learn from what humans do, and then substitute algorithms for humans in these tasks.

An elitist vision boosts this focus on automation. Most humans, according to proponents of this view, are error-prone and not very good at the tasks they perform. As one AI website states, "Humans are naturally prone to making mistakes." On the other hand, there are some very talented programmers who could design sophisticated algorithms. As Mark Zuckerberg puts it, "Someone who is exceptional in their role is not just a little better than someone who is pretty good. They are 100 times better." Or in the words of Netscape cofounder Marc Andreessen, "Five great programmers can completely outperform 1,000 mediocre programmers." Based on this worldview, it is desirable to use top-down design of technology by exceptional talent to limit human mistakes and their costs in workplaces. Replacing workers with machines and algorithms then becomes acceptable, and collecting massive amounts of data about people comes to be viewed as tolerable. This approach further justifies reaching human parity, rather than complementing humans, as the criterion for progress and comfortably fits with the emphasis of corporations on cutting labor costs.

The Underappreciated Human

Even with displacement and massive data collection, productivity growth from new technologies can sometimes increase demand for workers and boost their earnings. But benefits for workers appear only when new technologies substantially increase productivity. Today, this is a serious concern because AI has so far brought a lot of so-so automation, with limited productivity benefits.

When productivity increases substantially, this can undo some of the negative effects of automation—for example, by increasing demand for labor in nonautomated tasks or stimulating employment in other sectors that expand subsequently. However, if cost reductions and productivity gains are small, these beneficial effects will not take place. So-so automation is particularly troublesome because it displaces workers but fails to deliver in terms of productivity.

In the age of AI, there is a fundamental reason for so-so automation. Humans are good at most of what they do, and AI-based automation is not likely to have impressive results when it simply replaces humans in tasks for which we accumulated relevant skills over centuries. So-so automation is what we get, for example, when companies rush to install self-checkout kiosks that do not work well and do not improve service quality for customers. Or when skilled customer-service representatives, IT specialists, or financial advisers are sidelined by AI algorithms, which then perform badly.

Many of the productive tasks performed by humans are a mixture of routine and more complex activities that involve social communication, problem solving, flexibility, and creativity. In such activities, humans draw on tacit knowledge and expertise. Moreover, much of this expertise is highly context dependent and difficult to transfer to AI algorithms, thus likely to get lost once the relevant tasks are automated.

To illustrate the importance of accumulated knowledge, take the foraging societies we discussed in Chapter 4. Ethnographic studies show that hunter-gatherers consistently have a remarkable degree of adaptation to local conditions. For instance, cassava (also known as manioc) is a highly nutritious tuber plant originating in the American tropics. It is used for making cassava flour, breads, tapioca, and various alcoholic beverages. The plant is poisonous, however, because it contains two cyanide-producing sugars. If it is eaten raw or cooked without being properly processed, it can cause cyanide intoxication, with severe consequences in extreme cases, including death.

Indigenous peoples in the Yucatán figured out this problem and developed several practices to remove the poison, including peeling the plant and soaking it for a while before cooking it for a long time, and then disposing of the cooking water. Some Europeans at first did not understand these methods and sometimes interpreted them as primitive, nonscientific traditions, only to find out the perilous costs of not following them.

Human adaptability and ingenuity are no less important in the modern economy, though often ignored by technology-minded elites. There is a strong consensus among city planners and engineers that traffic lights are key to the safe and timely flow of cars. In September 2009 the coastal English town of Portishead turned off the traffic lights at one of its busiest intersections. Against the fears of many experts, drivers started using more common sense and responded adaptively to this new organization. At the end of four weeks, traffic flow had improved significantly at the intersection, with no increase in accidents or injuries. Portishead is no outlier. Several other experiments with such "naked streets" show similar results. There is a debate on the practicality of naked streets in large cities, and a complete lack of traffic lights is unlikely to be workable at the busiest intersections in megacities. Nevertheless, it is hard not to conclude from these experiments that technology, by taking away initiative and judgment from humans, sometimes makes things worse, not better.

The same is true when it comes to production tasks. Human intelligence derives its strength from being situational and social: The ability to understand and successfully respond to one's environment, enabling individuals to fluidly adapt to changing conditions. For example, people can be more alert when they are in an unfamiliar environment that provides subtle cues of danger, even while resting or in their sleep. In other environments they perceive as predictable, they can perform tasks faster using learned routines. It is also situational intelligence that helps people respond to changing circumstances more broadly and recognize faces and patterns, using inputs from multiple relevant contexts.

Human intelligence is also social in three important ways. First, a lot of the necessary information for successful problem solving and adaptation resides in the community. We acquire it via implicit and explicit communication—for example, by imitating others' behavior. Interpreting this type of external knowledge is a vital part of human cognition and is the basis of the emphasis on the "theory of the mind" in this area. Theory of the mind is what enables humans to reason about others' mental state and thus correctly understand their intentions and knowledge.

Second, our reasoning is based on social communication; we develop arguments and counterarguments in favor of different hypotheses and evaluate our understanding in light of this process. Humans would be terrible decision makers without this social dimension of intelligence. Yes, we make mistakes when placed into lab settings that prevent these aspects of intelligence from being activated, but we avoid some of the same mistakes in more natural settings.

Third, humans gain additional skills and capabilities from the empathy they have for others and the sharing of goals and objectives that this enables.

The central role of the situational and social dimensions of intelligence is related to the weak relationship between analytic aspects of human cognition, as measured by IQ tests, and various dimensions of success. Even in scientific and technical fields, individuals who are the most successful are those who combine moderately high IQ with social skills and other human capabilities.

In most work environments, situational and social intelligence enables not just flexible adaptation to circumstances but also communication with customers and other employees to improve service quality and reduce mistakes. It is therefore not surprising that despite the spread of AI technologies, many companies are increasingly seeking workers with social, rather than mathematical or technical, skills. At the root of this growing demand for social skills is the reality that neither traditional digital technologies nor AI can perform essential tasks that involve social interaction, adaptation, flexibility, and communication.

All the same, ignoring human capabilities can become a self-fulfilling prophecy because automation decisions can gradually reduce the scope for social interaction and human learning. Take customer service again as an example. Well-trained humans can be very effective in dealing with problems, precisely because they form a social bond with the person needing help (for instance, sympathizing with somebody who just had an accident and needs to file a claim). They can quickly understand the nature of their problem, partly because they are communicating with the customer, and come up with solutions that fit the needs on the basis of this communication. These interactions further enable customer-service representatives to get better at their job over time.

Now imagine the situation after the job of customer service is broken into narrower tasks and the front-end ones are assigned to algorithms, which will often fail to fully identify and deal with the complex problems they encounter. Humans are then brought in as troubleshooters, after a long series of menus. At this point, the customer is often frustrated, early opportunities for building a social bond have been lost, and the customer-service representative does not get the same extent of information from communication, limiting their ability to learn from and adapt to the specific circumstances. This makes the customer-service representative less effective and may encourage managers and technologists to seek additional ways of reducing the tasks allocated to them even further.

These lessons about human intelligence and adaptability are often ignored in the AI community, which rushes to automate a range of tasks, regardless of the role of human skill.

The triumph of AI in radiology is much trumpeted. In 2016 Geoffrey Hinton, cocreator of modern deep-learning methods, Turing Award winner, and Google scientist, suggested that "people should stop training radiologists now. It's just completely obvious that within five years deep learning is going to do better than radiologists."

Nothing of the sort has yet happened, and demand for radiologists has increased since 2016, for a very simple reason. Full radiological diagnosis requires even more situational and social intelligence than, for example, customer service, and it is currently beyond the capabilities of machines. In fact, recent research shows that combining human expertise with new technology tends to be much more effective. For example, state-of-the-art machine-learning algorithms can improve the diagnosis of diabetic retinopathy, which results from damage to blood vessels on the retina among diabetic patients. Nevertheless, accuracy increases significantly more when algorithms are used to identify difficult cases, which are then assigned to ophthalmologists for better diagnosis.

The chief technology officer of Google's self-driving car division confidently expected in 2015 that his then-eleven-year-old son would not need to get a driver's license by the time he turned sixteen. In 2019 Elon Musk predicted that Tesla would have one million fully automated, driverless taxicabs on the streets by the end of 2020. These predictions have not come to pass for the same reason. As the naked-streets experiment emphasized, driving in busy cities requires a tremendous amount of situational intelligence to adapt to changing circumstances, and even more social intelligence to respond to cues from other drivers and pedestrians.

General AI Illusion

The apogee of the current AI approach inspired by Turing's ideas is the quest for general, human-level intelligence.

Despite tremendous advances such as GPT-3 and recommendation systems, the current approach to AI is unlikely to soon crack human intelligence or even achieve very high levels of productivity in many of the decision-making tasks humans engage in. Tasks that involve social and situational aspects of human cognition will continue to pose formidable challenges for machine intelligence. Once we look at the details of what has been achieved, the difficulty of translating existing successes to most human tasks becomes clear.

Take the most vaunted successes of AI, such as the AlphaZero chess program, discussed in Chapter 1. AlphaZero is even argued to be "creative" because it has come up with moves that human chess masters had not considered or seen. Nevertheless, this is not true intelligence. To start with, AlphaZero is an extremely specialized program and can play only chess and other similar games. Even the simplest tasks beyond chess, such as simple arithmetic or playing games with more social interaction, are beyond AlphaZero's capabilities. Worse, there is no obvious way in which AlphaZero's architecture can be adapted to do many of the simple things humans do, such as drawing analogies, playing games that have less-strict rules, or learning a language, which is done masterfully by hundreds of millions of one-year-olds every year.

AlphaZero's intelligence within chess is also very specific. Although AlphaZero's chess moves within the rules of the game are impressive, they do not involve the type of creativity that humans regularly engage in—such as drawing analogies across unstructured, disparate environments and coming up with solutions to new and varied problems.

Even GPT-3, though more versatile and impressive than AlphaZero, shows the same limitations. It cannot perform tasks beyond those for which it has been pretrained and shows no judgment, so conflicting or unusual instructions can stump it. Worse, this technology has no element of the social or situational intelligence of humans. GPT-3 cannot reason about the context in which the tasks it is performing are situated and draw on causal relationships that exist between actions and effects. As a result, it sometimes misunderstands even simple instructions and has little hope of responding adequately to changing or completely new environments.

Indeed, this discussion illustrates a broader problem. Statistical approaches used for pattern recognition and prediction are ill-suited to capturing the essence of many human skills. To start with, these approaches will have difficulty with the situational nature of intelligence because the exact situation is difficult to define and codify.

Another perennial challenge for statistical approaches is "overfitting," which is typically defined as using more parameters than justified for fitting some empirical relationship. The concern is that overfitting will make a statistical model account for irrelevant aspects of the data and then lead to inaccurate predictions and conclusions. Statisticians have devised many methods to prevent overfitting—for example, developing algorithms on a different sample than the one in which they are deployed. Nevertheless, overfitting remains a thorn in the side of statistical approaches because it is fundamentally linked to the shortcomings of the current approach to AI: lack of a theory of the phenomena being modeled.

To explain this problem, it is useful to have a broader understanding of the overfitting problem, based on using irrelevant or nonpermanent features of an application. Consider the task of distinguishing wolves from huskies. Although humans are excellent at this task, it turns out to be a difficult one for AI. When some algorithms managed to achieve good performance, it was later understood that this was thanks to overfitting: huskies were recognized from urban backgrounds, such as nice lawns and fire hydrants, and wolves from natural backgrounds, such as snowy mountains. These are irrelevant characteristics in two fundamental senses. First, humans do not rely on these backgrounds for defining or distinguishing the animals. Second, and more troublingly, as the climate warms, wolves' habitats may change, or wolves may need to be identified in different settings. In other words, because the background is not a defining characteristic of wolves, any approach that relies on it will lead to mistaken predictions as the world evolves or the context changes.

Overfitting is particularly troublesome for machine intelligence because it creates a false sense of success, when the machine is in reality performing badly. For instance, a statistical association between two variables, say temperature and GDP per capita across countries, does not necessarily indicate that climate has a sizable impact on economic development. It may simply result from how European colonialism impacted areas with different climatic conditions and in different parts of the globe during a specific historical process. But without the right theory, it is easy to confuse causation and correlation, and machine learning often does this.

The overfitting problem becomes much worse when algorithms are dealing with an inherently social situation where humans react to new information. Human responses will mean that the relevant context evolves frequently, or it may change *because* of the actions they take on the basis of the information that algorithms provide. Let us give an economic illustration. An algorithm may observe the mistakes a person makes when looking for a job—for example, seeking occupations that have few job postings relative to the number of people applying to them—and may try to correct them. Procedures developed against overfitting, such as separating the training and the testing samples, do not remove the relevant overfitting problem: both samples may be adapted to a particular environment in which there are many unfilled vacancies in the retail sector. But this might change over time precisely because we are dealing with a social situation where humans respond to the available evidence. For instance, as individuals are encouraged by algorithms to apply to them, retail jobs may become oversubscribed and no longer as attractive. Without fully understanding this situational and social aspect of human cognition and how behavior changes dynamically, overfitting will continue to bedevil machine intelligence.

There are other troubling implications of AI's lack of social intelligence. Although it uses data from a large community of users and thus can embed the social dimension of data, with existing approaches it does not leverage the fact that human understanding is founded on selective imitation, communication, and argumentation between people. As a result, many automation attempts appear to reduce, rather than increase, flexibility, which well-trained workers can achieve by rapidly and fluidly responding to changing circumstances, often leveraging skills and perspectives they learn from their coworkers.

Of course, these arguments do not rule out the possibility that a completely new approach can crack the problem of AGI in the near future. Yet there is so far no indication that we are close to coming up with such an approach. Nor is this the main area in which AI dollars are being invested. Industry's focus continues to be on extensive data collection and the automation of narrow tasks based on machine-learning techniques.

The economic problem from this business strategy is clear: when humans are not as useless as sometimes presumed and intelligent machines not as intelligent as typically assumed, we get so-so automation—all the displacement and little of the promised productivity gains. In fact, even the companies themselves do not benefit much from this automation, and some of the AI adoption may be because of hype, as the former AI scientist Alberto Romero, whom we quoted earlier, noted: "The marketing power of AI is such that many companies use it without knowing why. Everyone wanted to get on the AI bandwagon."

The Modern Panopticon

Another popular use of modern AI illustrates how enthusiasm for autonomous technology, together with massive data collection, has forged a very specific direction for digital technologies and how it has again caused modest gains for corporations and significant losses for society and workers.

The use of digital tools for worker monitoring is nothing new. When the social psychologist and business scholar Shoshana Zuboff interviewed workers experiencing the introduction of digital technologies in the early 1980s, a common refrain was about the intensification of monitoring by management. As one office worker put it, "The ETS [digital expense tracking system] has become a vehicle for management to check up on us. They can pick up any changes on a minute-by-minute basis if they want to."

But earlier efforts pale in comparison to what we see today. Amazon, for example, collects a huge amount of data about its delivery workers and warehouse employees, which are then combined with algorithms for restructuring work in a way that increases throughput and minimizes disruptions.

The company, which is the second-largest private-sector employer in the United States, pays higher minimum wages than several other retailers, such as Walmart. But there is a fundamental sense in which Amazon jobs are not good jobs. Workers must abide by strict, fast-paced work routines and are continuously monitored to make sure that they are not taking longer or more frequent breaks and are exerting the required effort at all times. Recent news reports reveal that a sizable proportion of workers from many facilities are fired for not meeting these work expectations, and some of these terminations are automatic, based on the data collected (although Amazon disputes that there are automatic terminations). In the words of a labor advocate, "One of the things we hear consistently from workers is that they are treated like robots in effect because they're monitored and supervised by these automated systems."

Jeremy Bentham's panopticon was meant to be a model not only for prisons but also for early British factories. But eighteenth- and nineteenth-century bosses did not have the technology for constant surveillance. Amazon does. In the words of one New Jersey employee, "They basically can see everything you do, and it's all to their benefit. They don't value you as a human being. It's demeaning."

These high-monitoring environments are not just demeaning but also dangerous. A recent OSHA report finds that in 2020, Amazon warehouse workers suffered about 6 serious injuries per 200,000 hours worked, nearly twice as high as the average in the warehousing industry, and other studies find even higher injury rates, especially in peak business times such as the Christmas season, periods during which monitoring of workers intensifies. Amazon additionally requires its delivery employees and contractors to download and continuously run a data tracking app called "Mentor," which enables closer monitoring. The company recently announced additional AI tools for tracking delivery workers. FedEx and other delivery services also collect a lot of data from their employees and use these for imposing strict scheduling constraints, which explains why many delivery workers are perpetually in a race against time.

Extensive data collection is now spreading to white-collar occupations, with employers tracking how employees use their time on computers and various communication devices.

Some amount of monitoring is part of the prerogative of an employer, who needs to ensure that workers perform the tasks assigned to them and do not damage or misuse machinery. Traditionally, however, workers used to be motivated not just by monitoring but also by the goodwill that developed between them and their employer because of high wages and the general amenities of the workplace. For instance, an employer or a supervisor could recognize that the worker might not be feeling fully well on a given day and cut them some slack, or, conversely, employees could be willing to work harder than usual when the need arose on an occasional basis. Monitoring enables employers to cut wages and get more work out of the workers. In this way, monitoring is a "rent-shifting activity," meaning that it can be used to prevent sharing of productivity gains and to shift rents away from workers, without improving their productivity much or at all.

Another domain in which AI methods have been deployed for rent shifting is work scheduling. A key source of autonomy for workers is a clear separation between work and leisure time, and predictable scheduling. Take employees at fast-food restaurants. If they know that they must come to work at 8 a.m. and leave at 4 p.m., this gives them a high degree of predictability and some amount of autonomy beyond this eight-hour window. But what happens if the manager suddenly finds out that there will be many more customers coming after 4 p.m.? She may have an incentive to reduce this autonomy and order employees to stay past 4 p.m. Can she do that?

The answer depends on countervailing powers—for example, collective agreements preventing such impositions; on goodwill and norms of what is acceptable in workplaces; and on technology, which determines whether companies can predict demand in advance and arrange real-time scheduling.

Countervailing powers were already absent, especially in the service industry, and goodwill and norms of respect for worker autonomy had long subsided. The remaining barrier, technology, has now been overcome with AI and massive data collection, paving the way to "flexible scheduling."

Many customer-facing industries have abandoned predictable schedules, such as 8 a.m.–4 p.m. work hours, adopting instead a combination of "zero-hour contracts" and real-time schedule changes. Zero-hour contracts mean that the company rescinds the commitment that it will employ and pay the worker for regular hours every week. Real-time scheduling, on the other hand, allows companies to call employees on their cell phone the night before asking them to be at work early the following morning or extend their regular hours into longer workdays. It also includes canceling shifts at short notice, which reduces worker income.

Both are rooted in data-crunching and AI technologies—for example, scheduling software offered by tech companies such as Kronos—that enable employers to predict the demand they are going to face and then compel the workers to adapt to it. An extreme version of these practices is "clopening," the name given to the practice of the same employee closing late one evening and then opening the store early the following morning. This is once again imposed on workers, often at the last moment, as the managers, empowered by AI tools, see it fitting their needs.

There are many parallels between flexible scheduling practices and worker monitoring. The most important is that they are both examples of so-so technologies: they create little productivity gains, despite substantial costs for workers. With additional monitoring, companies can abandon efforts to build goodwill and cut wages. But this does not increase productivity by much: workers do not become better at their job because they are being paid less, and in fact may lose motivation and become less productive. With flexible scheduling, companies can increase their revenues a little by having more employees when the demand is high and fewer when the store is less busy. In both cases, the burden on workers is more substantial than the productivity benefits. In the words of a British worker employed with a zero-hour contract, "There is no career progression. . . . [I've] been in the job for six and a half years. Since then the role hasn't changed, no promotion. I've got no promotional prospects at all. I asked if I could perhaps go on a course, and I got an absolute no for that one." No matter the costs on workers and the small, ephemeral productivity gains, companies intent on cost cutting and increasing control over workers are continuing to demand AI technologies, and in response researchers beholden to the AI illusion are supplying them.

But is there another way than using digital technologies in the service of ceaseless automation and worker monitoring? The answer is yes. When digital technologies are steered toward helping and complementing humans, the results can be, and have been, much better.

A Road Not Taken

When interpreting both recent and distant history, there is often a deterministic fallacy: what happened had to happen. Often, this is not accurate. There are many possible paths along which history could have evolved. The same is true for technology. The current approach that dominates the third wave of AI based on massive data harvesting and ceaseless automation is a choice. It is in fact a costly choice, not just because it is following the bias of elites toward automation and surveillance, and damaging the economic livelihood of workers. It is also diverting energy and research away from other, socially more beneficial directions for general-purpose digital technologies. We will next see that paradigms prioritizing machine usefulness have had some remarkable successes in the past when tried and offer many fruitful opportunities for the future.

Even before the Dartmouth conference, MIT polymath Norbert Wiener had articulated a different vision, one that positioned machines as complements to humans. Although Wiener did not use the term, MU (machine usefulness) is inspired by his ideas. What we want from machines is not some amorphous notion of intelligence or "high-level capabilities" but their use for human objectives. Focusing on MU, rather than AI, is more likely to get us there.

Wiener identified three critical issues that have stymied dreams of autonomous machine intelligence since Turing. First, surpassing and replacing humans is difficult because machines are always imperfect at imitating living organisms. As Wiener put it in a slightly different context, "The best material model for a cat is another, or preferably the same cat."

Second, automation had an immediate negative effect on working people: "Let us remember that the automatic machine, whatever we think of any feelings it may have or may not have, is the precise economic equivalent of slave labor. Any labor which competes with slave labor must accept the economic consequences of slave labor."

And finally, the drive for automation also meant that scientists and technologists could lose control over the path of technology. "It is necessary to realize that human action is a feedback action" means that we adjust what we do based on information about what is happening around us. But "when a machine constructed by us is capable of operating on its incoming data at a pace which we cannot keep, we may not know, until it is too late, when to turn it off." None of this was inevitable, however: machines could be harnessed to the service of humans as a complement to our skills. As Wiener wrote in an article drafted in 1949 for the *New York Times* (parts of it were published posthumously in 2013), "We can be humble and live a good life with the aid of the machines, or we can be arrogant and die."

Two visionaries picked up Wiener's torch. The first was J. C. R. Licklider, who focused on encouraging others to adopt and develop this approach in productive ways. Originally trained as a psychologist, Licklider subsequently moved into information technology, and he proposed ideas that would become critical for networked computers and interactive computing systems. A clear articulation of this vision is contained in his 1960 pathbreaking article, "Man-Computer Symbiosis." Licklider's analysis is still relevant today, more than sixty years after its publication, especially in his emphasis that "relative to men, computing machines are very fast and very accurate, but they are constrained to perform only one or a few elementary operations at a time. Men are flexible, capable of 'programming themselves contingently' on the basis of newly received information."

The second proponent of this alternative vision, Douglas Engelbart, also articulated ideas that are precursors to our notion of machine usefulness. Engelbart strove to make computers more usable and easier to operate for nonprogrammers, based on his belief that they would be most transformative when they were "boosting mankind's capability for coping with complex, urgent problems."

Engelbart's most important innovations were revealed in spectacular fashion in a show that was later christened as the "Mother of All Demos." At a conference organized by the Association for Computer Machinery, jointly with the Institute for Electrical and Electronics Engineers on December 9, 1968, Engelbart introduced the prototypical computer mouse. This contraption, consisting of a big roller, a wooden-carved frame, and a single button, looked nothing like the computer mouse we are used to today, but with wires sticking from its back, it did look enough like a rodent to get the name. It transformed what most users could do with computers at one fell swoop. It was also the innovation that propelled Steve Jobs and Steve Wozniak's Macintosh computers ahead of PCs and operating systems based on Microsoft. Other Engelbart innovations, some of them also showcased at the Mother of All Demos, include hypertext (which now powers the internet), bitmapped screens (which made various other interfaces feasible), and early forms of the graphical user interface. Engelbart's ideas continued to generate several other advances, especially under the auspices of the Xerox company (and many of these ideas were again critical for Macintosh and other computers).

Wiener, Licklider, and Engelbart's alternative vision laid the foundations of some of the most fruitful developments in digital technology, even if today this vision is overshadowed by the AI illusion. To understand these achievements, and why they have not received as much attention as the successes of the dominant paradigm, we first need to discuss how MU works in practice.

Machine Usefulness in Action

We can distinguish four related but distinct ways in which digital technologies can be steered in the direction of MU, helping and empowering humans.

First, machines and algorithms can increase worker productivity in tasks they are already performing. When a skilled artisan is given a better chisel or an architect has access to computer-aided design software, their productivity can increase significantly. Such productivity increases need not just come from new tools and can also be accomplished by improving machine design. This is the aspiration of the fields known as human-computer interaction and human-centered design. These approaches recognize that all machines, and in particular computers, need to have certain features to be most productively used by people, and they prioritize designing new technologies that increase human convenience and usability. When successful, as was Engelbart's mouse and graphic user interface, new digital technologies can be what Steve Jobs referred to as "a bicycle for our minds" and expand human skills. Because this approach puts machine capabilities at the service of people, it tends to complement human intelligence.

Although this approach has already generated notable benefits, much more can be done. Virtual- and augmented-reality tools hold tremendous promise to increase human capabilities in tasks such as planning, design, inspection, and training. But applications can go beyond technical and engineering jobs.

The current consensus in the technology and engineering community is summarized by Kai-Fu Lee: "Robots and AI will take over the manufacturing, delivery, design and marketing of most goods." Such claims notwithstanding, as we saw in Chapter 8, efforts to deploy new software tools have been an important source of productivity growth in the context of the German Industry 4.0 program, which has enabled greater flexibility in the face of changing circumstances or demands.

This potential is even better illustrated by Japanese manufacturing, where many companies have prioritized flexibility and worker participation in decision making, even as advanced and sometimes automated machinery has been introduced. This approach was pioneered by W. Edwards Deming, another engineer following the same vision as Wiener, Licklider, and Engelbart. Deming was instrumental in setting up a quality-centered, flexible production approach in Japanese manufacturing. In return, he received the highest honors in Japan, and the Deming Prize has been established in his name. Augmented and virtual reality currently provides many new avenues for this type of human-machine collaboration, including improved capabilities for precision work by humans, more adaptive designs, and greater flexibility in responding to changing circumstances.

The second type of MU is even more important and was our focus in chapters 7 and 8: the creation of new tasks for workers. These tasks were critical for expanding the demand for both skilled and unskilled workers even as manufacturers such as Ford automated parts of the production process, reorganized work, and transitioned to mass production. Digital technologies have also created various new technical and design tasks over the last half century (even if most companies have prioritized digital automation). Augmented and virtual reality can also generate more new tasks in the future. Education and health care provide a vivid illustration of how algorithmic advances can introduce new tasks. More than four decades ago, Isaac Asimov noted the problem of our current system of education: "Today, what people call learning is forced on you. Everyone is forced to learn the same thing on the same day at the same speed in class. But everyone is different. For some, class goes too fast, for some too slow, for some in the wrong direction." When Asimov wrote these words, his proposal for personalized teaching was purely aspirational. Short of one-on-one teaching for all students, there was little possibility for such personalization. Today, we have the tools for making personalization a reality in many classrooms. Indeed, it should be possible to reconfigure existing digital technologies for this purpose. The same statistical techniques used for task automation can also be used for identifying in real time groups of students who have difficulties with similar problems, as well as students who can be exposed to more advanced material. The relevant content can then be adjusted for small groups of students. Evidence from the field of education research indicates that such personalization has considerable return and is most useful where exactly society has the greatest need: improving the cognitive and social skills of students from low socioeconomic backgrounds.

The situation in health care is similar: the right type of MU can significantly empower nurses and other health care professionals, and this would be most useful in primary health, prevention, and low-tech medical applications.

The third contribution of machines to human capabilities may be even more relevant in the near future. Decision making is almost always constrained by accurate information, and even human creativity relies on accessing accurate information in a timely fashion. Most creative tasks require drawing analogies, finding new combinations of existing methods and designs. People doing this work then come up with previously untried schemes that are confronted with evidence and reasoning, and are subsequently further refined. All these human tasks can be helped by accurate filtering and the provision of useful information.

The World Wide Web, often associated with the British computer scientist Tim Berners-Lee, is a quintessential example of this type of aid to human cognition. By the late 1980s, the internet, the global network of computers communicating with one another, had been around for about two decades, but there was no easy way of accessing the trove of information that existed in this network. Berners-Lee, together with Belgian computer scientist Robert Cailliau, extended Engelbart's hypertext idea and introduced hyperlinks to allow information on one site to be linked to the relevant information on other parts of the internet. The two scientists wrote the first web browser to retrieve this information, and named it the World Wide Web or simply the Web. The Web is a milestone in human-machine complementarity: it enables people to access information and wisdom that other humans have produced to a degree essentially unparalleled in the past.

MU can enable many more applications that provide better information to people in their capacities as workers, consumers, and citizens. Recommendation systems, at their best, have this ability: they can aggregate masses of information from others and present relevant aspects to users to aid in their decision making.

The fourth category, based on the use of digital technologies to create new platforms and markets, may turn out to be the most important application of the Wiener-Licklider-Engelbart vision. Economic productivity is inseparable from cooperation and trading. Bringing together people with different skills and endowments has always been a major aspect of economic dynamism and can be powerfully expanded by digital technologies.

A brilliant illustration of this phenomenon is provided by the fishing industry in the southern Indian state of Kerala, which was revolutionized by the use of mobile phones. In some local beach markets in Kerala, fishermen would come in with a good catch of fish but would encounter insufficient demand, driving the price to zero and leaving a lot of the fish to rot. A few kilometers away, a different beach market would have few fish for sale and many buyers, leading to high prices, unmet demand, and widespread inefficiencies. Beginning in 1997, mobile phone service was introduced throughout Kerala. Fishermen and wholesalers started using mobile phones to acquire information about the distribution of supply and demand across beach markets. Subsequently, price dispersion and fish wastage dropped sharply. The basic economics of this story is clear: communication technology enabled the creation of a unified fish market, and a careful study of this episode documents that both fishermen and consumers benefited significantly.

Opportunities for new connections and market creation are potentially greater with digital technologies, and some platforms have already made use of them. An inspiring example is the mobile currency and money-transfer system M-Pesa, which was introduced in Kenya in 2007 and provides cheap and fast banking services using mobile phones. This system spread to 65 percent of the Kenyan population two years after its introduction and has since been adopted by several other developing countries.

It is estimated to have generated broad-based benefits to these economies. As another example, Airbnb has created a new market where people can rent accommodations, expanding choice for consumers and generating competition with hotel chains.

Even in areas such as translation where AI-based automation has been quite successful, there are complementary alternatives based on the creation of new platforms. For example, rather than just relying on fully automated and often low-quality translation, one could also build platforms that bring together people needing higher-quality language service and qualified multilingual people around the world.

New platforms need not be confined to those specializing in monetary transactions. Decentralized digital structures can be used to build platforms for broader forms of collaboration, sharing of expertise, and collective action, as we will discuss in Chapter 11.

The successes of MU we have mentioned are among the most productive applications of digital technologies and have paved the way to myriad other innovations. Nevertheless, they are, overall, marginal to the current direction of AI. For 2016, McKinsey Global Institute estimated that $20–$30 billion out of the total global AI spending of $26–$39 billion comes from a handful of big tech companies in the United States and China. Unfortunately, as far as we can tell, most of this spending appears to go toward massive data collection that is targeted at automation and surveillance.

So why are tech companies not developing tools that help humans and at the same time boost productivity? There are several reasons for this, all of them informative about the broader forces we are confronted with. Consider the teaching example, and recall that new tasks, as in this example, are useful in part because they increase productivity by generating meaningful and high-paying jobs for humans—in this instance, for teachers. Yet new teaching tasks imply greater costs for schools already strapped for cash. Most public schools, like other modern organizations, have to focus on containing labor costs and may struggle to hire additional teachers. Consequently, new algorithms for automated grading or automated teaching could appear more attractive to them.

The same is true in health care. Despite the $4 trillion that the United States spends on health care, hospitals also face budget pressure, and a shortage of nurses became painfully evident during the COVID-19 pandemic. New technologies that increase nurses' capabilities and responsibilities would mean hiring more nurses for higher-quality health care. This observation reiterates a key point: human-complementary machines are not attractive to organizations when they are intent on cost cutting.

Another challenge is that new platforms and methods of aggregating and providing information to users also open up possibilities for novel exploitative uses. The World Wide Web, for instance, has become as much a platform for digital advertisement and propagation of misinformation as a source of useful information for people. Recommendation systems are often used for steering customers to specific products, depending on the platform's financial incentives. Digital tools can provide information to managers not just for better decision making but also for the better monitoring of workers. Some of the AI-powered recommendation systems have incorporated and reintensified existing biases—for example, toward race in hiring or toward race in the justice system. Platforms for ride sharing and delivery have imposed exploitative arrangements on workers lacking protection or job security. Hence, the way in which even the most promising applications of human-machine complementarity are used is still dependent on market incentives, the vision and priorities of tech leaders, and countervailing powers.

Besides, there is an equally insurmountable barrier to human-machine complementarity. Under the shadow of the Turing test and the AI illusion, top researchers in the field are motivated to reach human parity, and the field tends to value and respect such achievements ahead of MU. This then biases innovation toward finding ways of taking tasks away from workers and allocating them to AI programs. This problem is, of course, amplified by financial incentives coming from large organizations intent on cost cutting by using algorithms.

Mother of All Inappropriate Technologies

It is not only workers and citizens in the industrialized world who will pay the price for the AI illusion.

Despite economic growth in many poorer nations over the last five decades, more than three billion people in the developing world still live on less than $6 per day, making it difficult for them to achieve three square meals each day, together with money for housing, clothing, and health care. Many pin their hopes on technology to alleviate this poverty. New technologies, introduced and perfected in Europe, the United States, or China, can be transferred to and adopted by developing nations and power their economic growth. International trade and globalization are also argued to be critical ingredients in this process, for low-income nations can export the products they produce with advanced technologies.

Success stories of very rapid economic growth, including South Korea, Taiwan, Mauritius, and more recently China, seem to bear this out. Each country achieved per-capita average growth rates of over 5 percent a year for periods of more than thirty years. In all these cases, industrial technologies played a major role in growth, as did exports to world markets.

But how and whether developing countries benefited from technology imports is more nuanced than typically presumed. A few economists, such as Frances Stewart, realized in the 1970s that technology imports may not work, and in fact may make things worse in terms of inequality and poverty, because the West's technologies are often "inappropriate" for the needs of developing nations. African agriculture illustrates the problem. High- and middle-income countries account for almost all the research spending on agricultural technologies, and a significant fraction has been targeted at the most perennial problem of agriculture: crop pests and pathogens, which are estimated to destroy perhaps as much as 40 percent of the world's agricultural output. For example, the European maize borer, which affects corn in Western Europe and North America, has received a lot of attention, and resistant strains of crops have been developed (including more than five thousand biotech patents and various genetically modified varieties). The same is true for the western corn rootworm, also affecting corn in the United States and parts of Western Europe, and the cotton bollworm, once a key threat against US cotton.

But these crops and chemicals are not very useful for African and South Asian agriculture, which faces different pests and pathogens. The African maize stalk borer, which afflicts the same crops in Africa, and the desert locust, which ravages almost all crops in Africa and much of South Asia, are phenomenal barriers to agricultural productivity in these regions. But these have received much less attention (very few patents and no genetically modified varieties). The overall amount of research dollars and new innovations targeted at the problems of the low-income developing world have been pitiful in general. Estimates suggest that global agricultural productivity could be increased by as much as 42 percent if biotech research effort was redirected away from Western pests and pathogens toward those afflicting the developing world.

New crops and agricultural chemicals targeted predominantly to Western agriculture are an example of inappropriate technology. Stewart's emphasis was not so much on pests and pathogens, but on how capital-intensive new production methods were. For instance, complex industrial machinery in manufacturing and combine harvesters in agriculture may be mismatched to the needs of the developing world, where capital is scarce and creating jobs—good jobs—for their population during the growth process is a major imperative.

Such mismatches are costly for economic development. Developing nations may end up not using new technologies because they are ill-suited to their needs or are too capital-intensive. Indeed, crop varieties developed in the United States are rarely exported to poorer nations, unless they happen to have a very similar climate and pathogens. Even when new technologies developed in advanced economies are introduced in the developing world, the benefits are often limited because the receiving countries may lack the highly skilled labor required to maintain and operate the latest machines. Additionally, technologies imported from the rich world tend to create a dual structure, with a highly capital- and skill-intensive sector paying decent wages alongside a much larger sector with few good jobs. In sum, inappropriate technologies fail to reduce world poverty and may instead increase inequality both between the West and the rest, and within developing nations.

Many in the developing world were already aware of these imperatives. Some of the most transformative innovations of the twentieth century were developed in what is now referred to as the "Green Revolution," which was spearheaded by researchers in Mexico, the Philippines, and India. New rice varieties invented in the West were not suitable to the soil and climatic conditions in these countries. A breakthrough came in 1966 with the breeding of a new hybrid rice variety, IR8 rice, which rapidly doubled rice production in the Philippines. IR8 and related cultivars developed in collaboration with Indian research institutes were soon being adopted in India as well and revolutionized that country's agriculture, in some places increasing yields by as much as tenfold. International funding from the Rockefeller Foundation and the leadership of scientists, especially the agronomist Norman Borlaug, who was later awarded the Nobel Peace Prize for saving more than a billion people from starvation, were instrumental as well.

Today, we are confronted with the mother of all inappropriate technologies, in the form of AI, but there are no efforts analogous to the Green Revolution (nor are many AI researchers attempting to fill Borlaug's shoes).

Poverty reduction and rapid economic growth in cases such as South Korea, Taiwan, and China did not just come from the import of Western production methods. Economic success resulted from new technologies enabling the human resources of these countries to be used more effectively. In all these cases, the technologies created new employment opportunities for most of the workforce, and the countries themselves also increased investment in education in order to improve the match between the technologies and their population's skills.

The current trajectory of AI is precluding this pathway. Digital technologies, robotics, and other automation equipment have already increased the skill requirements of global production and started remaking the international division of labor—for example, contributing to a process of deindustrialization in many developing nations that have workforces primarily consisting of people with low education.

AI is again the next act in this process. Rather than creating jobs and opportunities for the majority of the population in poor and middle-income countries, the current path of AI is raising the demand for capital, highly skilled production workers, and even higher-skilled services, such as from management-consulting and tech companies. These are exactly the resources that are most lacking in the developing world. As in the examples of export-led growth and the Green Revolution, many of these economies have abundant resources that can be used for spearheading economic growth and reducing poverty. But these are the resources that will remain unused if the future of technology moves in the pathways that the AI illusion dictates.

Rebirth of the Two-Tiered Society

The Industrial Revolution started in eighteenth-century Britain, where most of the population had little political or social power. Predictably, the direction of progress and productivity growth in such a two-tiered society initially worsened the living conditions of millions. This began changing only when the distribution of social power shifted, altering technology's course so that it raised the marginal productivity of workers. Also critical were institutions and norms for robust rent sharing in workplaces, ensuring that higher productivity translated into wage growth. This struggle over technology and worker power started to transform the highly hierarchical nature of British society in the second half of the nineteenth century.

In chapters 6 and 7, we followed this process from Britain to the United States, as technological leadership shifted. Twentieth-century US technology moved even more decisively in the direction of raising worker marginal productivity. In this way it laid the foundations of shared prosperity, not just domestically but also in much of the world, as American techniques and innovations spread globally and enabled mass production and the rise of a middle class in scores of countries.

The United States has remained at the forefront of technology over the last fifty years, and its production methods and practices, especially its digital innovations, are still spreading throughout the world, but now with very different consequences. The US model of shared prosperity broke down as power became concentrated in the hands of big corporations, the institutions and norms of rent sharing unwound, and technology went in a predominantly automating direction starting around 1980.

All of this was underway, and the vision that animates the use of technology for automating work, monitoring, and squeezing out workers was firmly in place, before the latest wave of AI. We were already on our way back to a two-tiered society long before the 2010s. With a heightened AI illusion, we are seeing this process accelerate.

Modern AI amplifies the tools in the hands of tech elites, enabling them to create more ways of automating work, sidelining humans, and supposedly doing all sorts of good deeds such as increasing productivity and solving major problems facing humanity (they claim). Empowered by AI, these leaders feel even less need to consult the rest of the population. In fact, many of them think that most humans are not that wise and may not even understand what is good for them.

The marriage of digital technologies and big business had created a growing number of billionaires by the mid-2000s. Such fortunes multiplied once AI tools started spreading in the 2010s. But this was not because AI turned out to be anything as productive or amazing as its boosters have maintained. On the contrary, AI-based automation often fails to increase productivity by that much. Worse, it is no way to build shared prosperity. It nevertheless enthralls and enriches tycoons and top managers as it disempowers workers and opens up new ways of monetizing information about people, which we will discuss in Chapter 10.

That all of this gets ignored in a mad rush to use digital technologies to automate work and monitor humans is the reason why we have dubbed this new phase of the vision the AI illusion. This illusion is set to intensify in the next decade, as more powerful algorithms are developed, global online connectivity grows, and household appliances and other machines become permanently connected to the cloud, allowing more extensive data collection.

Today we are moving closer to H. G. Wells's *Time Machine* future dystopia. Our society has already become two-tiered. On top there are the big tycoons, who firmly believe they have earned their wealth because of their amazing genius. At the bottom we have regular people whom tech leaders view as error-prone and ripe for replacement. As AI penetrates more and more aspects of modern economies, it looks increasingly likely that the two tiers will grow further apart.

None of this had to be the case. Digital technologies did not have to be used for just automating work, and AI technologies did not have to be applied indiscriminately to amplify the same trend. The tech community did not have to be mesmerized by machine intelligence instead of working on machine usefulness. There is nothing foreordained about this path of technology, nor is there anything inevitable about the two-tiered society that our leaders are creating.

There are ways out of our current conundrum by reconfiguring the distribution of power in society and redirecting technological change. Such change will have to work through bottom-up, democratic processes. Ominously, however, AI is also breaking democracy.

10

Democracy Breaks

Social media's history is not yet written, and its effects are not neutral.

    —CHRIS COX, head of product, Facebook, 2019

If everybody always lies to you, the consequence is not that you believe the lies, but rather that nobody believes anything any longer.

    —HANNAH ARENDT, 1974 interview

On November 2, 2021, Chinese tennis star Peng Shuai posted on the social media site Weibo that she had been coerced into sex by a senior official. The message was removed within twenty minutes and never appeared again on Chinese social media. By the time it was removed, a number of users had taken screenshots of the post that were shared in foreign media. But access to these foreign outlets was also quickly censored. There was great interest in China about Peng Shuai, but few people were able to see the original post, and there was no public discussion.

Such swift removal of politically sensitive information is the rule and not the exception in China, where the internet and social media are under constant surveillance. The Chinese government reportedly spends an estimated $6.6 billion every year just on monitoring and censoring online content.

The government also invests massively in other digital tools and especially AI for surveillance. This is most visible in Xinjiang province, where systematic data collection on Uighur Muslims dates back to the immediate aftermath of the 2009 July riots there, but have multiplied since 2014. The Communist Party has instructed several leading technology companies to develop tools for collecting, aggregating, and analyzing data on individual and household habits, communication patterns, jobs, spending, and even hobbies to be used as inputs into "predictive policing" against the eleven million inhabitants of the province who are seen as potential dissidents.

Several of the major technology companies in China, including Ant Group (partly owned by Alibaba), the telecom giant Huawei, and some of the largest AI companies in the world such as SenseTime, CloudWalk, and Megvii, have cooperated with the government's efforts to develop surveillance tools and their rollout in Xinjiang. Efforts at tracking people using their DNA are underway. AI technologies that recognize Uighurs on the basis of their facial features are also used routinely.

What started in Xinjiang has since spread to the rest of China. Facial-recognition cameras are now widespread throughout the country, and the government has made steady progress toward introducing a national social credit system, which collects information on individuals and businesses to monitor their undesirable and untrustworthy activities. This, of course, includes dissent and subversive criticism of the government. According to the official planning document, a social credit system

> is founded on laws, regulations, standards and charters, it is based on a complete network covering the credit records of members of society and credit infrastructure, it is supported by the lawful application of credit information and a credit services system, its inherent requirements are establishing the idea of a sincerity culture, and carrying forward sincerity and traditional virtues, it uses encouragement to keep trust and

constraints against breaking trust as incentive mechanisms, and its objective is raising the honest mentality and credit levels of the entire society.

Early versions of the system were developed alongside private-sector firms, including Alibaba, Tencent, and the ride-sharing company Didi, with the purported aim of differentiating between acceptable (to the authorities) behavior and unacceptable behavior—and limiting the mobility and other actions of transgressors. Since 2017, prototypes of the social credit system have been implemented in dozens of major cities, including Hangzhou, Chengdu, and Nanjing. According to the Supreme People's Court, "Defaulters [on court orders] had been restrained from purchasing about 27.3 million plane tickets and nearly 6 million train tickets so far [July 9, 2019]." Some commentators have come to view the Chinese model and its social credit system as a prototype for a new kind of "digital dictatorship," in which authoritarian rule is maintained by intense surveillance and data collection.

Ironically, this is exactly the opposite of what many thought would be the effects of the internet and social media on political discourse and democracy. Online communication was promised to unleash the wisdom of the crowds, as different perspectives communicated and competed freely, enabling the truth to triumph. The internet was supposed to make democracies stronger and put dictatorships on the defensive as it revealed information on corruption, repression, and abuses. Wikis, such as the now infamous WikiLeaks, were viewed as steps toward democratizing journalism. Social media would do all the above and better by facilitating open political discourse and coordination among citizens.

Early evidence seemed to bear this out. On January 17, 2001, text messages were used to coordinate protests in the Philippines against its Congress, which had decided to disregard critical evidence against President Joseph Estrada in his impeachment trial. As messages moved from one user to another, more than a million people arrived in downtown Manila to object to congresspeople's complicity in Estrada's corruption and crimes. After the capital was brought to a standstill, the legislators reversed their decision, and Estrada was impeached.

Less than a decade later, it was social media's turn. Facebook and Twitter were used by protesters during the Arab Spring, helping to topple long-ruling autocrats Zine El Abidine Ben Ali in Tunisia and Hosni Mubarak in Egypt. One of the leaders of the Egyptian protests and a computer engineer at Google, Wael Ghonim, summarized both the mood among some of the protesters and the optimism in the tech world when he said in an interview: "I want to meet Mark Zuckerberg one day and thank him, actually. This revolution started—well, a lot of this revolution started on Facebook. If you want to liberate a society, just give them the Internet. If you want to have a free society, just give them Internet." A cofounder of Twitter adopted the same interpretation of its own role, claiming "Some Tweets may facilitate positive change in a repressed country. . . ."

Policy makers concurred. Secretary of State Hillary Rodham Clinton declared in 2010 that internet freedom would be a key pillar of her strategy for spreading democracy around the world.

With these hopes, how did we end up in a world in which digital tools are powerful weapons in the hands of autocrats for suppressing information and dissent, and social media has become a hotbed of misinformation, manipulated not just by authoritarian governments but also by extremists from both the Right and the Left?

In this chapter we argue that the pernicious effects of digital technologies and AI on politics and social discourse were not inevitable and resulted from the specific way in which these technologies were developed. Once these digital tools started being used primarily for massive data collection and processing, they became potent tools in the hands of both governments and companies interested in surveillance and manipulation. As people became more disempowered, top-down control intensified in both autocratic and democratic countries, and new business models based on monetizing and maximizing user engagement and outrage flourished.

A Politically Weaponized
System of Censorship

It was never easy to be in the opposition in Communist China. In what many interpreted as a partial relaxation of the repression that had already taken millions of lives, Chairman Mao declared in 1957, "Let a hundred flowers bloom," allowing criticism of the Communist Party. But hopes that this meant more accepting attitudes toward dissent were soon dashed. Mao initiated a vigorous "Anti-Rightist" campaign, and those who heeded his earlier invitation and attempted to express their critical views were rounded up, imprisoned, and tortured. At least five hundred thousand people were executed between 1957 and 1959.

But by the late 1970s and early 1980s, things were looking very different. Mao had died in 1976, and the hard-liners, including his wife, Jiang Qing, and three of her Communist Party associates, commonly known as the "Gang of Four," lost the ensuing power struggle and were sidelined. Deng Xiaoping, who was one of the revolution's leaders, a successful general during the civil war, the architect of the Anti-Rightist campaign, secretary-general and vice premier, and later purged by Mao, came back on the scene and took charge in 1978. Deng reinvented himself as a reformer and attempted a major economic restructuring of China.

This period witnessed a loosening of the power of the Communist Party. New independent media outlets sprang up, and some were openly critical of the party. Various grassroots movements also started during this period, including pro-democracy student movements and initiatives in the countryside to defend the rights of regular people against land grabs.

Hopes of a more open society crumbled, once more, during the Tiananmen Square massacre in 1989. In the relatively more permissive days of the 1980s, demands for greater freedoms and reforms had built up in the cities and especially among the students. A major wave of student demonstrations had already taken place in 1986, with demands for democracy, greater freedom of speech, and economic liberalization. The hard-liners blamed the pro-reform general secretary of the party, Hu Yaobang, for being soft on the protesters and removed him from power.

New protests broke out in April 1989 after Hu's death from a heart attack. Hundreds of students from Peking University marched to Tiananmen Square, at the center of Beijing, separated by the Gate of Heavenly Peace (*Tiananmen*) from the Forbidden City. As their ranks swelled over the next several hours, the students drafted the "Seven Demands," which included calls for affirming Hu Yaobang's views on democracy and freedom as correct, ending press censorship and restrictions on demonstrations, and curbing corruption by state leaders and their families.

As the government prevaricated on how to respond, support for the protests grew, especially after students began a hunger strike on May 13. As many as a million Beijing residents demonstrated in solidarity in the middle of May. Finally, Deng Xiaoping weighed in on the side of hard-liners and approved military action against the students. Martial law was declared on May 20, and in the next two weeks more than 250,000 troops were sent to Beijing to quell the unrest. By June 4, the protests were quashed, and the square was emptied. Independent sources estimate the death toll among protesters to have been as high as 10,000. Tiananmen Square was a turning point in the Communist Party's determination to clamp down on the freedoms that had emerged in the 1980s and to limit opposition activities.

All the same, the ability of the Communist Party to control dissent in the vast territories it controlled remained limited in the 1990s and most of the 2000s. The grassroots Weiquan movement, which brought together a large number of lawyers to defend victims of human rights abuses throughout China and advocate for environmental causes, housing rights, and freedom of speech, started in the early 2000s. One of the pro-democracy movements that gained highest visibility, Charter 08, led by the writer and activist Liu Xiaobo, published its platform in 2008 and proposed reforms that went far beyond the seven demands of the Tiananmen Square protests. They included a new constitution, election of all public officials, separation of powers, an independent judiciary, guarantees for basic human rights, and extensive freedom for association, assembly, and expression.

By 2010, public dissent had become much harder in China, with the internet a potent tool in the hands of the authorities to monitor and sanitize political discourse. The internet had arrived in China in 1994, and efforts at censorship started soon thereafter. The "Great Firewall," aimed at limiting what Chinese citizens could view and with whom they could communicate, originated in 2002, was completed in 2009, and has been extended periodically since then.

During the early 2010s, however, digital censorship had its limits. A major research effort collected and analyzed millions of social media posts across 1,382 Chinese websites and platforms in 2011 and then followed them up to see if they were removed by Chinese authorities. Results show that the Great Firewall was effective, but only up to a point. The authorities did not censor most of the (hundreds of thousands of) posts critical of the government or the party. Rather, they removed the much smaller subset of posts that were on sensitive topics, which posed a risk of large-scale response and the possibility of coalescing different opposition groups. For instance, the vast majority of posts about protests in Inner Mongolia or Zhengcheng province were removed quickly. Those about Bo Xilai (former mayor of Dalian, member of the Politburo, and being purged at the time) or Fang Binxing (father of the Great Firewall) were removed equally rapidly.

Another team of researchers found that, the Great Firewall and the systematic censorship notwithstanding, social media communication still acted as a trigger for protests. Messaging over Weibo enabled coordination and geographic spread of protest activities. Already during this period, however, social media–mediated dissident activities were short-lived.

The softer-touch censorship that allowed some critical messages to circulate ceased after 2014. Under the leadership of Xi Jinping, the government increased its demand for surveillance and related AI technologies first in Xinjiang and then throughout China. In 2017 it issued the "New Generation AI Development

Plan," with a goal of global leadership in AI and a clear focus on the use of AI for surveillance. Since 2014, China's spending on surveillance software and cameras and its share of global investment in AI have increased rapidly every year, now making up about 20 percent of worldwide AI spending. Researchers located in China now account for more AI-related patents than any other country.

With better AI technologies came more intense surveillance, and in the words of the founder of *China Digital Times*, Xiao Qiang, "China has a politically weaponized system of censorship; it is refined, organized, coordinated and supported by the state's resources. It's not just for deleting something. They also have a powerful apparatus to construct a narrative and aim it at any target with huge scale."

Today, very few dissenting posts escape censorship on any major social media platforms, the Great Firewall covers almost all politically sensitive foreign websites, and there is little evidence of protests being coordinated on social media. The Chinese can no longer access most independent foreign media, including the *New York Times*, CNN, BBC, the *Guardian*, and the *Wall Street Journal*. Major Western social media outlets and search engines, including Google, YouTube, Facebook, Twitter, Instagram, and various video-sharing blog sites, are also blocked.

AI has significantly amplified the Chinese government's ability to suppress dissent and circumvent political discourse and information, especially in the context of multimedia content and live chats.

A Braver New World

By the 2010s, Chinese political discourse was already looking like George Orwell's *1984*. By suppressing information and using systemic propaganda, the government attempted to tightly control the political narrative. When corruption investigations that touched upon high-level politicians or their families were prominently reported in the foreign press, government censorship ensured that the Chinese people did not see these details and were instead bombarded with propaganda about the virtuousness of their leaders.

Many people appeared to be at least partially convinced by indoctrination or at the very least did not dare admit that they thought this was propaganda. The Communist Party initiated a major reform of high school curriculums in 2001. The aim was to politically educate the nation's youths. A 2004 memo on the reform was titled "Suggestions on Strengthening the Ideological and Moral Construction of Our Youths." The new textbooks, which started being rolled out in 2004, had a more national-istic account of history and stressed the authority and virtues of the Communist Party. They criticized Western democracies and argued that China's political system was superior.

Students exposed to the new textbooks professed to hold very different opinions than their peers in the same province who grad-uated before the textbooks were introduced. They also reported higher levels of trust in government officials and deemed the Chi-nese system to be more democratic than did the students who were not indoctrinated by the same textbooks. Whether they truly believed these things or simply internalized the fact that they were expected to offer these opinions when asked is harder to establish. Nevertheless, it is clear that their reported views were strongly affected by the propaganda they were subjected to.

By the late 2010s, these tendencies were all significantly inten-sified. Digital censorship and propaganda meant that nationalism, unquestioning support for the government, and unwillingness to listen to critical news and opinions had become much more widespread among Chinese youths. After the massive investments in AI, the Great Firewall was also complemented by constant surveillance using data gathered on all Chinese platforms and workplaces. In such an environment, would Chinese university students even want to access foreign media sources if they could? This is the question that two researchers set out to explore in an ambitious study. The answer they found was surprising, even to themselves.

The Great Firewall had one weakness in the mid-2010s. It blocked Chinese users' access to foreign media and websites by using their IP addresses, which indicated whether they were located in mainland China. But VPNs (virtual private networks) could be used to hide IP addresses, allowing users in mainland China to access censored websites. The government had not explicitly banned VPNs, and information on websites visited by using them was unavailable to the authorities, making such a work-around fairly safe. (Things have changed since then, though, with private use of VPNs banned and all VPN providers having to register with the government.)

In a cleverly designed experiment, the two researchers offered university students in Beijing free VPN access (and sometimes extra encouragements via newsletters and other means) so that they could access Western news outlets for a period of eighteen months between 2015 and 2017. Students receiving these additional encouragements visited Western media, were interested in the news, and, once having started doing so, continued to get news from foreign sources. Their survey responses indicate that they understood and believed the information, changed their political opinions, and became more critical of the Chinese government. They also expressed much more sympathy for democratic institutions.

Nonetheless, without the extra encouragements, the vast majority of the students had no interest in visiting foreign websites and did not even want free VPN access. They were so convinced by the propaganda in schools and in the Chinese media that there was no relevant or reliable information about China in Western sources that they did not really need to be actively censored. They had already internalized the censorship.

The researchers interpreted this finding as Aldous Huxley's *Brave New World* rather than George Orwell's *1984*. In the words of the social critic Neil Postman, "What Orwell feared were those who would ban books. What Huxley feared was that there would be no reason to ban a book, for there would be no one who would want to read one."

In Huxley's dystopia, society is divided between rigidly segmented castes, going from alphas at the top to betas, gammas, deltas, and all the way down to epsilons. But there is no more need for constant censorship and surveillance because "under a scientific dictator education will really work—with the result that most men and women will grow up to love their servitude and will never dream of revolution. There seems to be no good reason why a thoroughly scientific dictatorship should ever be overthrown."

From Prometheus to Pegasus

The use of digital tools to suppress dissent is not unique to China. Iran and Russia, among other dictatorships, have also used them to track and punish dissenters and stifle access to free information.

Even before the Arab Spring, social media use in pro-democracy protests had come to international attention during Iran's ultimately unsuccessful Green Revolution. Huge crowds (by some estimated to be as many as three million) poured into the streets to bring down President Mahmoud Ahmadinejad, who was believed to have rigged the 2009 election to remain in power. Many social media tools, including text messages and Facebook, were used in coordinating the protests.

The protests were soon suppressed, and a large number of opposition figures and students were arrested. In the aftermath, Iranian internet censorship intensified. In 2012 the Supreme Council of Cyberspace was launched to oversee the internet and social media, and today almost all Western social media, various streaming services (including Netflix), and most Western news media are blocked in Iran.

The evolution of social media's role in politics and the resulting government crackdown in Russia are similar. The site VK (*VKontakte*) emerged as the most popular social media platform in the country and was already widely used by 2011. Electoral fraud in the December 4, 2011, parliamentary elections, documented on the internet with pictures of ballot stuffing and multiple votes cast by government supporters, ignited massive protests. Subsequent research found that protests were coordinated on the platform and there were significantly larger antigovernment actions in cities where VK was more widely used.

As in China and Iran, the protests acted as a trigger for increased government control and censorship of online activity in Russia. Systematic censorship has since intensified. The System for Operative Investigative Activities compels all telecommunication operators to install hardware provided by the Federal Security Service (FSB), which enables the FSB to monitor metadata or even content, and also block access, without a need for a warrant. After another round of protests in 2020, more dissident and news websites were blocked, VPN tools and the encrypted browser Tor were banned, and new, astronomical fines were introduced as a way of coercing companies to prevent access to illegal content, including social media posts and websites critical of the government. Although AI technologies are less important to Russian censorship efforts, their role has recently grown as well.

Abuse of digital tools directed against opposition groups is not confined to dictatorships. In 2020 a list of about fifty thousand phone numbers was leaked to Forbidden Stories, an international organization striving to publish stories from and about journalists under repression around the globe. The numbers belonged to opposition politicians, human rights activists, journalists, and dissidents who were reportedly being hacked using the Pegasus spyware developed by the Israeli tech company NSO Group (named after the first names of its founders, Niv Karmi, Shalev Hulio, and Omri Lavie). (NSO denies any wrongdoing, saying that the software is provided only to "vetted government customers" and that these customers decide how to use it.)

Pegasus is a "zero-click" software, meaning that it can be installed on mobile phones remotely and without requiring a user to click on any links—in other words, it can be installed without the user's knowledge or consent. Its name comes from the winged horse, Pegasus, in Greek mythology, with reference to the broad class of software it belongs to (Trojan horse) and the fact that it flies rather than being manually installed. As we saw in Chapter 1, today's tech leaders are keen to stress the fire-like powers of

AI and portray themselves as the current-day Prometheus, gifting technology's powers to humanity. But Pegasus, not Prometheus, is what we seem to have gotten from modern digital technologies.

Pegasus can read text messages, listen to calls, determine location, remotely collect passwords, monitor online activity, and even take control of a phone's camera and microphone. It is allegedly used routinely in many countries with authoritarian rulers, including Saudi Arabia, the United Arab Emirates, and Hungary. The journalist Jamal Khashoggi, who was later brutally murdered and dismembered, was allegedly under surveillance by Saudi Arabian agents using Pegasus. (The Saudi authorities have said this was a "rogue operation.")

Investigation of the numbers obtained by Forbidden Stories reveals systematic abuse of the software by many democratically elected governments as well. In Mexico, spyware was originally acquired as a weapon against drug cartels and deployed in the operation that led to the capture of the head of the Sinaloa cartel, El Chapo. But it was subsequently turned against journalists, lawyers investigating the massacre of forty-three students, and opposition parties, including one of the opposition's leaders, Andrés Manuel López Obrador, who later became the country's president. In India, Prime Minister Narendra Modi's government uses the software even more extensively and has placed many important opposition leaders, student activists, journalists, election commissioners, and even heads of the country's Central Bureau of Investigation under surveillance.

Abuses using Pegasus went beyond developing-country governments. French president Emmanuel Macron's phone was on the list, as were the numbers of several high-ranking US State Department officials.

The United States does not need Pegasus for high-tech transgressions (although some of its security agencies did experiment with the software and also acted as an intermediary in its sale to the Djibouti government). On June 5, 2013, the world was awakened to revelations from Edward Snowden, first published in the *Guardian* newspaper, about illegal data collection by the National

Security Agency (NSA). The NSA cooperated with Google, Microsoft, Facebook, Yahoo!, various other internet service providers, and telephone companies such as AT&T and Verizon to scoop up huge amounts of data about American citizens' internet searches, online communications, and phone calls. It also tapped communication by leaders of American allies, including Germany and Brazil. It collected data from satellites and underwater fiber-optic cables. Snowden described the reach of these programs by saying that when he was a contractor for the NSA, "I, sitting at my desk, certainly had the authorities to wiretap anyone, from you or your accountant, to a federal judge or even the President, if I had a personal e-mail." Though unconstitutional and taking place without the knowledge or oversight of Congress, some of these activities were sanctioned by FISA (the Foreign Intelligence Surveillance Court).

The United States is not China, and these activities had to be hidden from the news media and even from most lawmakers. When Snowden's revelations broke out, there was a powerful reaction against the NSA and other agencies' abusive data-collection strategies. But this was not enough to put a stop to most surveillance. Perhaps even worse, private companies such as Clearview AI have started collecting facial images from hundreds of millions of users and selling this information to law-enforcement agencies, with essentially no oversight from civil society or other institutions. There is nothing wrong in this, according to Clearview's founder and CEO, who states, "Our belief is that this is the best use of the technology."

Pegasus spyware, snooping by NSA, and Clearview's facial-recognition technology illustrate a deeper problem. Once out there, digital tools for extensive data collection will be adopted by many, if not most, governments to suppress opposition and better monitor their citizens. They will strengthen nondemocratic regimes and enable them to withstand opposition much more effectively. They could even create a slippery slope for democratic governments to become more authoritarian over time.

Democracy dies in darkness. But it also struggles under the light provided by modern artificial intelligence.

Surveillance and the
Direction of Technology

From the initial euphoria about the democratizing potential of the internet and social media, some have jumped to the polar opposite conclusion: digital tools are inherently antidemocratic. In the words of historian Yuval Noah Harari, "Technology favors tyranny."

Both of these binary perspectives are wrong. Digital technology is not pro-democratic or antidemocratic. Nor was there any necessity for AI technologies to be developed to empower governments to monitor media, censor information, and repress their citizens. All of this was a choice of direction for technology.

We saw in Chapter 9 that digital technologies, which are almost by their nature highly general purpose, could have been used to further machine usefulness—for example, by creating new worker tasks or new platforms that multiplied human capabilities. It was the vision and the business model of large tech companies that pushed toward a primary focus on worker monitoring and job destruction through automation. The same is true when it comes to the use of AI as a tool in the hands of authoritarian governments and some purportedly democratic ones.

The dreams of the internet and digital technologies empowering citizens against dictatorship were not completely surreal. Digital technologies can be used for encryption, making it impossible for authorities to snoop in private communications. Services such as VPNs can be used to thwart censorship. Search engines such as Tor are currently impossible for governments to decrypt (so far as we know) and hence offer greater levels of privacy and security. Nevertheless, early hopes of digital democratization have been dashed because the tech world put its effort where the money and power lie—with government censorship.

It is thus a specific path—a low road—chosen by the tech community that intensifies data collection and surveillance. Although advances in large-scale processing of data using tools from machine learning have been important in these efforts, the real secret sauce in surveillance by governments and companies is massive amounts of data.

Once AI technologies strengthen authoritarian impulses, they create a vicious circle. As governments become more authoritarian, their demand for AI to track and control their population increases, and this pushes AI further in the direction of becoming a fully fledged monitoring technology.

Since 2014, there has been, for example, a huge increase in demand from local Chinese governments for AI technologies that provide facial recognition and other types of monitoring. This demand seems to be triggered, in part, by local political unrest. Politicians want to increase policing and surveillance when they see discontent or protest activity brewing in their region. In the second half of the 2010s, massive protests, especially targeted against the national government, were all but impossible, although local protests were still taking place, and for a while, as we have seen earlier in this chapter, they were even coordinated on social media.

By this point, however, AI tools were firmly on the side of the crackdowns, not of the protesters. Once empowered by AI technologies, local authorities become better at putting down and avoiding protests. Incidentally, even though the Chinese central government and local authorities are willing to hire large numbers of police officers, the increase in AI investments appears to reduce the need for using manpower to do the surveillance and even the actual repression of protesters.

More strikingly, this demand from local governments affects the direction of innovation. Data on the universe of AI start-ups in China show that government demand for monitoring technologies fundamentally transforms subsequent innovation. AI firms contracting with Chinese local governments start shifting their research more and more toward facial recognition and other tracking technologies. Perhaps as a result of these incentives, China has emerged as a global leader in surveillance technologies, such as facial recognition, but lags in other areas, including natural-language processing, language-reasoning skills, and abstract reasoning.

International experts rank the quality of AI research in China to be still significantly behind that of the United States in all dimensions. There is one aspect in which China has an advantage, however: data.

Chinese researchers work with much larger quantities of data and without the privacy restrictions that often limit the type of data Western researchers can access. The impact of local government contracts on the direction of AI research is particularly pronounced when local governments share vast amounts of data in their procurement contracts. With abundant data available to them without any strings and strong demand for surveillance technologies, AI start-ups were able to test and develop powerful applications that could track, monitor, and control citizens.

There is a surveillance technology trap here: powerful and cash-rich governments that are intent on suppressing dissent demand AI technologies to control their population. The more they demand them, the more researchers produce them. The more AI moves in this repressive direction, the more attractive it becomes to authoritarian (or wannabe authoritarian) governments.

Indeed, Chinese start-ups are now exporting their AI products targeted at monitoring and repression to other nondemocratic governments. The Chinese tech giant Huawei, one of the main beneficiaries of unrestricted access to data and financial incentives to develop snooping technologies, exported these tools to fifty other countries. In Chapter 9 we saw how AI-based automation developed in technologically advanced countries will affect the rest of the world, with significant potential downsides for most workers. The same is true for AI-based surveillance: most citizens, wherever they are around the world, are finding it harder and harder to escape repression.

Social Media and Paper Clips

Internet censorship and even high-tech spyware may say nothing about the potential of social media as a tool to improve political discourse and coordinate opposition to the worst regimes in the world. That several dictatorships have used new technologies to repress their populations should not surprise anybody. That the United States has done the same is also understandable when you think about its security services' long tradition of lawless behavior that has only been amplified with the "War on Terror." Perhaps the solution is to double down on social media and allow more connectivity and unencumbered messaging to shine a brighter light on abuses. Alas, AI-powered social media's current path appears almost as pernicious for democracy and human rights as top-down internet censorship.

The paper-clip parable is a favorite tool of computer scientists and philosophers for emphasizing the dangers that super-intelligent AI will pose if its objectives are not perfectly aligned with those of humanity. The thought experiment presupposes an unstoppably powerful intelligent machine that gets instructions to produce more paper clips and then uses its considerable capabilities to excel in meeting this objective by coming up with new methods to transform the entire world into paper clips. When it comes to the effects of AI on politics, it may be turning our institutions into paper clips, not thanks to its superior capabilities but because of its mediocrity.

By 2017, Facebook was so popular in Myanmar that it came to be identified with the internet itself. The twenty-two million users, out of a population of fifty-three million, were fertile ground for misinformation and hate speech. One of the most ethnically diverse countries in the world, Myanmar is home to 135 officially recognized distinct ethnicities. Its military, which has ruled the country with an iron fist since 1962, with a brief period of parliamentary democracy under military tutelage between 2015 and 2020, has often stoked ethnic hatred among the majority-Buddhist population. No other group has been as often targeted as the Rohingya Muslims, whom the government propaganda portrays as foreigners, even though they have lived there for centuries. Hate speech against the Rohingya has been commonplace in government-controlled media.

Facebook arrived into this combustible mix of ethnic tension and incendiary propaganda in 2010. From then on it expanded rapidly. Consistent with Silicon Valley's belief in the superiority of algorithms to humans and despite its huge user base, Facebook employed only one person who monitored Myanmar and spoke

Burmese but not most of the other hundred or so languages used in the country.

In Myanmar, hate speech and incitement were rife on Facebook from the beginning. In June 2012 a senior official close to the country's president, Thein Sein, posted this on his Facebook page:

> It is heard that Rohingya Terrorists of the so-called Rohingya Solidarity Organization are crossing the border and getting into the country with the weapons. That is Rohingyas from other countries are coming into the country. Since our Military has got the news in advance, we will eradicate them until the end! I believe we are already doing it.

The post continued: "We don't want to hear any humanitarian issues or human rights from others." The post was not only whipping up hatred against the Muslim minority but also amplifying the false narrative that Rohingya were coming into the country from the outside.

In 2013 the Buddhist monk Ashin Wirathu, dubbed as the face of Buddhist terror by *Time* magazine that same year, was posting Facebook messages calling the Rohingya invading foreigners, murderers, and a danger to the country. He would eventually say, "I accept the term extremist with pride."

Calls from activists and international organizations to Facebook to clamp down on misleading information and incendiary posts kept growing. A Facebook executive admitted, "We agree that we can and should do more." Yet by August 2017, whatever Facebook was doing was far from enough to monitor hate speech. The platform had become the chief medium for organizing what the United States would eventually call a genocide.

The popularity of hate speech on Facebook in Myanmar should not have been a surprise. Facebook's business model was based on maximizing user engagement, and any messages that garnered strong emotions, including of course hate speech and provocative misinformation, were favored by the platform's algorithms because they triggered intense engagement from thousands, sometimes hundreds of thousands, of users.

Human rights groups and activists brought up these concerns about mounting hate speech and the resulting atrocities to Facebook's leadership as early as 2014, with little success. The problem was at first ignored and activists were stonewalled, while the amount of false, incendiary information against the Rohingya continued to balloon. So did evidence that hate crimes, including murders of the Muslim minority, were being organized on the platform. Although the company was reluctant to do much on the hate-crime problem, this was not because it did not care about Myanmar. When the country's government closed down Facebook, its executives immediately jumped into action, fearing that the shutdown might drive away some of its twenty-two million users in the country.

Facebook also accommodated the government's demands in 2019 to label four ethnic organizations as "dangerous" and ban them from the platform. These websites, though associated with ethnic separatist groups, such as the Arakan Army, the Kachin Independence Army, and the Myanmar National Democratic Alliance Army, were some of the main repositories of photos and other proofs of murders and other atrocities by the army and extremist Buddhist monks.

When Facebook finally responded to earlier pressure, its solution was to create "stickers" to identify potential hate speech. The stickers would allow users to post messages that included harmful or questionable content but would warn them, "Think before you share" or "Don't be the cause of violence." However, it turned out that just like a dumb version of the paper clip–obsessed AI program, Facebook's algorithm was so intent on maximizing engagement that it registered harmful posts as being more popular because people were engaging with the content to flag it as harmful. The algorithm then recommended this content more widely in Myanmar, further exacerbating the spread of hate speech.

Myanmar's lessons do not seem to have been well learned by Facebook. In 2018 similar dynamics started being played out in Sri Lanka, with posts on Facebook inciting violence against Muslims. Human rights groups reported the hate speech, but to no avail. In the assessment of a researcher and activist, "There's incitements to violence against entire communities and Facebook says it doesn't violate community standards."

Two years later, in 2020, it was India's turn. Facebook executives overrode calls from their employees and refused to remove Indian politician T. Raja Singh, who was calling for Rohingya Muslim immigrants to be shot and encouraging the destruction of mosques. Many were indeed destroyed in anti-Muslim riots in Delhi that year, which also killed more than fifty people.

Misinformation Machine

The problems of hate speech and misinformation in Myanmar are paralleled by how Facebook has been used in the United States, and for the same reason: hate speech, extremism, and misinformation generate strong emotions and increase engagement and time spent on the platform. This enables Facebook to sell more individualized digital ads.

During the US presidential election of 2016, there was a remarkable increase in the number of posts with either misleading information or demonstrably false content. Nevertheless, by 2020, 14 percent of Americans viewed social media as their primary source of news, and 70 percent reported receiving at least some of their news from Facebook and other social media outlets.

These stories were not just a sideshow. A study of misinformation on the platform concluded that "falsehood diffused significantly farther, faster, deeper, and more broadly than the truth in all categories of information." Many of the blatantly misleading posts went viral because they kept being shared. But it was not just users propagating falsehoods. Facebook's algorithms were elevating these sensational articles ahead of both less politically relevant posts and information from trusted media sources.

During the 2016 presidential election, Facebook was a major conduit for misinformation, especially for right-leaning users. Trump supporters often reached sites propagating misinformation from Facebook. There was less traffic going from social media to traditional media. Worse, recent research documents that people tend to believe posts with misinformation because they are bad at remembering where they saw a piece of news. This may be particularly significant because users often receive unreliable and sometimes downright false information from their like-minded friends and acquaintances. They are also unlikely to be exposed to contrarian voices in these echo chamber–like environments.

Echo chambers may be an inevitable by-product of social media. But it has been known for more than a decade that they are exacerbated by platform algorithms. Eli Pariser, internet activist and executive director of MoveOn.org, reported in a TED talk in 2010 that although he followed many liberal and conservative news sites, after a while he noticed he was directed more and more to liberal sites because the algorithm had noticed he was a little more likely to click on them. He coined the term *filter bubble* to describe how algorithm filters were creating an artificial space in which people heard only voices that were already aligned with their political views.

Filter bubbles have pernicious effects. Facebook's algorithm is more likely to show right-wing content to users who have a right-leaning ideology, and vice versa for left-wingers. Researchers have documented that the resulting filter bubbles exacerbate the spread of misinformation on the social media site because people are influenced by the news items they see. These filter-bubble effects go beyond social media. Recent research that incentivized some regular Fox News users to watch CNN found that exposure to CNN content had a moderating effect on their beliefs and political attitudes across a range of issues. The main reason for this effect appears to be that Fox News was giving a slanted presentation of some facts and concealing others, pushing users in a more right-wing direction. There is growing evidence that these effects are even stronger on social media.

Although there were hearings and media reaction to Facebook's role in the 2016 election, not much had changed by 2020. Misinformation multiplied on the platform, some of it propagated by President Donald Trump, who frequently claimed that mail-in ballots were fraudulent and that noncitizen immigrants were voting in droves. He repeatedly used social media to call a stop to the vote count.

In the run-up to the election, Facebook was also mired in controversy because of a doctored video of House Speaker Nancy Pelosi, giving the impression that she was drunk or ill, slurring her words and sounding unwell in general. The fake video was promoted by Trump allies, including Rudy Giuliani, and the hashtag #DrunkNancy began to trend. It soon went viral and attracted more than two million views. Crazy conspiracy theories, such as those that came from QAnon, circulated uninterrupted in the platform's filter bubbles as well. Documents provided to the US Congress and the Securities and Exchange Commission by former Facebook employee Frances Haugen reveal that Facebook executives were often informed of these developments.

As Facebook came under increasing pressure, its vice president of global affairs and communications, former British deputy prime minister Nick Clegg, defended the company's policies, stating that a social media platform should be viewed as a tennis court: "Our job is to make sure the court is ready—the surface is flat, the lines painted, the net at the correct height. But we don't pick up a racket and start playing. How the players play the game is up to them, not us."

In the week following the election, Facebook introduced an emergency measure, altering its algorithms to stop the spread of right-wing conspiracy theories claiming that the election was in reality won by Trump but stolen because of illegal votes and irregularities at ballot boxes. By the end of December, however, Facebook's algorithm was back to its usual self, and the "tennis court" was open for a rematch of the 2016 fiasco.

Several extremist right-wing groups as well as Donald Trump continued to propagate falsehoods, and we now know that the January 6, 2021, insurrection was partly organized using Facebook and other social media sites. For example, members of the far-right militia group Oath Keepers used Facebook to discuss how and where they would meet, and several other extremist groups live-messaged each other over the platform on January 6. One of the Oath Keepers' leaders, Thomas Caldwell, is alleged to have posted updates as he entered the Capitol and to have received information over the platform on how to navigate the building as well as to incite violence toward lawmakers and police.

Misinformation and hate speech are not confined to Facebook. Around 2016, YouTube emerged as one of the most potent recruitment grounds for the Far Right. In 2019, Caleb Cain, a twenty-six-year-old college dropout, made a video about YouTube explaining how he had been radicalized on the platform. As he said, "I fell down the alt-right rabbit hole." Cain explained how he "kept falling deeper and deeper into this" as he watched more and more radical content recommended by YouTube's algorithms.

Journalist Robert Evans studied how scores of ordinary people around the country were recruited by these groups, and he concluded that the groups themselves mentioned YouTube most often on their website: "15 out of 75 fascist activists we studied credited YouTube videos with their red-pilling." ("Red-pilling" refers to the lingo that these groups used, with reference to the movie *The Matrix*: accepting the truths propagated by these far-right groups was the equivalent of taking the red pill in the movie.)

YouTube's algorithmic choices and intent to boost watch time on the platform were critical for these outcomes. To increase watch time, in 2012 the company modified its algorithm to give more weight to the time that users spend watching rather than just clicking on content. This algorithmic tweak started favoring videos that people became glued to, including some of the more incendiary extremist content, the sort that Cain became hooked on.

In 2015 YouTube engaged a research team from its parent company's AI division, Google Brain, to improve the platform's algorithm. New algorithms then led to more pathways for users to become radicalized—while, of course, spending more time on the platform. One of Google Brain's researchers, Minmin Chen, boasted in an AI conference that the new algorithm was successfully altering user behavior: "We can really lead the users towards a different state, versus recommending content that is familiar." This was ideal for fringe groups trying to radicalize people. It meant that users watching a video on 9/11 would be quickly recommended content on 9/11 conspiracies. With about 70 percent of all videos watched on the platform coming from algorithm recommendations, this meant plenty of room for misinformation and manipulation to pull users into the rabbit hole.

Twitter was no different. As the favorite communication medium of former president Trump, it became an important tool for communication between right-wingers (and separately among left-wingers as well). Trump's anti-Muslim tweets were widely disseminated and subsequently caused not just more anti-Muslim and xenophobic posts on the platform but also actual hate crimes against Muslims, especially in states where the president had more followers.

Some of the worst language and consistent hate speech were propagated on other platforms, such as 4chan, 8chan, and Reddit, including its various sub-Reddits such as The_Donald (where conspiracy theories and misinformation related to Donald Trump originate and circulate), Physical_Removal (advocating the elimination of liberals), and several others with explicitly racist names that we prefer not to print here. In 2015 the Southern Poverty Law Center named Reddit as the platform hosting "the most violent racist" content on the internet.

Was it unavoidable that social media should have become such a cesspool? Or was it some of the decisions that leading tech companies made that brought us to this sorry state? The truth is much closer to the latter, and in fact also answers the question posed in Chapter 9: Why has AI become so popular even if it is not massively increasing productivity and outperforming humans?

The answer—and the reason for the specific path that digital technologies took—is the revenues that companies that collect vast amounts of data can generate using individually targeted digital advertising. But digital ads are only as good as the attention that people pay to them, so this business model meant that platforms strove to increase user engagement with online content. The most effective way of doing this turned out to be cultivating strong emotions such as outrage or indignation.

The Ad Bargain

To understand the roots of misinformation on social media, we must turn to Google's origin story.

The internet was flourishing before Google, but the available search engines were not helping. What makes the internet so special is its amazing size, with the number of websites estimated at 1.88 billion in 2021. Sifting through these many websites and finding the relevant information or products was bound to be a challenge.

The idea of early search engines was familiar to anyone who has used a book index: find all the occurrences of a given search word. If you wanted to find where the Neolithic Age was discussed in a book, you would look at the index and see the list of pages where the word *Neolithic* appeared. It worked well because a given word appeared a limited number of times, making the method of "exhaustive search" among all of the indicated pages feasible and quite effective. But imagine that you are looking into the index of an enormously large book, such as the internet. If you get the list of the instances in which the word *Neolithic* is mentioned in this humongous book, it may be hundreds of thousands of times. Good luck with exhaustive search!

Of course, the problem is that many of these mentions are not that relevant, and only one or two websites would be the authoritative sources in which one can obtain the necessary information about the Neolithic Age and how, say, humans transitioned to settled life and permanent agriculture. Only a way of prioritizing the more important mentions would enable the relevant information to be quickly retrieved. But this is not what the early search engines were capable of doing.

Enter two brash, smart young men, Larry Page and Sergey Brin. Page was a graduate student, working with the famous computer scientist Terry Winograd at Stanford, and Sergey Brin was his friend. Winograd, an early enthusiast for the currently dominant paradigm of AI, had by that point changed his mind and was working on problems in which human and machine knowledge could be combined, very much as Wiener, Licklider, and

Engelbart had envisaged. The internet, as we have seen, was an obvious domain for such a combination because its raw material was content and knowledge created by humans but it needed to be navigated by algorithms.

Page and Brin came up with a better way of achieving this combination, a true human-machine interaction in some sense: Humans were the best judge of which websites were more relevant, and search algorithms were excellent at collecting and processing link information. Why not let human choices about linkages guide how search algorithms should prioritize relevant websites?

This was at first a theoretical idea—the realization that this could be done. Then came the algorithmic solution of how to do it. This was the basis of their revolutionary PageRank algorithm ("Page" here reputedly refers both to Larry Page and the fact that pages are being ranked). Among the relevant pages, the idea was to prioritize those that received more links. So rather than using some ad hoc rules to decide which ones of the pages that have the word *Neolithic* should be suggested, the algorithm would rank these pages according to how many incoming links they received. The more popular pages would be more highly ranked. But why stop there? If a page received links from other highly ranked pages, that would be more informative about its relevance. To encapsulate this insight, Brin and Page developed a recursive algorithm where each page has a rank, and this rank is determined by how many other highly ranked pages were linking to it ("recursive" means that each page's rank depends on the ranks of all others). With millions of websites, calculating these ranks is no trivial matter, but it was already feasible by the 1990s.

Ultimately, how the algorithm computes the results is secondary. The important breakthrough here was that Page and Brin had come up with a way of using human insights and knowledge, as encapsulated in their subjective evaluations of which other pages were relevant, to improve a key machine task: ranking search outcomes. Brin and Page's 1998 paper, titled "The Anatomy of a Large-Scale Hypertextual Web Search Engine," starts with this sentence: "In this paper, we present Google, a prototype of a large-scale search engine which makes heavy use of the structure present in hypertext. Google is designed to crawl and index the Web efficiently and produce much more satisfying search results than existing systems."

Page and Brin understood that this was a major breakthrough but did not have a clear plan for commercializing it. Larry Page is quoted as saying that "amazingly, I had no thought of building a search engine. The idea wasn't even on the radar." But by the end of the project, it was clear that they had a winner on their hands. If they could build this search engine, it would tremendously improve how the World Wide Web functioned.

Thus arrived Google, as a company. Page and Brin's first idea was to sell or license their software to others. But their initial attempts did not gain traction, partly because other major tech companies were already locked into their own approaches or were prioritizing other areas: at that point search was not seen as a major moneymaker. Yahoo!, the leading platform at that time, showed no interest in Page and Brin's algorithm.

This changed in 1998, when a tech investor, Andy Bechtolsheim, entered the scene. Bechtolsheim met with Page and Brin and immediately got the promise of the new technology, if they had the right way of monetizing it. Bechtolsheim knew what that would be—ads.

Selling ads was not what Page and Brin were planning or even considering. But right away Bechtolsheim changed the game with a check for $100,000 to Google Inc., even though Google was not yet "incorporated." Soon the company was incorporated, the ad potential of the new technology became clear, and a lot more money poured in. A new business model was born.

The company introduced AdWords in 2000, a platform that sold advertising to be displayed to users searching for websites using Google. The platform was based on an extension of well-known auction models used in economics, and it rapidly auctioned off the most valuable (highly visible) places on the search screen. Prices depended on how much potential advertisers bid and how many clicks their ads received.

In 1998, or even in 2000, almost nobody was thinking about big data and AI. However, AI tools applied to large amounts of data would soon mean a lot of information for companies to target ads to users according to what they were interested in. AI quickly revolutionized Google's already-successful monetization model. This meant, in particular, that Google could track which websites were visited from the user's unique IP address, and thus direct individualized ads for this specific user. Hence, users looking at Caribbean beaches would receive ads from airlines, travel agencies, and hotels, and those browsing clothes or shoes would be bombarded with ads from relevant retailers.

The value of targeting in advertising cannot be overstated. The perennial problem of the ad industry is encapsulated in a saying that dates to the late 1800s: "I know half of my advertising is wasted, but I just don't know which half." Early internet advertising was afflicted by this problem. Ads from a menswear retailer would be shown to all users on a particular platform, say the Pandora music program, but half of the users would be women, and even most men would have no interest in buying clothes online at that point. With targeting, ads could be sent only to those who have demonstrated interest in making a purchase—for example, having visited a clothing store website or browsed some fashion items elsewhere. Targeting revolutionized digital advertising, but as with many revolutions, there would be plenty of collateral damage.

Google soon accelerated its data collection by offering a range of sophisticated free products, such as Gmail and Google Maps, which enabled the company to learn much more about users' preferences beyond the items they were searching for and their exact location. It also acquired YouTube. Now ads could be catered much more specifically for each user depending on their entire profile of purchases, activities, and location, boosting profitability. The results were striking, and in 2021 the vast majority of Google's (or its parent company Alphabet's) $65.1 billion revenue came from ads.

Google and other companies figured out how to make a lot of money from ads, and this not only explains the emergence of a new business model. It also answers a fundamental question we posed in Chapter 9: If it often leads to so-so automation, why is there so much enthusiasm about AI? The answer is largely about massive data collection and targeted ads, and there was much more of both to come.

The Socially Bankrupt Web

What Google can learn about its users from the metadata of their email activity and location pales in comparison with what some people are willing to share with their friends and acquaintances about their activities, intentions, desires, and views. Social media was bound to put the targeted-ad business model into a higher gear.

Mark Zuckerberg saw from the very beginning that key to Facebook's success would be its ability to be a vehicle, or in fact even a manufacturer, of a "social web," in which people would engage in a range of social activities. To accomplish this, he prioritized the growth of the platform above all else.

But monetizing this information was always going to be a challenge, even with Google's successful business model out there to be emulated. Facebook's first few forays into data collection as a means of improving its ability to target ads were failures. In 2007 the company introduced a program called Beacon as a way of scooping up information about Facebook users' purchases on other sites and then sharing it with their friends on their newsfeed. The initiative was immediately seen to be a colossal violation of user privacy and was discontinued. The company needed to forge an approach that combined a massive amount of data collection for digital advertisements and at least some amount of user control.

The person who made this a reality was Sheryl Sandberg, who had been in charge of Google's AdWords and had been instrumental in the transformation of that company into a targeted-ad machine. In 2008 she was hired at Facebook as chief operating officer. Sandberg understood how to make this combination work and also the potential that Facebook had in this space: the company could create new demand for products, and thus for advertising, by leveraging its knowledge about users' social circles and preferences. Already in November 2008, Sandberg summarized this combination as foundational to the company's growth, stating that "what we believe we've done is we've taken the power of real trust, real user privacy controls, and made it possible for people to be their authentic selves online." If people were their authentic selves, then they would reveal more about themselves, and there would be more information to be used for generating ad revenue.

The first important innovation in this effort was the "Like" button, which not only revealed much more about user preferences but also would act as an emotional cue to encourage greater engagement. Several other architectural changes—for example, concerning how the newsfeed works and how users can give feedback—were also introduced. Most importantly, AI algorithms started organizing each user's newsfeed to attract and retain their attention and, of course, place ads in the most profitable manner.

Facebook also began offering new tools to advertisers, again based on basic AI technologies. These included the ability to build custom audiences so that ads could be sent to users with certain specific demographics, and capabilities to form look-alike audiences, which Facebook itself describes as "a way your ads can reach new people who are likely to be interested in your business because they share similar characteristics to your existing customers."

Social media's great advantage over search engines when it came to ads was intense engagement. People sometimes pay attention to ads when they search for a product or shop using an engine such as Google, but this is a short engagement, and the amount that the company can make by selling ads is correspondingly limited. If people were to spend more time watching what pops up on their screens, that would mean greater ad revenues. Working to increase Likes for posts from friends and acquaintances proved a great way of boosting such engagement.

From its early days, Facebook played with people's psychology for achieving these objectives, and in fact engaged in systematic testing and experiments with its users to determine which types of posts and which ways of presenting them would generate more emotion and reaction.

Social relations, especially within groups, are always fraught with feelings of disapproval, rejection, and envy. There is now abundant evidence that Facebook triggers not just outrage at political content but also strong negative emotions in other social contexts. It then exploits all these emotions to encourage people to spend more time on the platform. Sensational content makes people spend more time on the platform, as does anxiety. Several social psychology studies show that social media use is intertwined with feelings of envy and inadequacy, and often leads to concerns about self-esteem.

The expansion of Facebook across US college campuses, for example, had a powerful negative impact on mental health, often leading to feelings of depression. Students whose campus gained access to the platform also started reporting significantly worse academic performance, indicating that the effects are not confined to emotions but affect off-line behavior as well. Facebook powerfully monetizes these feelings because both anxieties and efforts to gain greater approval increase the time that people spend on the platform.

An ambitious research project is revealing about this issue. Researchers incentivized some people on Facebook to (temporarily) give up using the platform and then compared their time use and emotional states to members of a control group who were given no such inducement and continued to use Facebook intensively. Those who were encouraged to stop using Facebook spent more time doing other social activities and were significantly happier. But, reflecting the social pressure that they might have felt from peers and from the platform trying to reengage them, when the study was over, they went back to Facebook—worse mental state and all.

In the service of increasing user engagement, many new features and algorithms at Facebook were introduced rapidly and without much study of how they would affect user psychology and misinformation on the platform. The general approach of the company and its engineers toward introducing new features aimed at increasing user engagement is summarized by "Fuck it, ship it," an expression used frequently by its employees.

But it was not just a case of unintentional damages on the way to achieve greater engagement. Facebook leadership was intent on maximizing user engagement and did not want other considerations to stand in the way. Sandberg repeatedly insisted that there should be more ads on Instagram, which had been acquired by Facebook in 2012 with promises that the app would remain independent from Facebook and make its own business decisions, including about the design of the app and about advertisements.

When Facebook had decided to change its algorithm so that it would not promote misleading stories and untrustworthy websites after the 2020 US presidential election, the results were striking. Hateful content and misinformation stopped going viral. But a short while later, the changes were reversed, and the platform was back to business as usual, largely because when the company tested the effect of the change on engagement, it found that when people were getting less enraged and triggered, they were spending less time there.

Throughout, Zuckerberg and Sandberg, later joined by Clegg, defended these decisions on the basis that the platform should not limit anyone's free speech. In response, the British comedian Sacha Baron Cohen summarized what many thought was the problem: "This is about giving people, including some of the most reprehensible people on earth, the biggest platform in history to reach a third of the planet."

The Antidemocratic Turn

We cannot understand the political mess that social media has created without recognizing the profit motive based on targeted ads, which makes these companies prioritize maximizing user engagement and sometimes rage. Targeted advertisements, in turn, would not have been possible without the collection and processing of massive amounts of data.

The profit motive is not the only factor that has pushed the tech industry in this antidemocratic direction. These companies' founding vision, which we dubbed the AI illusion, has played an equally important role.

As we discussed in Chapter 1, democracy, above all else, is about a multitude of voices, critically including those of ordinary people, being heard and becoming significant in public-policy directions. The notion of the "public sphere," proposed by the German philosopher Jürgen Habermas, captures some of the essential features of healthy democratic discourse. Habermas argued that the public sphere, defined as forums where individuals form new associations and discuss social issues and policy, is pivotal for democratic politics. Using British coffeehouses or French salons of the nineteenth century as the model, Habermas suggested that the critical ingredient of the public sphere is the ability that it offers to people to freely participate in debates on issues of general interest without a strict hierarchy based on preexisting status. In this way, the public sphere generates both a forum for diverse opinions to be heard and a springboard for these opinions to influence policy. It can be particularly effective when it allows people to interact with others on a range of cross-cutting issues.

Early on, there was even a hope that online communications could generate a new public sphere, one where people from even more diverse backgrounds than in local politics could freely interact and exchange opinions.

Unfortunately, online democracy is not in line with the business models of leading tech companies and the AI illusion. In fact, it is diametrically opposed to a technocratic approach, which maintains that many important decisions are too complex for regular people. The vibe in the corridors of most tech companies is that men (and sometimes, but not that often, women) of genius are at work, striving for the common good. It is only natural that they should be the ones making the important decisions. When approached this way, the political discourse of the masses becomes something to be manipulated and harvested, not something to be encouraged and protected.

The AI illusion thus favors an antidemocratic impulse, even as many of its executives view themselves to be on the center-left

17. A decisive moment in the development of countervailing power in modern America: United Auto Workers sit comfortably while stopping production at the General Motors plant in Flint, Michigan, 1937.

18. A rally of Professional Air Traffic Controllers Organization employees in 1981. Their strike was broken by President Ronald Reagan.

19. Dockworkers loading one bag at a time at the Royal Albert Docks, London, 1885.

20. Dock work today: one worker, one crane, many containers.

21. An IBM computer, 1959.

22. Robots at a Porsche plant, 2022. A worker watches, wearing gloves.

23. A reconstruction of the Bombe, designed by Alan Turing to speed up the decryption of German signals during World War II.

24. MIT math professor Norbert Wiener brilliantly warned in 1949 about a new "industrial revolution of unmitigated cruelty."

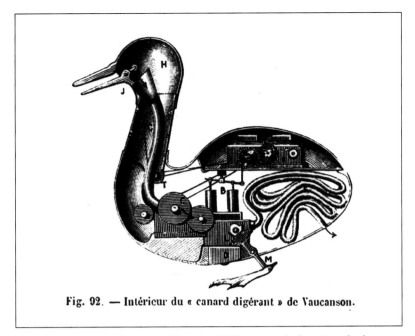

Fig. 92. — Intérieur du « canard digérant » de Vaucanson.

25. An imaginative drawing of Jacques de Vaucanson's digesting duck.

26. Human-complementary technology: Douglas Engelbart's mouse to control a computer, introduced at the "Mother of All Demos" in 1968.

27. So-so automation: customers trying to do the work, and sometimes failing, at self-checkout kiosks.

28. Facebook deciding what is and what is not fit for people to see.

29. Monitoring workflow inside an Amazon fulfillment center.

30. Digital surveillance with Chinese characteristics: a machine for checking social credit scores in China.

31. Milton Friedman: "The social responsibility of business is to increase its profits."

32. Ralph Nader: "The unconstrained behavior of big business is subordinating our democracy to the control of a corporate plutocracy that knows few self-imposed limits. . . ."

33. Ted Nelson: "COMPUTER POWER TO THE PEOPLE!"

34. Elon Musk: "Robots will be able to do everything better than us."

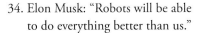

and supporters of democratic institutions and even the Demo-cratic Party. Their support is often rooted in cultural issues and conveniently bypasses the vital building block of democracy: people's active participation in politics. Such participation is especially discouraged when it comes to AI because most entre-preneurs and venture capitalists believe that people do not under-stand the technology and unnecessarily worry about its intrusive effects. As one venture capitalist put it, "Most of the fears of arti-ficial intelligence are overblown if not altogether unfounded." The solution is to ignore these concerns, forge ahead, and inte-grate AI into every aspect of our lives because "perhaps it's only when a technology is fully integrated into daily life, and recedes into the background of our imagination, that people stop fearing it." This was essentially the same approach advocated by Mark Zuckerberg when he told *Time* magazine, "Whenever any tech-nology or innovation comes along and it changes the nature of something, there are always people who lament the change and wish to go back to the previous time. But, I mean, I think that it's so clearly positive for people in terms of their ability to stay connected to folks."

Another aspect of the AI illusion, the elevation of disruption as a virtue encapsulated by "move fast and break things," has accelerated this antidemocratic turn. *Disruption* came to mean any negative effects on others, including workers, civil society organizations, traditional media, or even democracy. It was all fair game, in fact encouraged, so long as it was a consequence of exciting new technologies and consistent with bigger market share and moneymaking.

A reflection of this antidemocratic impulse can be seen in Facebook's own research on how users respond to negative and positive emotions from friends in their newsfeed. In 2014 the company undertook a massive internal study, manipulating the newsfeed of nearly seven hundred thousand users by reducing their exposure to either positive or negative expressions for a week. Unsurprisingly, greater exposure to negative emotions and lower exposure to positive emotions impacted users, with lasting adverse effects.

The company did not ask for any permission for this massive study from the users or even attempt to adhere to commonly accepted standards in scientific research, where informed consent from subjects is necessary. After some of the results of the study were published by Facebook researchers and others in the *Proceedings of the National Academy of Sciences*, the editor in chief published an Editorial Expression of Concern because the study was done without informed consent and did not meet accepted standards of academic research. Google followed the same playbook in its efforts to expand the amount of information that it collected with Google Books and Google Maps. The company ignored privacy concerns and acted first, without permission or consultation, hoping that things would get sorted out or, at the very least, its fait accompli would be accepted. That worked out, at least for Google.

Facebook and Google are not exceptional in the industry. It is now routine for tech companies to collect vast amounts of data without any consent from the people whose information or photos are being harnessed. In the area of image recognition, for example, many AI algorithms are trained and sometimes take part in competitions on the ImageNet data set, initiated by the computer scientist and later chief scientist of Google Cloud, Fei-Fei Li. The data set, which contains more than 15 million images sorted into more than 22,000 categories, was built by collecting private photos uploaded to various applications on the internet, with no permission from the people who took or appear in these pictures. This was generally viewed as acceptable in the tech industry. In Li's assessment, "In the age of the Internet, we are suddenly faced with an explosion in terms of imagery data."

According to reporting in the *New York Times*, Clearview has systematically collected facial images without consent, aiming to build predictive tools that identify illegal immigrants and people likely to commit crimes. Such strategies are justified by arguing that large-scale data collection is necessary for technological advancement. As an investor in a facial-recognition start-up summed up, the defense for massive data collection is that "laws have to determine what's legal, but you can't ban technology. Sure, that might lead to a dystopian future or something, but you can't ban it."

The truth is more nuanced. Imposing massive surveillance and data collection is not the only path of technological advance, and limiting it does not mean banning technology. What we are experiencing instead is an antidemocratic trajectory charted by the profit motive and the AI illusion, which involves authoritarian governments and tech companies foisting their vision on everybody else.

Radio Days

Perhaps all of these issues are not specific to digital technologies and AI. Every great new communication technology contains the potential for abuse.

Consider another one of the transformative communication technologies of the twentieth century: radio. Radio is also a general-purpose technology and, in its way, was as revolutionary as social media, allowing for the first time in history different forms of entertainment, mass broadcasting of information, and, of course, propaganda. The technology was developed soon after the German physicist Heinrich Hertz proved the existence of radio waves in 1886, and the first radio transmitters were built by the Italian physicist Guglielmo Marconi a decade later. By the early 1900s, there were radio broadcasts, and during the 1920s commercial radio became widespread in many Western nations. Propaganda and misinformation started almost immediately. President Franklin D. Roosevelt understood the importance of the technology and made his fireside chats on live radio a key part of the efforts to explain his New Deal policies to the American public.

An early supporter of FDR came to be identified with radio propaganda in the United States: Father Charles Coughlin, a Roman Catholic priest with great oratory skills. By the mid-1930s, however, Father Coughlin had turned against New Deal policies and founded the National Union for Social Justice. His radio speeches, initially broadcast on the CBS network, focused as much on anti-Semitic propaganda as on his policy ideas. Father Coughlin was soon supporting Benito Mussolini and Adolf Hitler on the airways.

Coughlin's blend of anti-FDR, fascist, and anti-Semitic broadcasts had major effects on US politics in the 1930s. Recent research used differences across US counties in the strength of radio signals, determined by geographic and topological obstacles to transmission, to investigate this issue. It finds that Father Coughlin's radio propaganda reduced support for New Deal policies and depressed FDR's vote in the 1936 presidential election by several percentage points (even if it could not prevent his landslide victory). It was not just presidential votes that Coughlin influenced. Counties receiving his broadcasts uninterrupted were more likely to open a local branch of the pro-Nazi German-American Bund and to lend less support to America's World War II effort. Several decades later, they still exhibited more anti-Jewish feelings.

What Father Coughlin effectively exploited in the United States was perfected in Germany at the same time. The Nazis, once in power, heavily relied on radio propaganda. Hitler's propaganda minister, Joseph Goebbels, became an expert at using the airways to whip up support for Nazi policies and hatred against Jewish people and "Bolsheviks." Goebbels himself said that "our way of taking power and using it would have been inconceivable without the radio and the airplane."

Nazis were indeed quite effective at manipulating sentiments with radio broadcasts. Exploring once again variations in the strength of radio signals across different parts of Germany, as well as changes in radio broadcast content over time, a team of researchers found powerful effects from Nazi propaganda. These radio broadcasts increased anti-Semitic activities and denunciations of Jews to authorities.

Radio propaganda by extremists was ultimately brought under control in the United States and Germany, and how this was done is revealing about the differences between social media and radio.

It also suggests some lessons for how new communication technologies can be best used.

The problem in the 1930s was that Father Coughlin had a national platform to reach millions with inflammatory rhetoric. The problem today is that misinformation is propagated by Facebook and other social media sites' algorithms to reach potentially billions of people.

Coughlin's pernicious effects were neutralized when FDR's administration decided that the First Amendment protected free speech but not the right to broadcast. It argued that radio spectrum was a publicly owned commons that must be regulated. With new regulations requiring broadcasting permits, Father Coughlin's programs were forced off the air. Coughlin continued to write and soon started broadcasting again, though with more limited access and only through individual stations. His antiwar, pro-German propaganda was further curtailed after the outbreak of World War II.

Today, there is plenty of misinformation and hate speech on AM talk shows, but they do not have the reach that Father Coughlin's national broadcasts achieved or the kind of platform that Facebook's algorithms provide for online misinformation.

The postwar German reaction to radio propaganda was even more comprehensive. The German Constitution bans speech classified as *Volksverhetzung*, meaning "incitement to hatred," as well as incitement to violence or acts denying the dignity of certain segments of the population. Under this law, denying the Holocaust and spreading incendiary anti-Jewish propaganda are outlawed.

Digital Choices

AI technologies did not have to focus on automating work and monitoring employees in workplaces. Nor did they have to be developed to empower government censorship. There is also nothing inherently antidemocratic in digital technologies, and social media certainly does not have to focus on maximizing outrage, extremism, and indignation. It was a matter of choice—choice by tech companies, AI researchers, and governments—that got us into our current predicament.

As we mentioned earlier in this chapter, YouTube and Reddit were initially as much afflicted by far-right extremism, misinformation, and hate speech as Facebook was. But over the last five years, these two platforms took some steps to lessen the problem.

As public pressure mounted on YouTube and its parent company, Google, after insider accounts such as Caleb Cain's and exposés in the *New York Times* and the *New Yorker* came out, the platform started modifying its algorithms to reduce the spread of the most malicious content. Google now claims that it promotes videos from "authoritative sources," which are less likely to be used for radicalization or contain misinformation. It also states that these algorithmic adjustments have reduced the viewing of "borderline content" by 70 percent ("borderline" here refers to the fact that the company maintains that hate speech is already vetted out).

The story of Reddit is similar. Home to some of the worst extremist and incendiary material, initially defended by one of its founders, Steve Huffman, as being completely consistent with the "open and honest discussion" philosophy of the site, the platform has since responded to public pressure and tightened its moderation standards. After the 2017 white supremacist Unite the Right rally in Charlottesville, Virginia, fueled and organized on the platform, turned violent and killed a counterprotester and injured dozens of others, Reddit's founders and the platform had an about-face. The platform started removing scores of sub-Reddits advocating hate speech, racist language, and blatant misinformation. In 2019 it removed The_Donald.

Improvements resulting from self-regulation by the platforms should not be exaggerated. There is still plenty of misinformation and manipulation, often aided by algorithms on YouTube, and plenty of hateful content on Reddit. Neither platform has changed its business model, and for the most part both platforms continue to rely on maximizing engagement and targeted-ad revenues. Platforms that have different business models, such as Uber and Airbnb, have been much more proactive in banning hate speech from their websites.

But the best demonstration of the viability of alternative models comes from Wikipedia. The platform is one of the most visited services on the web, having received more than 5.5 billion unique annual visitors over the last few years. Wikipedia does not try to monopolize user attention because it does not finance itself by advertisements.

This has allowed the platform to develop a very different approach to misinformation. Entries in this online encyclopedia are written by anonymous volunteers, and any volunteer editor can start a new entry or modify an existing one. The platform has several layers of administrators, promoted from frequent users with good track records. Among the volunteer contributors, there are experienced editors with additional privileges and responsibilities, such as maintenance or dispute resolution. At a higher level, "stewards" have greater authority to deal with disagreements. According to the platform itself, stewards are "tasked with technical implementation of community consensus, dealing with emergencies, intervening against cross-wiki vandalism." Above stewards is the "Arbitration Committee," consisting of "volunteer editors who act in concert or in subgroups imposing binding solutions on conduct disputes the community has been unable to resolve." "Administrators" have the ability to protect and delete pages, and block editing in case of disputed content or past occurrence of vandalism or misinformation. Administrators themselves are overseen and promoted by "bureaucrats."

This administrative structure is instrumental in the site's ability to prevent the propagation of misinformation and the type of polarization that has been all too common on other sites. Wikipedia's experience suggests that the wisdom of the crowd, so dearly admired by early techno-optimists of social media, can work, but only when underpinned and monitored by the right organizational structure and when appropriate choices are made on the use and direction of technology.

Alternatives to the targeted-ad business model are not confined to nonprofits such as Wikipedia. Netflix, based on a subscription model, also collects information about users and invests heavily in AI in order to make individual-specific recommendations. But there is little misinformation or political outrage on the platform, for its goal is to improve user experience to encourage subscriptions, not to ensure maximal engagement.

Social media platforms can work with and make money from a subscription model as well. Such a model will not remedy all of the problems of social media. People can create their own echo chambers in a subscription-based platform, and new ways of monetizing misinformation and insecurities may arise. Nevertheless, alternative business models can move away from seeking intense user engagement, which has proved to be conducive to the worst type of social interaction, damaging both to mental health and democratic discourse.

A "social web" can have myriad positive effects if its pernicious impact on misinformation, polarization, and mental health can be contained. Recent research tracks the start of Facebook's service in new languages and shows that small businesses in the affected countries get access to information from foreign markets and expand their sales as a result. There is no reason to believe that a company could not make money on the basis of these types of services rather than on its ability to manipulate users. Social media and digital tools can also provide greater protection to individuals against surveillance and can even play a pro-democratic role, as we will discuss in Chapter 11. Pushing buttons for emotional reactions and targeting ads to users when they are thus triggered were never the only options for social media.

Democracy Undermined When We Most Need It

The tragedy is that AI is further undermining democracy when we need it most. Unless the direction of digital technologies is altered fundamentally, they will continue to fuel inequality and marginalize large segments of the labor force, both in the West and increasingly around the world. AI technologies are also being used to more intensively monitor workers and, through this channel, create even more downward pressure on wages.

You can pin your hopes on the productivity bandwagon if you like. But there is no indication that shared productivity gains will be forthcoming soon. As we have seen, managers and entrepreneurs often have a bias to use new technologies to automate work and disempower people, unless reined in by countervailing powers. Massive data collection has exacerbated this bias.

Countervailing powers are hard to come by without democracy, however. When an elite completely controls politics and can use tools of repression and propaganda effectively, it is hard to build any meaningful, well-organized opposition. So robust dissent will not rise in China anytime soon, especially under the increasingly effective system of censorship and AI-based surveillance that the Communist Party has established. But it is also becoming increasingly difficult to hope for the resurgence of countervailing powers in the United States and much of the rest of the Western world. AI is choking democracy while also providing the tools for repression and manipulation to both authoritarian and democratically elected governments.

As George Orwell asked in *1984*, "For, after all, how do we know that two and two make four? Or that the force of gravity works? Or that the past is unchangeable? If both the past and the external world exist only in the mind, and if the mind itself is controllable, what then?" This question is even more relevant today because, as philosopher Hannah Arendt anticipated, when bombarded with falsehoods and propaganda, people both in democratic and nondemocratic countries stop believing any news. It may be even worse than that. Glued to their social media and frequently outraged and very often absorbed by strong emotions, people may become divorced from their community and democratic discourse because an alternative, segregated reality has been created online, where extremist voices are loudest, artificial echo chambers abound, all information is suspect or partisan, and compromise has been forgotten or even condemned.

Some are optimistic that new technologies, such as Web 3.0 or the metaverse, can provide different dynamics. But as long as the current business model of tech companies and the surveillance obsession of governments prevail, they are more likely to further exacerbate these trends, creating even more powerful filter bubbles and a wider wedge with reality.

It is late, but perhaps not too late. Chapter 11 outlines how the tide can be turned and which specific policy proposals hold promise for such a transformation.

11

Redirecting Technology

Computers are mostly
used against people instead of for people
used to control people instead of to *free* them
time to change all that—
we need a . . .
PEOPLE'S COMPUTER COMPANY

—first newsletter of the
PEOPLE'S COMPUTER COMPANY,
October 1972 (italics in original)

Most of the things worth doing in the world had been
declared impossible before they were done.

—ATTORNEY LOUIS BRANDEIS, arbitration proceedings,
New York Cloak Industry, October 13, 1913

The Gilded Age of the late nineteenth century was a period of rapid technological change and alarming inequalities in America, like today. The first people and companies to invest in new technologies and grab opportunities, especially in the most dynamic sectors of the economy, such as railways, steel, machinery, oil, and banking, prospered and made phenomenal profits.

Businesses of unprecedented size emerged during this era. Some companies employed more than a hundred thousand people, significantly more than did the US military at the time. Although real wages rose as the economy expanded, inequality skyrocketed, and working conditions were abysmal for millions who had no protection against their economically and politically powerful bosses. The robber barons, as the most famous and unscrupulous of these tycoons were known, made vast fortunes not only because of ingenuity in introducing new technologies but also from consolidation with rival businesses. Political connections were also important in the quest to dominate their sectors.

Emblematic of the age were the massive "trusts" these men built, such as Standard Oil, which controlled key inputs and eliminated rivals. In 1850 British chemist James Young discovered how petroleum could be refined. Within a few years, scores of oil refineries were operating around the world. In 1859 oil reserves were discovered in Titusville, Pennsylvania, and oil became the engine of industrialization in the United States. The sector was soon defined by Standard Oil, founded and run by John D. Rockefeller, who symbolizes both the opportunities of the age and their abuse. Born into poverty, Rockefeller understood the importance of oil and of becoming the dominant player in a sector, and rapidly turned his company into a monopoly. By the early 1890s, Standard Oil controlled around 90 percent of oil-refining facilities and pipelines in the country, and it developed a reputation for predatory pricing, questionable side deals—for example, with railways that barred its competitors from shipping their oil—and the intimidation of rivals and workers.

The track record of other dominant firms, such as Andrew Carnegie's steel company, Cornelius Vanderbilt's railway conglomerate, DuPont in chemicals, International Harvester in farm machinery, and J.P. Morgan in banking, was similar.

There was a clear sense that the institutional fabric of the United States was ill-suited to contain the heft of these companies. They wielded growing political power, both because several US presidents sided with them, and even more because they had great sway over the US Senate, whose members in that era were not directly elected but selected by state legislatures. The general feeling (and indeed the reality) was that senatorial positions were "bought and sold," and the robber barons were heavily involved in the buying. It was not just the Senate. President William McKinley's campaigns in 1896 and 1900 were generously funded by businesses, organized in part by Senator Mark Hanna, who summed up the system this way: "There are two things that are important in politics. The first is money and I can't remember what the second one is." There were few effective laws to prevent the companies owned by robber barons from controlling their sector and thwarting competition by using the power conferred on them by their size.

When workers organized to ask for higher wages or better working conditions, they were often harshly suppressed, including in the Great Railroad strike of 1877, the Great Southwest Railroad strike of 1886, the Carnegie Steel strike of 1892, the 1894 Pullman strike, and the coal strike of 1902. In the 1913–1914 United Mine Workers strike at the Colorado Fuel and Iron Company, controlled by Rockefeller, the altercations between strikers and mine guards, troops, and strikebreakers hired by the company escalated and led to the deaths of twenty-one people, including women and children.

The United States today would be a very different place if the economic and social conditions of the Gilded Age had endured. But a broad Progressive movement formed to oppose the trusts' power and demand institutional change. Although the movement had its roots in earlier rural organizations, such as the National Grange of the Order of Patrons of Husbandry and later the Populist Party, Progressives built a much broader coalition around urban middle classes and had a momentous impact on the history of the United States.

Central to their success was a change in the views and norms of the American public, especially the middle classes. The transformation was in large part the result of the work of a group of journalists who came to be known as the muckrakers, as well as the writings of other reformers, such as the lawyer and later Supreme Court justice Louis Brandeis. Upton Sinclair's *The Jungle* revealed horrible working conditions in the meatpacking industry, and Lincoln Steffens reported on political corruption in many major cities.

Perhaps most influential was the work of another muckraker, Ida Tarbell, on Standard Oil. In a series of articles in *McClure's Magazine* starting in 1902, she exposed the company's and Rockefeller's alleged intimidation, price-fixing, illegal practices, and political shenanigans. Tarbell had personal knowledge of Rockefeller's business practices. Her father had been an oil producer in western Pennsylvania and was driven out of business by Standard Oil when Rockefeller made a secret deal with local railroads to raise the prices they charged on the oil shipments of the company's rivals. Tarbell's articles, collected in her 1904 book *The History of the Standard Oil Company*, did as much as any other to transform the American public's perception of the trusts' and robber barons' pernicious effects on society.

Where Tarbell led, other muckrakers followed. In a series of articles titled "The Treason of the Senate" in *Cosmopolitan* magazine in 1906, David Graham Phillips shone the light on shady deals and corruption in the Senate. Brandeis's *Other People's Money and How Bankers Use It* did the same for the banking industry, and especially for J.P. Morgan.

Also important was the work of community activists such as Mary Harris Jones (known as Mother Jones), who played a leading role in the organization of the United Mine Workers and the more radical Knights of Labor. Mother Jones was the key instigator of the 1903 Children's Crusade, a march of children working in mines and mills. They carried banners such as "We want to go to school and not to the mines!" to President Teddy Roosevelt's summer home in order to protest the lack of enforcement of laws banning child labor.

Progressives did not just change the public's mind; they also organized politically. The Populists had already provided a template of how a protest movement could coalesce into a nationally influential party. In the 1892 election the Progressive Party won 8.5 percent of the total votes. Urban middle classes built on this early success, and a wide variety of politicians such as William Jennings Bryan, Teddy Roosevelt, Robert La Follette, William Taft, and then Woodrow Wilson brought Progressive politics into mainstream parties, winning elections and paving the way to reform.

Progressives had an ambitious reform agenda, including the regulation and breakup of trusts, new financial regulations, political reform directed at cleaning up corruption in the cities and the Senate, and tax reform. Their policy proposals were not just slogans. Progressives deeply believed in the role of expertise in policy making and were instrumental in triggering new professional associations and systematic inquiries of many of the key social issues of the age.

Key policy reforms of the age were the outgrowth of the ideas that muckrakers, activists, and reformers had popularized. For example, Sinclair's exposé directly led to the Pure Food and Drug Act and the Meat Inspection Act. Ida Tarbell's research and writings inspired the application of the 1890 Sherman Antitrust Act to industrial and railway conglomerates. This was reinforced by passage of the Clayton Act in 1914 and the creation of the Federal Trade Commission for further regulation of monopoly and antitrust action. Progressive pressure was also instrumental in the formation of the Pujo Committee, which investigated misdeeds in the financial industry.

Even more consequential institutional changes included the Tillman Act of 1907, banning corporate contributions to federal political candidates; the Sixteenth Amendment, ratified in 1913, which introduced the federal income tax; the Seventeenth Amendment of 1913, which required the direct election of all US senators by popular vote; and the Nineteenth Amendment of 1920, giving women the right to vote.

Progressive reforms did not change American political economy in one fell swoop. Large corporations remained powerful, and inequality stayed high. Nevertheless, as we saw in Chapter 7, the Progressives laid the foundations for the New Deal reforms and for post–World War II shared prosperity.

Progressivism was a bottom-up movement, full of a diverse set of voices, which was critical for its success in building its populist coalition and generating new policy ideas. But this also led to some of its unappealing elements, including the overt or covert racism of some of its leading lights (including Woodrow Wilson), ideas of eugenics that gained prominence among some

Progressives, and Prohibition, established by the Eighteenth Amendment in 1919. All of these failings notwithstanding, the Progressive movement completely remade American institutions.

THE PROGRESSIVE MOVEMENT provides a historical perspective on the three prongs of a critical formula necessary for escaping our current predicament.

The first is altering the narrative and changing norms. The Progressives enabled individual Americans to have an informed view about troubles in the economy and society—rather than just accepting the line coming from lawmakers, business tycoons, and the yellow journalists allied with them. For example, Tarbell never presented herself as a political candidate or even committed to one cause. Instead, she honed her craft of investigative journalism to expose the main facts about Standard Oil and its boss, Rockefeller. Critically, Progressives transformed what was viewed as acceptable for companies to do and what ordinary citizens thought they could do about injustices.

The second is cultivating countervailing powers. Building on the change in the narrative and social norms, Progressives helped organize people into a broad movement that could oppose robber barons and push politicians to reform, including via labor unions.

The third prong is policy solutions, which Progressives articulated based on the new narrative, research, and expertise.

Redirecting Technological Change

Although the challenges facing us today are digital and global, the lessons of the Progressive Era are still relevant. The contemporary environmental movement confronting the existential threat of climate change demonstrates that the formula's three prongs can work to redirect technological change today. Despite the continued reliance of most large energy companies on fossil fuels and the failure of most policy makers to act, there have been remarkable advances in renewable energy technologies.

Fossil-fuel emissions are first and foremost a technology problem. Industrialization was built on fossil-fuel energy, and technological investments since the middle of the eighteenth century have focused on improving and expanding these conventional energy sources. As early as the 1980s, it was clear that fossil-fuel emissions could not be reduced to levels that would prevent continuous warming of the climate by just relying on small tweaks to coal and oil production and consumption. New sources of energy were needed, which meant a major redirection of technology. Little of this happened for several decades. As late as the mid-2000s, solar energy was more than twenty times as expensive as energy from fossil fuels. For wind, the factor was about ten. Although hydroelectric energy was already cheaper in the 1990s, capacity was limited.

Today, solar, wind, and hydro energy are cheaper to operate than fossil-fuel power stations. For example, the International Renewable Energy Agency estimates that fossil fuels cost between $50 and $150 per 100 kWh (kilowatt hours), photovoltaic solar power costs between $40 and $54, and onshore wind power comes in at less than $40. Although there are some activities for which renewables cannot be used effectively (such as jet fuel) and there are important storage challenges, most of the electric grid of the world could be powered by renewables, when policy makers decide to move in that direction.

How did this impressive achievement take place? First came the change in narrative on climate. Rachel Carson's *Silent Spring*, published in 1962, was one of the first steps. Several organizations, most prominently Greenpeace, were campaigning strongly to protect the environment by the 1970s. Greenpeace initiated a program on global warming in the early 1990s, attempting to be a counterweight against the tactics used by big oil companies to hide the environmental damages that fossil fuels were causing.

The 2006 documentary film *An Inconvenient Truth*, focusing on former vice president and presidential candidate Al Gore's efforts to inform the public about global warming, played a major role. The movie was watched by millions all over the world.

Around the same time, new organizations focused on climate change, such as 350.org, were launched. As 350.org's founder, Bill McKibben, put it, the environment was the main issue, and all else paled in comparison: "In 50 years, no one will care about the fiscal cliff or the Euro crisis. They'll just ask, 'So the Arctic melted, and then what did you do?'"

The change in narrative coalesced into a more organized political movement with Green parties, which made global warming the centerpiece of their agenda. The German Green Party became a powerful electoral force and entered the government a number of times. Environmentalists played a similar role in other Western European nations as well. A show of force by the environmental movement came during a series of climate strikes in September 2019, with protests and walkouts in schools and workplaces across 4,500 cities around the globe.

Two major consequences followed from the second prong. These movements put pressure on the corporate sector. As the population in many Western nations became informed about the dangers of climate change, they demanded cleaner products, such as electric cars and renewable energy, and employees at many large corporations insisted on reductions in their companies' carbon footprint. Equally, they pushed (some) policy makers into taking global warming seriously.

These developments activated the third prong, technical and policy solutions. Economic and environmental analyses identified three critical levers for combating climate change: a carbon tax to reduce fossil-fuel emissions, support for innovation and research in renewable energy and other clean technologies, and regulation against the worst polluting technologies.

Although carbon taxes face stiff opposition in many countries, not least in the United States, Britain, and Australia, they have been introduced in several European countries. The levels of carbon tax adopted around the world are still inadequate given global warming trends, but some nations are gradually increasing this tax. The Swedish rate now stands above $120 per metric ton of carbon dioxide, amounting to a significant increase in the price of coal-powered energy.

A carbon tax is a powerful tool to curb carbon emissions. Because it reduces the profitability of fossil-fuel production, it can spur investment in alternative energy sources. But at the current levels, it makes only a small dent in the profits of fossil-fuel companies and would not lead to a major redirection of technology. Much more potent are schemes that directly incentivize innovations and investment in clean energy. The US government recently provided annual tax credits of more than $10 billion for renewables and almost $3 billion for improving energy efficiency. Some funds are also directly targeted at new technologies—for example, under the auspices of the National Renewable Energy Laboratory, NASA, and the Department of Defense. Subsidies to renewables research have been even more generous in Germany and the Nordic countries.

Regulations, such as emissions standards by the state of California, first adopted in 2002, played a role in directly discouraging the most inefficient uses of fossil fuel—for instance, by forcing older models of vehicles that have much higher gas consumption off the road. These regulations have also encouraged further research in electric cars.

The three policy levers (carbon taxes, research subsidies, and regulations), together with pressure from consumers and civil society, led to a boost both in innovations in renewables and much larger levels of production of solar panels and wind energy. The basic technology that generates energy through the photovoltaic effect, by using the sun's photons, has been known since the late nineteenth century, and viable solar panels were first produced in the 1950s in the Bell Labs. Important breakthroughs followed, starting in the 2000s, as the number of patents related to clean energy increased dramatically in the United States, France, Germany, and Britain. As production expanded, the costs of solar panels plummeted. As a result of these rapid improvements, renewables already make up over 20 percent of total energy consumption in Europe, although the United States is lagging behind.

Remarkably, China followed the European and American redirection of this technology. The country began producing solar panels in response to growing demand in Europe, especially Germany, in the late 1990s, following European climate-mitigation policies. Motivated by a desire to play a leading role in this sector and also to deal with the country's own severe pollution problem, the Chinese government provided generous subsidies and loans to producers, rapidly boosting productive capacity. Costs of photovoltaic panels and other solar equipment started declining thanks to "learning by doing" (meaning that as volumes rose, firms became better and better at producing cost-effective and energy-efficient solar panels). Chinese producers introduced new machinery and improved techniques for cutting polysilicon wafers more thinly, enabling them to produce more solar cells from the same amount of material, reducing costs and increasing production. The country is now the largest producer of solar panels and polysilicon in the world (even if many of the solar panel factories are powered by electricity generated from coal). According to the Chinese government's statistics, renewable energy accounted for about 29 percent of electricity consumption in 2020.

Of course, success to date should not be exaggerated. There are still many areas, such as cost-effective energy storage, in which breakthroughs are needed, and several sectors, such as air transport and agriculture, have not reduced their carbon emissions. Emissions in the developing world, including in China and India, are still growing, despite advances in renewable technology. There is little prospect for a global carbon tax that can powerfully reduce consumption in the immediate future.

Nevertheless, from the point of view of the challenge of digital technologies, there is much we can learn from how technology is being redirected in the energy sector. The same combination—altering the narrative, building countervailing powers, and developing and implementing specific policies to deal with the most important issues—can work in redirecting digital technology.

Remaking Digital Technologies

Our current problems are rooted in the enormous economic, political, and social power of corporations, especially in the tech industry. The concentrated power of business undercuts shared prosperity because it limits the sharing of gains from technological change. But its most pernicious impact is via the direction of technology, which is moving excessively toward automation, surveillance, data collection, and advertising. To regain shared prosperity, we must redirect technology, and this means activating a version of the same approach that worked more than a century ago for the Progressives.

This can start only by altering the narrative and the norms. The necessary steps are truly fundamental. Society and its powerful gatekeepers need to stop being mesmerized by tech billionaires and their agenda. Debates on new technology ought to center not just on the brilliance of new products and algorithms but also on whether they are working for the people or against the people. Whether digital technologies should be used for automating work and empowering large companies and nondemocratic governments must not be the sole decision of a handful of entrepreneurs and engineers. One does not need to be an AI expert to have a say about the direction of progress and the future of our society forged by these technologies. One does not need to be a tech investor or venture capitalist to hold tech entrepreneurs and engineers accountable for what their inventions do.

Choices over the direction of technology should be part of the criteria that investors use for evaluating companies and their effects. Large investors can demand transparency on whether new technologies will automate work or create new tasks, whether they will monitor or empower workers, and how they will affect political discourse and other social outcomes. These are not decisions investors should care about only because of the profits they generate. A two-tiered society with a small elite and a dwindling middle class is not a foundation for prosperity or democracy. Nevertheless, it is possible to make digital technologies useful to humans and boost productivity so that investing in technologies that help humans can also be good business.

As with the Progressive Era reforms and redirection in the energy sector, a new narrative is critical for building countervailing powers in the digital age. Such a narrative and public pressure can trigger more responsible behavior among some decision makers. For example, we saw in Chapter 8 that managers with business-school educations tend to reduce wages and cut labor costs, presumably because of the lingering influence of the Friedman doctrine—the idea that the only purpose and responsibility of business is to make profits. A powerful new narrative about shared prosperity can be a counterweight, influencing the priorities of some managers and even swaying the prevailing paradigm in business schools. Equally, it can help reshape the thinking of tens of thousands of bright young people wishing to work in the tech sector—even if it is unlikely to have much impact on tech tycoons.

More fundamentally, these efforts must formulate and support specific policies to rechart the course of technology. As we explained in Chapter 9, digital technologies can complement humans by:

- improving the productivity of workers in their current jobs

- creating new tasks with the help of machine intelligence augmenting human capabilities

- providing better, more usable information for human decision making

- building new platforms that bring together people with different skills and needs

For example, digital and AI technologies can increase effectiveness of classroom instruction by providing new tools and better information to teachers. They can enable personalized instruction by identifying in real time areas of difficulty or strength for each student, thus generating a plethora of new, productive tasks for teachers. They can also build platforms that bring teachers and teaching resources more effectively together. Similar avenues are open in health care, entertainment, and production work, as we have already discussed.

An approach that complements workers, rather than sidelining and attempting to eliminate them, is more likely when diverse human skills, based on the situational and social aspects of human cognition, are recognized. Yet such diverse objectives for technological change necessitate a plurality of innovation strategies, and they become less likely to be realized when a few tech firms dominate the future of technology.

Diverse innovation strategies are also important because automation is not harmful in and of itself. Technologies that replace tasks performed by people with machines and algorithms are as old as industry itself, and they will continue to be part of our future. Similarly, data collection is not bad per se, but it becomes inconsistent both with shared prosperity and democratic governance when it is centralized in the hands of unaccountable companies and governments that use these data to disempower people. The problem is an unbalanced portfolio of innovations that excessively prioritize automation and surveillance, failing to create new tasks and opportunities for workers. Redirecting technology need not involve the blocking of automation or banning data collection; it can instead encourage the development of technologies that complement and help human capabilities.

Society and government must work together to achieve this objective. Pressure from civil society, as in the case of successful major reforms of the past, is key. Government regulation and incentives are critical too, as they were in the case of energy. However, the government cannot be the nerve center of innovation, and bureaucrats are not going to design algorithms or come up with new products. What is needed is the right institutional framework and incentives shaped by government policies, bolstered by a constructive narrative, to induce the private sector to move away from excessive automation and surveillance, and toward more worker-friendly technologies.

A central question is whether efforts toward redirecting technology in the West would be of any use if China continues to pursue automation and surveillance. The answer is likely yes. China is still a follower in most frontier technologies, and redirection efforts in the United States and Europe will have a major impact on global technology. As in the case of energy innovations, a serious redirection in the West can powerfully influence Chinese investments as well.

How to foster countervailing powers that influence the path of future technologies and incentivize socially beneficial technological change is our focus in the rest of this chapter.

Remaking Countervailing Powers

We cannot redirect technology without building new countervailing powers. And we cannot build countervailing powers without relying on civil-society organizations that bring people together around shared issues and cultivate norms of self-governance and political action.

Worker Organization. Labor unions have been a mainstay of countervailing powers since the beginning of the industrial age. They are a key vehicle for supporting the sharing of productivity gains between employers and workers. In workplaces where labor has a voice (either in the form of unions or work councils, as in many German companies), workers are consulted in technology and organizational decisions, and they have at times been successful in acting as a counterweight to excessive automation.

In their heyday, labor unions succeeded because they formed bonds among their members. They provided camaraderie for people working together and on similar tasks. They were a nexus of cooperation along common economic interests, centered on better working conditions and higher wages. And they cultivated political objectives aligned with their membership's beliefs and interests, such as the right to vote. These ingredients are unlikely to work as synergistically today.

Workplaces have become much less concentrated and more diverse, so camaraderie is harder to achieve. With the rise of more highly educated and white-collar employees in most workplaces, economic interests among workers are more divergent as well. Blue-collar production workers are now a smaller fraction of the US labor force (about 13.7 percent as of 2016), and organizational forms centered on them are unlikely to speak for the entire workforce. There are also fewer common-interest political aims within the working population, which is now more divided between Right and Left than was the case half a century ago.

Nevertheless, new methods for organizing workers may succeed where older approaches have failed, and some of this can already be seen in successful unionization drives in companies such as Amazon and Starbucks in 2021–2022. The worker-initiated union election of Amazon's Staten Island warehouse used very different tactics to succeed in an environment unlike those where traditional labor movements once thrived. Turnover rates were enormous in the company's warehouses, and the workforce was diverse in every respect, coming from myriad backgrounds and speaking dozens of different languages. The movement was organized by workers on the shop floor, not by professional union personnel. It funded itself over the social media platform GoFundMe rather than receiving centralized union money. It appears to have succeeded by developing a less rigid and less ideological approach, focusing on issues relevant to most Amazon warehouse employees, such as excessive monitoring, insufficient breaks, and high injury rates. Although its strategy is very different from the iconic "sit-down strike" of GM workers in 1936, which was a turning point for the US labor movement, it is reminiscent in terms of developing new methods of organizing from the ground up.

The other problem with labor unions in the United States and Britain is, as we have seen, their traditional structure is organized at the level of individual plants, which breeds a more conflictual relationship with management. More broad-based organizations, rather than just at the level of plants or firms, will be necessary moving into the future. These may take the form of multilayer organizations, whereby some decisions are taken at the workplace level and others are made at the industry level. The dual-track German system is a relevant example: work councils are engaged in communication and coordination in workplaces and can have a say in technology and training decisions, whereas industry unions are more focused on wage setting. Of course, it is possible that future labor movements may end up looking more like other civil-society organizations or more loosely associated industry-level confederations. This suggests that experimenting with new organizational forms is an important step moving forward.

Civil-Society Action, Alone and Together. The West is now a consumer society, and consumer preferences and action are important levers for influencing companies and technologies. Consumer reaction was vital in the case of renewable energy and electric cars. It was also pressure from consumers, together with media coverage, that forced YouTube and Reddit to take some steps to limit extremism on their platforms.

However, collective action requires a large group of people acting together to achieve an objective—for example, pushing companies toward reducing their carbon footprint. Such action is costly for most people, who will have to spend time to become informed, to attend meetings, to change their consumption habits, and to occasionally go out and protest. These costs multiply when there is a counterpush from companies and sometimes, even worse, from state security services. In authoritarian and even semi-democratic regimes, authorities can clamp down on protests and civil-society organizations.

These dynamics produce the "free-rider" problem: people who share the same values may nevertheless be tempted not to take part in collective action in order to avoid paying the costs. This tendency of course intensifies when punishments against dissidents increase. For example, recent research on Hong Kong protests shows that when pro-democracy university students expect others to take part in rallies against antidemocratic measures, they themselves become less likely to join the protests, free riding on others' efforts. Free riding is at the root of the collective-action dilemma: without coordination, only a minority of people who desire social change take part in collective action.

Consumer choice, the ultimate uncoordinated, individual action, suffers mightily from the collective-action dilemma. Only a fraction of those who want to reduce carbon emissions will give up air travel or fossil-fuel energy, for example. Civil-society organizations that coordinate consumers and make them act more as citizens rather than individual decision makers in the marketplace are vital.

In addition to providing a forum for debate and trusted information dissemination, civil-society organizations can create both carrots and sticks to coordinate protests and public pressure on misbehaving companies. On the carrot side, they cultivate an ethos for taking part in activities that are good for the public interest and additionally develop links between different people, who then encourage one another to participate. For sticks, they can sometimes shame individuals for free riding on the efforts of others.

Although other organizations, such as labor unions, can also play these roles, civil-society organizations are important, especially when the main issues, such as climate change or digital technologies, have effects on large numbers of people and cut across traditional groups. For instance, although labor unions can contribute to climate-change activism and mitigation, they are not ideally situated to solve climate-related collective-action problems, compared to, say, Greenpeace or 350.org, which can organize people from different walks of life and backgrounds. The same considerations apply to action on digital technologies and business regulation. In both cases, the effects are wide-ranging, necessitating broad coalitions that can best be constructed by civil-society organizations.

Can online organizations help rather than hinder these efforts? Is broad-based civil society even possible in the digital age? Although the optimism of the early 2000s about social media and the internet providing a forum for a digital "public sphere" has been dashed, new and better online communities can be built.

Periodic elections to select representatives are not the only aspect of democratic politics. Self-governance, both at workplaces and more broadly, may be as important as elections. Indeed, successful democratic periods in the West often coincide with other institutional vehicles for people to participate in political decision making, express and develop their opinions, and pressure public policy. These include local politics, town hall–type arrangements, and most importantly various civil-society clubs and organizations.

Meanwhile, in some non-Western societies, bottom-up political participation has taken place without such elections—for instance, in the context of village councils and the election of traditional chiefs in parts of sub-Saharan Africa. Among others, this type of participation (via traditional assemblies called *kgotla*) played a defining role in the economic and political development of Botswana, one of the most economically successful countries of the last fifty years.

Pathways for democratic institutions to cultivate new and better online communities are critical. There are some digital technologies that can play a helpful role rather than a harmful one, and finding ways to encourage their development is crucial. For example, digital tools are well suited for creating new forums in which debate and exchange of opinions can be carried out in real time and within a prespecified set of rules. Online meetings and communication can reduce participation costs, enabling larger-scale cross-cutting associations. Digital tools can also ensure that even in large meetings, individuals can participate in debate by making comments or recording their approval or disapproval. If these tools are well designed, they can help empower and amplify diverse voices—an imperative for successful democratic governance. Efforts in this direction include the New_Public project, founded by internet activist Eli Pariser and Professor Talia Stroud, which seeks to develop a platform and tools for deliberation and bottom-up participation, especially on issues relevant to the future of technology. The project advocates a richer view of technology as "what we can learn to do" (as articulated by the science-fiction writer Ursula Le Guin) and calls for a more decentralized approach to its development.

The new democracy initiative led by Audrey Tang, a former activist and currently the digital minister of Taiwan, is particularly noteworthy. Tang entered politics as part of the student-led Sunflower Movement, which occupied the Taiwanese Parliament to protest against the 2013 trade deal with China that was being signed by the ruling Kuomintang Party without sufficient public review or consultation.

Tang, previously a software entrepreneur and programmer, volunteered to help the movement communicate its message to the broader public. After the Democratic Progressive Party came to power in the 2016 general election, Tang was appointed as a minister, with a focus on digital communication and transparency. She has built a variety of digital tools for providing transparency in government decision making and for increasing deliberation and consultation with the public. This digital-democracy approach was used for a number of key decisions, including the regulation of the ride-sharing platform Uber and of liquor sales. It involves a "presidential hackathon," which allows citizens to make proposals to the executive. Another platform, g0v, provides open data from several Taiwanese ministries, which civic hackers can use to develop alternative versions of bureaucratic services. These technologies helped Taiwan's early and effective response to COVID-19, in which the private sector and civil society collaborated with the government to develop tools for testing and contact tracing.

New forums for virtual participation can of course repeat the same mistakes that social media commits today, exacerbating echo chambers and extremism. Once such tools start being used extensively, some parties will come up with strategies to spread disinformation, whereas others might use such platforms for demagoguery. Sensational, misleading content may start spreading, and rival viewpoints can begin shouting each other down rather than deliberating constructively. The best way of avoiding such mistakes is to view pro-democratic online tools as a work in progress that needs to be updated continuously as new challenges develop, and also as a complement, rather than as a full substitute, to traditional, in-person civic engagement.

These solutions have a technical aspect as well as a social dimension. The algorithmic architecture of online systems can be designed to aid deliberation and dialogue, instead of attention grabbing and provocation. Because algorithms need to come from the private sector, improved market incentives for technology development remain critical, as we discuss next.

Civil-society action also depends on information on deals and decisions in the corridors of power. Digital technologies can help shed light on the influence of large corporations and corporate money in politics. Online tools can track links and flows of money and favors between companies and politicians and bureaucrats. We certainly do not agree with former US Supreme Court justice Anthony Kennedy's excessively optimistic outlook: "With the advent of the Internet, prompt disclosure of expenditures can provide shareholders and citizens with the information needed to hold corporations and elected officials accountable for their positions and supporters." This can happen only when there are other traditional safeguards. Ensuring transparency should thus be viewed as complementary to more traditional types of civil-society action. For example, it could take the form of automatic detection and public posting of all meetings and interactions of politicians and top bureaucrats with lobbies and private-sector managers.

It is important to strike the right balance in transparency. The public does not need to be informed about every policy debate and all the negotiations that politicians undertake in order to build coalitions. However, with the amount spent on lobbying in the Western world reaching astronomical levels, the public has a right to know about deals reached by lobbyists, politicians, and firms, and these connections need to be regulated.

Policies for Redirecting Technology

The existence of countervailing powers and even new institutions by themselves will not redirect technology. Specific policies that change incentives and encourage socially beneficial innovations are necessary. Complementary policies—including subsidies and support for more worker-friendly technologies, tax reform, worker-training programs, data-ownership and data-protection schemes, breaking up of tech giants, and digital advertisement taxes—can help initiate a major redirection of technology.

Market Incentives for Redirection. Government subsidies for developing more socially beneficial technologies are one of the most powerful means of redirecting technology in a market economy. Subsidies are more potent when they are bolstered by changes in social norms and consumer preferences that push in the same direction, as demonstrated by our experience with renewable energy.

There are important differences between green and digital technologies, however. When environmental concerns first surfaced, activists did not have a full understanding of how energy consumption was affecting climate or of how the carbon content of energy could be consistently measured. All the same, the scientific understanding and the measurement framework developed rapidly and were in place as early as the 1980s. It then became straightforward to estimate how much greenhouse gases different energy sources were emitting. This knowledge now underpins most carbon taxes, cap-and-trade schemes, and subsidies for renewable energy and electric cars.

Determining how different digital technologies are used and their impact on wages, inequality, and surveillance is much harder. For example, new digital technologies that enable managers to more efficiently track the performance of their subordinates could be viewed as human complementary, for they are enabling managers to do new tasks and expand their capabilities. Simultaneously, they may intensify surveillance or eliminate tasks that used to be performed by other white-collar workers.

Nevertheless, there are a number of principles useful for creating a framework for measuring the impact of digital technology. First, whether new technologies are used for monitoring and surveillance is fairly straightforward to ascertain. Both the development and deployment of these technologies should be discouraged. A government agency such as OSHA could develop clear guidelines that prevent the most intrusive forms of surveillance and data collection on employees, and other agencies could similarly regulate data collection on consumers and citizens. As an additional step, the federal government could also decide not to enforce patents on technologies that are aimed at worker or citizen surveillance—including those filed in China. Conversely, technologies that provide tools for worker and user privacy can also be identified and subsidized.

Second, there is a telltale sign of automation technologies: reducing the labor share of value added, meaning that once these technologies are introduced, how much of value added goes to capital increases and how much gets captured by labor decreases. Existing research documents that the introduction of robots and other automation technologies almost always leads to significantly lower labor share. Likewise, technologies that create new tasks for workers tend to increase the labor share. On this basis, technologies that raise the labor share can be encouraged via subsidies for their use and their development. Such policies can be also useful in encouraging the sharing of productivity gains with workers because higher wage raises would increase the labor share and thus qualify companies for additional subsidies.

Third, subsidies for research directions that are human complementary can be provided on the basis of more detailed data about whether new methods are complementing humans or automating work when used in practice. We have already mentioned several examples where new digital technologies can complement humans by creating new tasks—for example, by providing better information for personalized teaching or health care, or by enabling improved design and production work on shop floors with the help of augmented- and virtual-reality capabilities. Although such a classification may be much easier after technologies are deployed, some of this information is available at the development stage and could be a first step toward a measurement framework for the extent of automation of new technologies. This measurement framework could then be used for providing subsidies to certain lines of innovation.

Some amount of ambiguity in the exact purpose and application of new technologies is not a major problem: preventing automation is not the objective. What policy makers should strive for is cultivating a plurality of approaches to encourage a greater focus on human-complementary and human-empowering technologies. This goal does not require a perfect metric for classifying whether a technology will automate work or create new tasks for workers. Rather, it requires a commitment to experiment with new technologies that try to help workers and citizens.

For the same reasons, we do not support automation taxes aimed at directly discouraging the development and adoption of automation technologies. Redirection should target a more balanced technology portfolio, and subsidies to new human-complementary technologies can achieve this more effectively. Moreover, given the difficulty of distinguishing automation from other uses of digital technologies, automation taxes are currently not practical. Just taxing the clear examples of automation technologies, such as industrial robots, would not be optimal either, for such a policy would leave out the much more pervasive algorithmic automation technologies. Nevertheless, if subsidies and other policies cannot succeed in redirecting technological efforts, automation taxes may need to be considered in the future.

Breaking Up Big Tech. Big businesses have become too powerful, and that is a problem in and of itself. Google dominates search, Facebook has few rivals in social networking, and Amazon is developing a lock on e-commerce. These overwhelming market shares remind us of Standard Oil, which had a 90 percent market share of oil and oil products when it was broken up in 1911, and AT&T, which had a near monopoly on telephone service when it was broken up in 1982.

High levels of market concentration and gargantuan monopolies can choke off innovation and distort its direction. For example, Netscape Navigator created a much better product than Microsoft's browser in the mid-1990s and altered the direction of browsers by spurring a series of follow-on innovations by other companies (it was selected as "the best tech product of all time" by *PC Magazine* in 2007). Unfortunately, Netscape ended up being crushed by Microsoft, despite a Department of Justice antitrust case.

These considerations may be more important today because a handful of companies are dominating the direction of digital technologies and especially AI. Their business models and priorities focus on automation and data collection. Hence, breaking up the largest tech giants to reduce their dominance and create room for greater diversity of innovations is an important part of redirecting technology.

Breakup by itself is not sufficient because it will not redirect technology away from automation, surveillance, or digital advertising. Take Facebook, which would likely be the first target of such antitrust action, owing to its controversial acquisitions of WhatsApp and Instagram. If the company were broken up and these two apps were separated from Facebook, data sharing between them would cease, but their business models would remain intact. Facebook itself would continue to seek the attention of its users and therefore continue to be a platform for the exploitation of insecurities, misinformation, and extremism. WhatsApp and Instagram would also adopt the same business model, unless pushed away from it by regulation or public pressure. The same is likely true of YouTube, even if it were separated from Google's parent company, Alphabet.

Therefore, breakup and more broadly antitrust should be considered as a complementary tool to the more fundamental aim of redirecting technology away from automation, surveillance, data collection, and digital advertising.

Tax Reform. The current tax system of many industrialized economies encourages automation. We saw in Chapter 8 that the US has taxed labor at an average rate of about 25 percent over the last four decades because of payroll and federal income taxes, while imposing much lower taxes on equipment and software capital. Moreover, taxes on these types of investments have fallen steadily since 2000 because of the broader reductions in corporate income taxes and federal income taxes on high earners, and increasingly generous allowances to write off tax obligations when firms invest in machinery and software.

A company investing in automation equipment or software today pays a tax of less than 5 percent—fully 20 percentage points lower than the taxes it faces when it hires workers to perform the same tasks. Specifically, what this means is that when a company hires more workers and pays them $100,000 per year, it and the workers will jointly owe $25,000 in payroll taxes. When it instead buys new equipment costing $100,000 to perform the same tasks, it pays less than $5,000 in taxes. This asymmetry is an impetus to additional automation and is present in similar forms, even if sometimes less pronounced, in the tax codes of several other Western economies.

Tax reform can remove this asymmetry and thus the incentives for excessive automation. A first step to achieve this would be to significantly reduce or even fully eliminate payroll taxes. The last thing we want today is to make it more expensive for people to work.

A second step would be a modest increase in taxes on capital. Eliminating provisions that reduce the effective taxation of capital, such as generous depreciation allowances and the advantageous tax status of private equity and carried interest, would be one way of achieving this. Additionally, moderately higher corporate income taxes would directly increase the marginal tax rates faced by capital owners, further closing the gap between the taxation of capital and labor. It is important to simultaneously close tax loopholes, including schemes that minimize tax liabilities of multinational corporations by shifting their accounting profits across jurisdictions; otherwise, corporate income taxes could be avoided and would not be fully effective.

Investing in Workers. The tax incentives for investments in equipment and software are not available to companies when it comes to investments in workers. Equalizing the rates at which capital and labor are taxed is an important step in removing the bias in favor of automation ahead of hiring and investing in workers.

But there is more that the tax code can do. Worker marginal productivity can be raised by post-schooling training. Even workers with college or post-college degrees learn most of the skills required for a given task or industry once they start working in a company. Some of these training investments take place in formal settings, such as vocational courses, whereas other relevant skills are learned on the job, from senior coworkers and supervisors, a process that is often aided by how jobs are designed and how much time employees are allowed to allocate to training activities. We have seen that the training of low-education workers was an important pillar of shared prosperity before the 1980s.

There are good reasons why the level of training investment that companies choose may be insufficient. Much of what a worker learns via training is "general" in the sense that they could use their skills productively with other employers as well. Investing in general training is less attractive for firms because competition from other employers implies that they would have to pay higher wages or may even lose the worker after training, without being able to recoup their investments. The Nobel Prize–winning economist Gary Becker pointed out how more efficient levels of training could be supported when workers indirectly pay for them by taking a pay cut during training, with the hope that they will be able to enjoy higher wages in the future. This solution is often imperfect, however. Workers may not be able to afford pay cuts and may not trust that firms would really devote enough care and time to training once they do take such cuts. Worse, when wages are negotiated, as they often are, neither the firm nor the worker receives the full returns from training investments, making it impossible even for pay cuts to support adequate levels of training.

Institutional solutions and government subsidies to training could rectify the resulting underinvestment problem. For example, the German apprenticeship system incentivizes firms to fund major training efforts. The programs in many industries last two, three, or sometimes even four years and are made feasible by the fact that workers develop close relations with their employer and do not immediately leave after apprenticeship. These schemes are often supported and supervised by labor unions. Similar apprenticeship programs exist in other countries but would be difficult to institute in the US and the UK, where unions are unlikely to play the same role and where quit rates for younger workers are much higher than in Germany. Government subsidies—for example, allowing companies to deduct training investments from taxable profits—should therefore play a more important role in the United States.

Government Leadership to Redirect Technological Change. The government is not the engine of innovation, yet it can play a central role in redirecting technological change through taxes, subsidies, regulation, and agenda setting. Indeed, in many frontier research areas, the identification of specific need, combined with government leadership, has been critical because it focuses the attention of researchers on specifying attainable goals or aspirations.

That was certainly the case with antibiotics, one of the most transformative technologies of the twentieth century. The importance of drugs that could fight bacteria was already well understood when Alexander Fleming serendipitously discovered the bacteria-killing properties of penicillin at St. Mary's Hospital in London in 1928. Ernst Chain, Howard Florey, and later other chemists built on Fleming's breakthrough to purify and produce penicillin that could be administered to patients. As important as the scientific advances, however, was the demand from the military, especially the US Army. The first successful application of the drug during World War II came in 1942. By D-Day on June 6, 1944, the US military had already procured 2.3 million doses of penicillin. Remarkably, financial incentives played relatively little role in this discovery and development process.

The same combination was important for many postwar scientific breakthroughs where the US government had articulated a strategic need, including for air defense, sensors, satellites, and computers. The recipe often brought together leading scientists to work on the problem and subsequently generated a sizable demand for these technologies, encouraging the private sector to jump in. A variant of this approach led to rapid vaccine development during the COVID-19 pandemic.

A similar combination could be effective in redirecting digital technology. When the social value of new research directions is established, it can draw many researchers in. Guaranteed demand for successful technologies can additionally incentivize private companies. For instance, the US government could convene and fund research teams to develop digital technologies that complement human skills to be used in education and health care, and commit to deploying them in US schools and Veterans Administration hospitals, provided that they meet the requisite technical standards.

This is not, we hasten to point out, traditional "industrial policy," which involves bureaucrats attempting to pick winners, either in terms of companies or specific technologies. Industrial policy's track record is mixed. When it succeeded, it took the form of government inducement for broad sectors, such as the chemical, metal, and machine-tool industries of South Korea in the 1970s or the metal industry of Finland between 1944 and 1952 (because of the in-kind war reparations that the country had to pay to the Soviet Union).

Instead of picking winners, redirecting technology is much more about identifying classes of technologies that have more socially beneficial consequences. In the energy sector, for example, technological redirection requires support for green technologies overall rather than attempts to determine whether wind or solar, let alone what type of photovoltaic panel, is more promising. The type of government leadership we advocate builds on the same approach and seeks to encourage the development of technologies that are more complementary to workers and citizen empowerment rather than trying to select specific technological trajectories.

Privacy Protection and Data Ownership. Controlling and redirecting the technology of the future is in large part about AI, and AI is mostly about incessant data collection from everyone. Two proposals in this realm are worth discussing.

The first is strengthening the protection of user privacy. Massive data collection about users and their friends and contacts has a variety of adverse effects. Platforms harvest these data to manipulate users (which is of course a core part of their ad-based business model). Such data collection also opens the way to nefarious collaboration between platforms and governments wishing to snoop on citizens. Relatedly, so much data in the hands of a few platforms cultivates an imbalance of power between them and their competitors and users.

Stronger privacy protection, requiring platforms to get explicit approval from users on what data they will collect and how they will use it, could be useful. But attempts to implement it—for example, with the European Union's General Data Protection Regulation (GDPR) in 2018—have not been very successful. Many users are not privacy conscious, even when prompted, because they do not understand how data will be utilized against them. Evidence suggests that GDPR disadvantaged smaller companies but was not effective in circumventing data collection and surveillance by large companies, such as Google, Facebook, and Microsoft.

There is another fundamental reason why privacy protection is difficult: platforms obtain information from users about others, either because they are indirectly revealing information about their friends or because they enable the platform to learn more about the specifics of their demographic groups, which can be used for targeting ads or products to others with similar characteristics. This type of "data externality" is often ignored by users.

A related idea, centered on providing ownership rights to users for their data, may be more effective than privacy regulations. Data ownership, originally proposed by the computer scientist Jaron Lanier, can simultaneously protect how user data are harvested and prevent large tech companies from corralling their data as a free input for their AI programs. It can also limit the ability of tech companies to collect vast amounts of data from the web and public records without the consent of the people involved. Data ownership may even, directly or indirectly, discourage ad-based business models.

Part of the goal of data ownership is to ensure that users receive income from their data. However, for many applications, the data of one user is highly substitutable with others' data. For example, from the viewpoint of a platform, there are hundreds of thousands of users who can identify cute cats, and who does so is of no importance. This implies that platforms will have all the bargaining power against users and that even when user data are valuable, the platforms will be able to buy the data on the cheap. This problem is made worse in the presence of data externalities. Lanier recognizes this issue and advocates "data unions," built on the model of the Writers Guild of America, which represents writers providing content for movies, television, and online shows. Data unions can negotiate prices and terms for all users or subgroups, thus circumventing "divide-and-conquer" strategies by platforms, which could otherwise obtain data from one subgroup and then use the data for getting better terms from others. Data unions can also prevent tech giants from using the data that they have collected in one part of their business in order to create an entry barrier in other activities—such as Uber using data from its ride-sharing app to gain an advantage in food delivery (a data-sharing practice that regulators in Vancouver recently tried to prevent).

Data unions could additionally provide models for other types of workplace organizations. They can become powerful civil-society associations and contribute to the emergence of a broader social movement, especially if combined with the other measures we are proposing.

Repeal Section 230 of the Communications Decency Act. Central to the regulation of the tech industry is Section 230 of the 1996 Communications Decency Act, which protects internet platforms against legal action or regulation because of the content they host. As Section 230 explicitly states, "No provider or user of an interactive computer service shall be treated as the publisher or speaker of any information provided by another information content provider." This passage has given protection to platforms such as Facebook and YouTube against accusations that they are featuring misinformation or even hate speech. This is often supplemented by arguments of executives defending freedom of speech on their platform. Mark Zuckerberg was quite categorical on this in a Fox News interview in 2020: "I just believe strongly that Facebook shouldn't be the arbiter of truth of everything that people say online."

Under public pressure, tech platforms have recently taken some steps to limit misinformation and extreme content. But they are unlikely to do much by themselves for a simple reason: their business model thrives on controversial and sensational material. This means that government regulation has to play a role, and a first step in this would be to repeal Section 230 and make platforms accountable when they *promote* such material.

The emphasis here is important. Even with much better monitoring, it would be unrealistic to expect that Facebook can eliminate all posts that include misinformation or hate speech. Yet it is not too much to expect that their algorithms should not give such material a much broader platform by "boosting" it and actively recommending it to other users, and this is what the repeal of Section 230 should target.

We should also add that such a relaxation of Section 230 protection would be most effective for platforms such as Facebook and YouTube that use algorithmic promotion of content, and is less relevant for other social media, such as Twitter, where direct promotion is less relevant. For Twitter, experimenting with different regulation strategies, requiring the monitoring of the most heavily subscribed accounts, may be necessary.

Digital Advertising Tax. Even getting rid of Section 230 is not enough, however, because it leaves unchanged the business model of internet platforms. We advocate a nontrivial digital advertising tax to encourage alternative business models, such as those based on subscription, instead of the currently prevailing model that largely relies on individualized targeted digital advertising. Some companies, such as YouTube, have taken some (albeit halfhearted) steps in that direction. But currently, without a digital ad tax, a subscription-based system is not as profitable. Since digital ads are the most important source of revenues from data collection and the surveillance of consumers, a change of business model can be a powerful tool to redirect technology as well.

Advertisement in general has an important element of an "arms race." Although some advertisements introduce consumers to brands or products they may not be aware of, expanding their choices, much of it simply tries to make their product more appealing than their competitors'. Coca-Cola advertises not to make consumers aware of its brand (it is safe to assume that everybody, at least in the United States, knows of Coca-Cola) but to convince them to buy Coke instead of Pepsi. Pepsi then responds by increasing its own advertisements. For arms race–type activities, when costs decline or potential impact increases, more wastefulness may follow. Digital advertising has taken us into this territory by individualizing ads and increasing their impact while also reducing the cost of advertisements to businesses. This bolsters the economic case for a digital advertisement tax.

Although we do not currently know how high such digital taxes need to be in order to have a meaningful impact on massively profitable business models, we suspect that they have to be significant. Recall that the point of such taxes is not to raise revenues or have a small influence on the volume of advertisement but to alter the business model of online platforms. In any case, some amount of policy experimentation will probably be needed to determine and set the right level of taxes.

Misinformation and manipulation are present off-line as well—for example, on Fox News. Although there may be reasons for extending advertisement taxes to TV, there is a big difference from online platforms: TV channels do not have access to the technology for individualized digital advertisements and do not collect and then use vast amounts of data about the audience.

Other Useful Policies

Policies that do not directly redirect technology are less well suited to the task at hand, but may still be worth considering, especially when they tackle the large inequalities and the excessive political power of companies and their bosses.

Wealth Taxes. Wealth taxes, imposed on those above a certain wealth threshold, have started to gain traction over the last decade. For example, in 1989 President Mitterrand introduced a tax in France on wealth levels above €1.3 million, which was reduced in scope by President Macron in 2017. In the US, Bernie Sanders and Elizabeth Warren, who both ran for president in 2020, have proposed wealth taxes. Sanders's 2020 plan was to impose a 2 percent wealth tax on households with wealth in excess of $50 million, rising gradually to 8 percent for those whose wealth exceeds $10 billion. Warren's most recent proposal is to impose a 2 percent wealth tax on households with wealth above $50 million and a 4 percent tax on those with wealth in excess of $1 billion. Given the vast fortunes that have been made over the last several decades, coupled with the need for additional tax revenues for bolstering the social safety net and other investments (as we detail below), well-administered wealth taxes can raise valuable revenue.

Although wealth taxes would not directly contribute to redirecting technological change, they would be helpful to reduce the wealth gaps that exist in many industrialized nations today. For example, a 3 percent wealth tax would, over time, significantly eat into the fortunes of tech tycoons such as Jeff Bezos, Bill Gates, and Mark Zuckerberg. An important question is whether smaller wealth gaps would also reduce their persuasion power. This would depend on other broader social changes, not just their exact wealth.

Wealth taxes are also difficult to estimate, and taxing in this fashion will multiply trickeries aimed at hiding wealth in trusts and other complex vehicles, sometimes offshore. For this purpose, wealth taxes should be combined with corporate income taxes, imposed directly on company profits, which are easier to assess and collect. At the very least, wealth taxes would have to be coupled with stronger international cooperation among tax authorities, including an overhaul of the rules for offshore tax havens and a concerted effort to close loopholes. Any wealth tax would also need to be embedded within the constraints imposed by the rule of law and democratic politics and clear constitutional guidelines to assuage concerns that such taxes could be used for expropriating certain groups.

On balance, we believe that wealth taxes, if coupled with significant efforts to close tax loopholes and change the accounting industry, could have benefits but are not a major part of the more systemic solutions we are seeking.

Redistribution and Strengthening the Social Safety Net. The US needs a better social safety net and better and more redistribution. Much evidence shows that social safety nets have gotten much weaker in the United States and Britain, and this deficiency contributes to poverty and reduced social mobility. Social mobility today is much lower in the US than in Western European countries.

For example, 85 percent of the income differences between families are eliminated within a generation in Denmark, where children of poor parents tend to get richer. The same number is only about 50 percent in the United States. Bolstering the safety net and improving schools in less advantaged areas have become urgent needs. Such policies need to be supplemented with broader redistributive measures.

Although robust redistribution and improved social safety nets, by themselves, are not going to affect the direction of technology or reduce the power of large tech companies, they can be an effective tool in reducing large inequalities that have emerged in the United States and other industrialized nations.

One specific proposal, popularized by Andrew Yang's Democratic primary campaign in 2020, deserves discussion: universal basic income. UBI, which promises an unconditional dollar amount for every adult, has emerged as a popular policy idea in some left-wing circles, among more libertarian scholars such as Milton Friedman and Charles Murray, and with tech billionaires such as Amazon's Jeff Bezos. Support for the idea is rooted, in part, in the clear inadequacies of the safety net in many countries, including the United States. But it also receives a powerful boost from the narrative that robots and AI are pushing us toward a jobless future. And so, the narrative goes, we need the UBI in order to provide income to most people (and to prevent the pitchfork uprising that some tech billionaires are fearing).

UBI is not ideal for bolstering the social safety net, however, because it transfers resources not just to those who need them but to everybody. In contrast, many of the programs that have formed the basis of twentieth-century welfare states around the world target transfers, including health spending and redistribution, to those in need. Because of this lack of targeting, UBI would be more expensive and less effective than the alternative proposals.

UBI is also likely the wrong type of solution to our current predicament, especially compared to measures aimed at creating new opportunities for workers. There is considerable evidence suggesting that people are more satisfied and are more engaged with their community when they feel that they are contributing value to society. In studies, people not only report improved psychological well-being when they work, compared to just receiving transfers, but are even willing to forgo a considerable amount of money rather than give up work and accept pure transfers.

The more fundamental problem with UBI is not related to psychological benefits of work but to the misguided narrative about the problems facing the world it propagates. UBI naturally lends itself to interpretations of our current predicament that are wrong and counterproductive. It implies that we are inexorably heading toward a world of little work for most people and growing inequality between the designers of more and more-advanced digital technologies and the rest, so major redistribution is the only thing we can do. In this way, it is also sometimes justified as the only way of quelling growing discontent in the population. As we have emphasized, this perspective is wrong. We are heading toward greater inequality not inevitably but because of faulty choices about who has power in society and the direction of technology. These are the fundamental issues to be addressed, whereas UBI is defeatist and accepts this fate.

In fact, UBI fully buys into the vision of the business and tech elite that they are the enlightened, talented people who should generously finance the rest. In this way, it pacifies the rest of the population and amplifies the status differences. Put differently, rather than addressing the emerging two-tiered nature of our society, it reaffirms these artificial divisions.

This all suggests that rather than search for fanciful transfer mechanisms, society should strengthen its existing social safety nets and crucially attempt to combine this with the creation of meaningful, well-paying jobs for all demographic groups—which means redirecting technology.

Education. Most economists' and policy makers' favorite tool for combating inequality is more investment in education. There is some wisdom to this conventional wisdom: schools are critical for worker skills and contribute to society by inculcating its core values among the young. There is also a sense in which schooling is deficient in many nations, especially for students from low socioeconomic backgrounds. In addition, as we have seen, schools are one of the areas where human-complementary AI can be most fruitfully deployed to improve outcomes and create meaningful new jobs. There are parts of the schooling system in the US, such as community colleges and vocational schools, that are due for a major revamp, particularly in order to focus more on skills that will be in greater demand in the future.

Although education by itself is not going to alter the trajectory of technology or reenergize countervailing powers, educational investments can help some of the most disadvantaged citizens who do not have access to good schooling opportunities.

Greater educational investments can help society produce more engineers and computer programmers, who will have higher earnings as a result of their upgraded skills, but we have to bear in mind that there is a limit to how many of these positions will be demanded by companies. Education has an indirect beneficial effect as well, which can help the rest. When there are more engineers and computer programmers, this may increase the demand for other, lower-skill occupations, and less educated workers may also benefit—even if they are not the ones receiving the education and obtaining the coveted programming and engineering jobs. This transmission of prosperity is related to the productivity bandwagon and sometimes works in the hoped-for fashion, but its reach depends on the nature of technology and the extent of worker power. Hence, these indirect effects from education can be more significant when there is some redirection of technology (so that not all lower-skill jobs are automated) and when institutions enable even lower-skill workers to bargain for decent wages.

Finally, we warn against the view that technology should adjust in its own way and the only thing society can do to counter its adverse effects is to educate more of the workforce. The direction of technology, its inequality implications, and the extent to which productivity gains are shared between capital and labor are not inescapable givens; they are societal choices. Once we accept this reality, the case that society should let technology go wherever powerful corporations and a small group of people want, then do its best by trying to catch up with education, seems less compelling. Rather, technology should be steered in a direction that best uses a workforce's skills, and education should of course simultaneously adapt to new skill requirements.

Minimum Wages. Minimum-wage floors can be a useful tool for economies where low-wage jobs are a persistent problem, as in the United States and the United Kingdom. Many economists once opposed minimum wages because of the fear that they would reduce employment: higher wage costs would discourage firms from hiring. The consensus among economists has been shifting, for evidence from many Western labor markets has indicated that moderate-level minimum wages do not reduce employment significantly. In the US the current federal minimum wage is $7.25 per hour, which is very low, especially for workers in urban areas. In fact, many states and cities have their own higher minimum wages. For instance, Massachusetts currently has a minimum wage of $14.25 for employees who do not receive tips.

The evidence also indicates that minimum wages reduce inequality because they increase wages for workers in the bottom quarter of the wage distribution. Modest increases in the federal minimum wage in the United States (for example, in line with the proposals to gradually raise it to $15 per hour) and similar increases in wage floors in other Western nations will be socially beneficial, and we support them.

Nevertheless, minimum-wage hikes are not a systemic solution to our problems. First, minimum wages have their biggest impact on the lowest-paid workers, whereas reducing overall inequality necessitates sharing productivity gains more equitably throughout the population. Second, minimum wages can have only a small role in countering the excessive power of big business and labor markets.

Most importantly, if the direction of technology remains distorted toward automation, higher minimum wages may backfire. As evidence from the COVID pandemic shows, when workers are not available to take jobs at relatively low wages in the hospitality and service sectors, companies have a powerful incentive to automate work. Hence, in the age of automation, minimum wages can have unintended consequences—unless accompanied by a broader redirection of technology.

This motivates our perspective that the minimum wage is most useful as part of a broader package aimed at redirecting technology away from automation. If technology can become more worker friendly, businesses would be less tempted to automate work as soon as they face higher wages. In such a scenario, when faced by higher wages, employers may also choose to invest in worker productivity—for example, with training or technological adjustments. This reiterates our overall conclusion that redirecting technological change and making corporations view workers as an important resource is critical. If this can be achieved, minimum wages can be more effective and less likely to backfire.

Reform of Academia. Last but not least is need for reform in academia. Technology depends on vision, and vision is rooted in social power, which is largely about convincing the public and decision makers of the virtues of a particular path of technology. Academia plays a central role in the cultivation and exercise of this type of social power because universities build the perspectives, interests, and skills of millions of talented young people who will work in the technology sector. In addition, top academics often work with leading tech firms and also directly influence public opinion. We would therefore benefit from a more independent academia. Over the last four decades, academics in the United States and other countries have started losing this independence because the amount of corporate money has skyrocketed. For example, many academics in computer science, engineering, statistics, economics, and physics departments—and, of course, business schools—in leading universities receive grants and consulting engagements from tech companies.

We believe that it is imperative to require greater transparency of such funding relationships and potentially establish some limits to restore greater independence and autonomy to academia. More government funding for basic research would also remove the dependence of academics on corporate sponsors. Nevertheless, obviously, academic reform will not redirect technology by itself and should be viewed as a complementary policy lever.

The Future Path of Technology Remains to Be Written

The reforms we have outlined are a tall order. The tech industry and large corporations are politically more influential today than they have been for much of the last hundred years. Despite scandals, tech titans are respected and socially influential, and they are rarely questioned about the future of technology—and the type of "progress"—they are imposing on the rest of society. A social movement to redirect technological change away from automation and surveillance is certainly not just around the corner.

All the same, we still think the path of technology remains unwritten.

The future looked bleak for HIV/AIDS patients in the late 1980s. In many quarters they were viewed as perpetrators of their own fate, not as innocent victims of a deadly disease, and they did not have any strong organizations or even any national politicians defending their cause. Although AIDS was already killing thousands of people around the world, there was very little research for a treatment or a vaccine against the virus.

This all changed during the subsequent decade. First there was a new narrative, showcasing the plight of tens of thousands of innocent people who were suffering from this debilitating, deadly infection. This was led by the activism of a few people, such as playwright, author, and film producer Larry Kramer and author Edmund White. Their campaigns were soon joined by journalists and other media personalities. The 1993 movie *Philadelphia* was one of the first big-screen depictions of the problems of HIV-positive gay Americans, and it had a major impact on the perceptions of moviegoing audiences. TV series tackling similar issues followed.

As the narrative changed, gay-rights and HIV activists started organizing. One of their demands was more research into cures and vaccines for HIV. This was initially resisted by US politicians and some leading scientists. But organizing paid off, and soon there was an about-face by lawmakers and the medical policy establishment. Millions of dollars started pouring into HIV research.

Once the money and societal pressure built up, the direction of medical research altered, and by the late 1990s, there were new drugs that could slow down AIDS infections, as well as novel therapeutics, including early stem-cell treatments, immunotherapies, and genome-editing strategies. By the early 2010s, an effective cocktail of drugs was available to contain the spread of the virus and provide more normal life conditions for most infected people. Several HIV vaccines are now in clinical trials.

What seemed impossible was achieved fairly rapidly in the fight against HIV/AIDS, as it was in renewable energy. Once the narrative changed and people became organized, societal pressure and financial incentives redirected the path of technological change.

The same can be done for the future direction of digital technologies.

Bibliographic Essay

Part I: General Sources and Background

In Part I of this essay, we explain how our approach relates to past work and theories. Detailed sources for data, facts, quotations, and other material are provided in Part II. Throughout Part II we also highlight work that has particularly inspired our approach to specific topics.

Our conceptual framework differs from conventional wisdom in economics and much of social sciences in four critical ways: first, how productivity increases affect wages and thus the validity of the productivity bandwagon; second, the malleability of technology and importance of choice over the direction of innovation; third, the role of bargaining and other noncompetitive factors in wage setting and how these affect the way in which productivity gains are or are not shared with workers; and fourth, the role of noneconomic factors—in particular, social and political power, ideas and vision—in technology choices. The first of these is explicitly discussed in Chapter 1, whereas the other three are more implicit. Here, we provide some additional background on these notions, emphasizing how they build on and differ from existing contributions. We also highlight how, based on these ideas, our interpretation of the major technological transitions in history differs from past work. Finally, we relate our approach to a few recent books on technology and inequality.

We start with the four building blocks that distinguish our conceptual framework from past approaches.

First, with competitive labor markets, wages are determined by the *marginal productivity of labor*, as we discuss in Chapter 1. Most common approaches in economics relate this marginal productivity to *average productivity* (output or value added per worker) and hence generate the prediction that the average wage varies with average productivity (or, simply, productivity). As a result, when productivity increases, average wages increase as well—what we dub the "productivity bandwagon."

Although the term *productivity bandwagon* is not used in standard textbooks, the ideas that it captures are common. Most of the models covered in textbooks on economic growth (including Barro and Sala-i-Martin 2004, Jones 1998, and Acemoglu 2009) imply that higher productivity directly translates into higher wages. Seminal contributions on technological progress, such as Solow (1956), Romer (1990), and Lucas (1988), maintain that technological progress will lift all living standards.

The most popular textbook for undergraduates today, Gregory Mankiw's *Principles of Economics*, states that "almost all variation in living standards is attributable to differences in countries' *productivity*—that is, the amount of goods and services produced by each unit of labor input" (Mankiw 2018, 13, italics in original). Mankiw then links productivity to technological change and gives a succinct statement of the productivity bandwagon. In a section called "Why Productivity Is Important," he explains that living standards are determined by productivity, which depends on technology, and writes that "Americans live better than Nigerians because American workers are more productive than Nigerian workers" (518–519). He also declares this observation to be one of the ten most important principles of economics. Mankiw recognizes the possibility of job losses but frames the issue this way: "It is also possible for technological change to reduce labor demand. The invention of a cheap industrial robot, for instance, could conceivably reduce the marginal product of labor, shifting the labor-demand curve to the left. Economists call this *labor-saving* technological change. History suggests, however, that most technological progress is instead *labor-augmenting*" (Mankiw 2018, 367, italics in original).

The rise in wages implied by the productivity bandwagon need not be one-to-one, so productivity growth can raise the capital share and reduce the labor share in national income. But in the standard view it will always benefit workers. When there are multiple types of labor (for example, skilled and unskilled), technological progress can raise inequality, but it will also increase the wage level of all types of labor. As a result, although technological change can bring inequality, it will be a tide that lifts all boats. For example, as explained in Acemoglu (2002b), in the most common framework used in economics, technological progress always increases the average wage, and even if it raises inequality, it also raises wages at the bottom of the distribution.

These results are a consequence of the type of model that most economists focus on, which assumes that technological changes directly raise the productivity of either capital or labor or both—in other words, in the terminology of economics, technological change is either "labor-augmenting" or "capital-augmenting" (see Barro and Sala-i-Martin 2004 and Acemoglu 2009 for an overview of standard growth models and the forms of technological change). With these types of technological change and under the assumption that there are "constant returns to scale" (so that doubling capital and labor doubles output), there is indeed a tight relationship between productivity and wages of all types of labor.

The fundamental problem is that automation, which we argue to have been critical during many stages of modern industrialization, does not correspond to an increase in the productivity of capital or labor. Rather, it involves the substitution of machines (or algorithms) for tasks previously performed by labor. Advances in automation technology can increase average productivity and at the same time reduce average real wages. Furthermore, technology's inequality implications can be much more amplified when automation encroaches on the tasks performed by low-skill workers, reducing their real wages while raising the returns to capital and the wages of higher-skilled labor (Acemoglu and Restrepo 2022).

It is important to emphasize that automation *can*—but does not *necessarily*—reduce wages. Theoretically, it displaces workers from the tasks they used to perform and thus is predicted to always reduce the labor share in value added (how much of total production value goes to labor as opposed to capital). This prediction is borne out empirically (see, for example, Acemoglu and Restrepo 2020a and Acemoglu, Lelarge, and Restrepo 2020). As mentioned briefly in Chapter 1, if automation raises productivity by enough, it can increase the demand for labor and real wages, even as it displaces workers and reduces the labor share. This can happen because lower costs (higher productivity) encourage automating firms to hire more workers into nonautomated tasks. This type of high-productivity automation also increases the demand for the products of other sectors, either through the demand for inputs from the firms installing automation technology or because the real incomes of consumers increase owing to the cheaper products of these firms. Critically, however, these benefits will not occur when automation is "so-so," meaning that it increases productivity only by a little (see our discussion below and in the context of Chapter 9). Another key part of our conceptual framework, the role of new tasks in generating opportunities for workers and counterbalancing automation, is also distinct from most approaches in economics.

Our overall approach builds on a number of prior contributions in the economics literature. Atkinson and Stiglitz (1969) proposed a model of technological change that differed from the conventional wisdom in allowing for innovations to affect productivity "locally"—meaning only at the prevailing capital-labor ratio. The first work that proposed a theory based on machines substituting for labor in certain activities was Zeira (1998). A related approach was developed in Acemoglu and Zilibotti (2001). This idea was further investigated and developed in the seminal work by Autor, Levy, and Murnane (2003), who proposed the mapping of tasks into routine and nonroutine categories and argued that it was routine activities that could be automated. Autor, Levy, and Murnane (2003) also undertook the first systematic empirical analysis of automation, demonstrating that it was closely related to the increase in inequality in the United States. Acemoglu and Autor (2011) developed a general task-based model and derived the wage and employment polarization implications of automation.

Our framework in this book most closely follows Acemoglu and Restrepo (2018 and 2022). The 2018 paper introduced a model in which economic growth takes place via a process of automation and new task creation, and identified conditions under which technological progress and productivity growth reduce wages. This paper also proposed the idea of new tasks as key elements potentially counterbalancing the effects of automation, and modeled how the simultaneous expansion of automation and new tasks affects the evolution of labor demand. This modeling clarifies that automation is not necessarily bad for wages or inequality but has adverse effects when the adoption of more worker-friendly types of technologies lags behind the rate of automation. The 2022 paper presents a general, multisector framework in which the distributional and wage implications of different types of technologies can be systematically measured. It also provides evidence showing that automation has been the major cause of the widening inequality trends in the US economy. This paper further underpins our discussion in Chapter 1 on how sufficiently large productivity increases can trigger employment and wage growth— for example, by inducing other sectors to expand.

This framework is also the basis of our discussion of "so-so automation" or "so-so technology" (a term introduced in Acemoglu and Restrepo 2019b). In particular, when some tasks that used to be performed by labor are automated but the cost reductions (productivity increases) are limited, this technological change generates significant worker displacement but little in the way of a productivity bandwagon. So-so automation is more likely to appear when human labor is fairly productive in the tasks that are being automated and machines and algorithms are not very productive. Excessive automation—which goes beyond what would be efficient from a pure production viewpoint and may thus even reduce correctly measured productivity—is then so-so by definition. The reference to "correctly measured productivity" is because automation always mechanically increases output per worker by reducing the need for labor in production, but it may reduce total factor productivity, which takes into account the contribution of both labor and capital, as explained in Chapter 7.

Second, most theories of economic growth either take the path of technological change as exogenous, as in Solow (1956), or endogenize the rate of innovations but assume that these take place along a given trajectory, as in Lucas (1988) or Romer (1990). Incidentally, both these lines of work introduce technology in the same way—as directly increasing labor's productivity—and this is the reason why they affirm the productivity bandwagon.

Our conceptual framework differs by emphasizing the malleability of technology and the fact that the direction of technological change—for example, how much new techniques will economize on different factors and how they will change their productivities—is a choice. Here we also build on a number of previous works. The first economist to discuss these issues was Hicks (1932), who conjectured that higher labor costs induce firms to adopt technologies that save on labor. Related ideas were developed by the "induced-innovation" literature of the 1960s, including among others Kennedy (1964), Samuelson (1965), and Drandakis and Phelps (1966), although those contributions mostly focused on whether there are natural reasons for technological change to keep the capital and labor shares in national income constant.

The first major empirical application of these ideas was by Habakkuk (1962), in the context of nineteenth-century American technology. Habakkuk's main argument was in line with Hicks's claim: scarcity of labor and especially skilled labor in America was a trigger for the rapid adoption and development of labor-saving machinery, as we discuss in Chapter 6. Robert Allen (2009a) proposed the related idea that the high cost of labor was a major cause of the onset of the British industrial revolution in the mid-eighteenth century. Our interpretation of late nineteenth-century technological developments in the United States draws heavily on Habakkuk's thesis, and we further argue that this induced direction of technology persisted into the first half of the twentieth century and spread to Britain and other industrializing nations as well.

Our theory also builds on the more recent literature on directed technological change, which starts with Acemoglu (1998, 2002a) and Kiley (1999). These papers focused on inequality implications, but subsequent work explored other dimensions of technological malleability, including general issues related to division of national income between labor and capital in Acemoglu (2003a), the effects of international trade and labor market institutions on inequality in Acemoglu (2003b), and the causes and consequences of inappropriate technology in Acemoglu and Zilibotti (2001) and Gancia and Zilibotti (2009). There is now a sizable empirical literature inspired by these ideas. Relevant works include those focusing on the direction of pharmaceutical research in Finkelstein (2004) and Acemoglu and Linn (2004); climate change and green technologies in Popp (2002) and Acemoglu, Aghion, Bursztyn, and Hemous (2012); textile innovations during the British industrial revolution in Hanlon (2015); and agriculture in Moscona and Sastry (2022). Whether the direction of technology saves on labor or complements labor is explored theoretically in Acemoglu (2010) and Acemoglu and Restrepo (2018).

We extend these approaches in conceptual, empirical, and historical directions. Conceptually, we emphasize the role of political and social factors in shaping the direction of technology, whereas the previous literature mostly focused on economic factors. In Acemoglu and Restrepo (2018), for example, the direction of technological change is determined by purely economic factors, such as the labor share in national income, the long-run price of capital, and labor market rents.

Another implication of these ideas, briefly mentioned in chapters 1 and 8, is worth emphasizing here: the malleability of technology opens the door to socially costly choices regarding the direction of innovation. In fact, when there are major decisions about the direction of technology, there is no guarantee that the market-based innovation process will select areas that are more beneficial for society as a whole or for workers. One reason for this is that some types of technologies may generate more profits for businesses than others, even if they do not contribute to or may even reduce social welfare. Examples include technologies that increase the productivity and dominance of monopolies or large oligopolies (which can charge higher prices and make greater profits), those that help companies to better monitor workers and thus increase profits by reducing wages, and those that are complementary to data collection and lock in the power of companies that monopolize data. An even more important reason for distortions in the direction of innovation, pointed out in Acemoglu and Restrepo (2018), is that firms may have an excessive demand for automation technologies, particularly when this enables them to economize on high wages. Innovation distortions potentially multiply when there are noneconomic factors influencing technology choices—for example, when the vision of influential individuals, entrepreneurs, and organizations determines major investments (as with the US tech sector at the moment) or when a powerful government demands and pushes innovators toward surveillance technologies (as with the policies of the Chinese government, discussed in Chapter 10).

From the empirical and historical viewpoint, we provide an account of the distributional consequences of economic growth for the last thousand years, focusing especially on the direction of industrial technologies from the middle of the eighteenth century to today. We are not aware of other precursors to our interpretation and historical evidence, which emphasize the following: how the balance between automation technologies and those that are more worker-friendly was first forged during early industrialization; then transformed in a more worker-friendly direction in the second half of the nineteenth century, persisting into the first eighty years of the twentieth century; and subsequently changed again since 1980, once more in an automation-focused direction. Partial exceptions are Acemoglu and Restrepo's (2019b) exploration of the extent of displacement and reinstatement of labor in the US economy since 1950, and Brynjolfsson and McAfee's (2014) book and Frey's (2019) more recent book, which we discuss below.

Third, most economic approaches, even when they recognize important deviations from the benchmark of competitive labor markets (for example, because of the power of firms to set wages, bargaining, or informational problems), do not emphasize these as central determinants of whether productivity increases will translate into wage growth. For example, the canonical approach in modern economics that incorporates labor market rents and frictions originates from the work by Diamond (1982), Mortensen (1982), and Pissarides (1985); as highlighted in Pissarides's (2000) leading treatise on the subject, *Equilibrium Unemployment Theory*, it predicts that productivity growth will translate into wage growth one-for-one.

In contrast to these approaches, we make the extent and nature of rent sharing an essential feature of how gains from productivity growth will be divided. Important precursors of our approach include Brenner's (1976) critique of neoclassical and neo-Malthusian theories of the collapse of feudalism. Brenner singled out the role of political power in the functioning and the end of feudalism. According to Brenner, demographic factors were secondary, and what mattered most was whether peasants had enough power to resist the demands of lords. Brenner's approach was a major inspiration for the theory of Acemoglu and Wolitzky (2011), on which we build. In their theory, productivity improvements can reduce rather than increase wages because employers can decide to intensify coercion (for example, hire more guards or make investments that prevent workers from quitting) instead of paying their employees more. Whether this happens or not is determined by the institutional context and the outside options of workers (for instance, whether despite employer coercive measures they can flee and find an alternative means of subsisting). Some of these implications can be extended to noncoercive environments. For example, when the balance of power in bargaining between firms and workers is held fixed, a new technology that increases productivity will raise wages. However, new technologies can also change the balance of power against labor, and if so, wages may decline. Alternatively, technological change can alter the trade-off between building goodwill and high morale among workers relative to monitoring them closely, and this can again break the link between high productivity and high wages.

Our present approach generalizes these perspectives, particularly in Chapter 4, which discusses agricultural economies. It then focuses on the role of technological change in such a framework, and in chapters 6, 7, and 8 it develops similar ideas that apply to rent sharing in modern economies. These ideas are then combined with two other notions that are also typically ignored in discussions of the effects of technology on wages. The first, proposed in Acemoglu (1997) and Acemoglu and Pischke (1999), is the possibility that in the presence of rent sharing, higher wages can sometimes increase investment in worker marginal productivity because firms find it more profitable to raise worker productivity. The second, proposed in Acemoglu (2001), points out that higher protection for workers can incentivize employers to create "good jobs" (with higher wages, greater job security, and career-building opportunities), and good jobs contribute to wage growth. These ideas help us understand why during certain episodes rent sharing went together with rapid wage growth and broadly shared prosperity (chapters 6 and 7), and how the weakening of worker power can be associated with less shared growth and less investment in worker-friendly technologies (Chapter 8).

Fourth, we offer a theory of vision of technology and the role of social power in shaping such visions. Specifically, we emphasize that once the malleability of technology and the lack of an automatic productivity bandwagon are recognized, the question of what determines the direction of technology, and thus who wins and who loses, becomes central. The key factors we focus on in this context include who has persuasion power and whose vision becomes influential.

Our emphasis on the role of economic and social power links us to the large and still growing literature on institutions, politics, and economic development. Here, we are building on the works by North and Thomas (1973), North (1982), North, Wallis, and Weingast (2009), and Besley and Persson (2011), as well as our own earlier work—Acemoglu, Johnson, and Robinson (2003, 2005b), Acemoglu and Johnson (2005), and Acemoglu and Robinson (2006b, 2012, and 2019)—and Brenner's (1976) ideas, already mentioned above. We add social factors related to visions and ideas, persuasion, and status to these theories, stressing the interplay of politics and economics. In this, we build on Mann's (1986) seminal book on the sources of social power and his distinction among economic, military, political, and ideological power. Relative to Mann, we stress the critical role of persuasion power, especially in modern societies, and also emphasize how persuasion power is shaped by institutions. In addition, our discussion of the sources of persuasion power is inspired by the social psychology literature on how persuasion works, summarized in Cialdini (2006) and Turner (1991).

Beyond these foundational differences, the way that we conceptualize the role of political and social factors in technological change is different from most existing approaches. In both economics and much of the rest of social science, because the malleability of technology is not considered, the main emphasis has been on whether institutions and social forces block technological change. This perspective was first articulated systematically in Mokyr (1990) and was modeled in economics by, among others, Krusell and Ríos-Rull (1996) and Acemoglu and Robinson (2006a).

An additional implication of these considerations is the greater room for agency and choice that they create among powerful actors. In the simplest political economy approaches, institutional factors work primarily by changing market incentives and technology, and wage policies of firms are largely dictated by profit maximization. This is no longer the case when ideas and visions matter. In this case, as influential visions shift, there can be major changes in the direction of innovation and rent-sharing patterns, altering how productivity gains are distributed within society.

Our framework combines these four building blocks. To the best of our knowledge, how political and social power shapes technological choices, and how institutions and technology choices together determine how much owners of capital, entrepreneurs, and workers of different skill levels benefit from new production methods, are original to this book. Using this framework, we reinterpret the major economic developments of the last thousand years.

Recent and important contributions in this context include Brynjolfsson and McAfee (2014) and Frey (2019). Brynjolfsson and McAfee (2014) discussed issues related to our focus almost a decade ago and anticipated many of the labor market disruptions that would follow from the next wave of AI technologies, although their interpretation is more optimistic than ours. Both their book and Frey's recognize the displacing effects of automation and some of the social and economic costs that these impose, and Frey vividly describes some of these costs in the context of the economic developments of the nineteenth and twentieth centuries, as we do. Specifically, Frey builds on the framework of Acemoglu and Restrepo (2018) and emphasizes the possibility that technology may either automate or increase worker productivity. However, he does not allow the direction of technology to be determined by institutions and social forces, and his main concern, like that of Brynjolfsson and McAfee (2014) and Mokyr (1990), remains the possibility that inequality and wage-level implications of automation technologies can lead to the blocking of progress.

In contrast, the framework in this book emphasizes that resistance to automation technologies is not always an impediment to economic growth; it can also be socially beneficial when it redirects innovation away from paths that have negative effects on workers and toward more worker-friendly directions (or away from those that disrupt democratic participation toward those that empower broader social groups). Because these positive effects of resistance and political reaction from workers and other segments of society are missing in Frey's framework, Frey views them as negatives, and his policy recommendations are likewise about preventing such resistance—for example, by redistributing gains resulting from automation or increasing education.

In this context, we should also relate our book to two other recent contributions, West (2018) and Susskind (2020). These authors also worry about the negative implications of automation, and especially AI, but do not recognize the directed nature of technology. Moreover, they stress, contrary to our emphasis, that AI is already a very capable technology that will quickly replace many jobs. This makes them view a future with fewer jobs as inevitable and thus favor measures such as universal basic income to combat the negative implications of these inexorable technological trends. This is sharply different from our perspective. Specifically, we emphasize (in chapters 9 and 10) that many uses of current AI are so-so, precisely because the capabilities of machine intelligence are more limited than sometimes presumed and because humans perform many tasks drawing on large amounts of accumulated expertise and social intelligence. Nevertheless, so-so automation technologies can still be adopted, and in this case tend to be damaging to workers, without generating major productivity gains or cost reductions for companies (see Acemoglu and Restrepo 2020c, Acemoglu 2021). As a result, and in contrast to West's and Susskind's emphasis, our book argues that the main issue is the redirection of technological change away from a singular focus on automation and data collection toward a more balanced portfolio of new innovations.

Part II: Sources and References, by Chapter

Epigraph

"If we combine . . ." is from Wiener (1949).

Prologue: What Is Progress?

Jeremy Bentham, "You will be surprised . . .," is from Steadman (2012), with details in his note 7. This is from a letter from Bentham to Charles Brown in December 1786. For context and details, see Bentham (1791).

"No man would like" appears in Select Committee (1834, 428, paragraph 5473), testimony of Richard Needham on July 18, 1834, and also appears in Thompson (1966, 307). "I am determined . . ." is from Select Committee (1835, 186, paragraph 2644), testimony of John Scott on April 11, 1835, and also appears in Thompson (1966, 307). "In consequence of better machinery . . ." is from Smith (1776 [1999], 350). "Laws of nature," is from Burke (1795, 30). The full sentence reads: "We, the people, ought to be made sensible, that it is not in breaking the laws of commerce, which are the laws of nature, and consequently the laws of God, that we are to place our hope of softening the Divine displeasure to remove any calamity under which we suffer, or which hangs over us."

"The fact is, that monopoly . . ." is from Thelwall (1796, 21), and a partial version is in Thompson (1966, 185).

Chapter 1: Control over Technology

It is useful to briefly review the historical debates surrounding the notion of technological unemployment and David Ricardo's views on machinery, which are discussed in this chapter.

The idea of technological unemployment resulting from improvements in production methods is often attributed to John Maynard Keynes (1930 [1966]). In reality, this idea significantly predates Keynes. Several authors in the eighteenth century worried about labor-displacing technological change. Thomas Mortimer wrote about this possibility in the early stages of the Industrial Revolution (Mortimer 1772). One of the leading economists of the era, James Steuart, studied these issues as well, recognizing that machinery may "force a man to be idle," although he viewed this as the less likely scenario (Steuart 1767, 122). Peter Gaskell emphasized these dangers more vividly in the early 1800s: "The adaptation of mechanical contrivances to nearly all the processes which have as yet wanted the delicate tact of the human hand, will soon either do away with the necessity for employing it, or it must be employed at a price that will enable it to compete with mechanism" (Gaskell 1833, 12).

Prominent economists were less worried, at least at first. In *An Inquiry into the Nature and Causes of the Wealth of Nations*, Adam Smith (1776 [1999]) viewed technological improvements to be beneficial broadly. For example, as we saw in the Prologue, he argued that "better machinery" tends to increase real wages "very considerably."

As we discuss at the beginning of Chapter 1, this optimism was initially shared by the other foundational figure of the discipline of economics from this era, David Ricardo. In his *Principles of Political Economy*, first published in 1817, Ricardo drew a parallel between machinery and foreign trade, viewing both as beneficial. He wrote, for example, that "the natural price of all commodities, excepting raw produce and labour, has a tendency to fall, in the progress of wealth and population; for though, on the one hand, they are enhanced in real value, from the rise in the natural price of the raw material of which they are made, this is more than counterbalanced by the improvements in machinery, by the better division and distribution of labour, and by the increasing skill, both in science and art, of the producers" (Ricardo 1821 [2001], 95).

However, Ricardo later changed his mind. He added a chapter, "On Machinery," to the third edition of the *Principles*, articulating a first version of the theory of technological unemployment. Here he wrote that "all I wish to prove, is, that the discovery and use of machinery may be attended with a definition of gross produce; and whenever that is the case, it will be injurious to the labouring class, as some of their number will be thrown out of employment, and population will become redundant, compared with the funds which are to employees" (Ricardo 1821 [2001], 286). But his ideas did not sway most of his followers. Even when economists noted the possibility of such negative effects on laborers or unskilled workers, they concluded that these were unlikely or could be at most temporary. For example, as John Stuart Mill stated, "I do not believe that . . . improvements in production are often, if ever, injurious, even temporarily, to the labouring classes in the aggregate" (Mill 1848, 97).

Similar fears about technological unemployment were expressed by a number of other prominent economists, most importantly by Wassily Leontief, whom we cite in Chapter 8. The history of these early debates is covered in Berg (1980) and Hollander (2019). Frey (2019) and Mokyr, Vickers, and Ziebarth (2015) also include detailed discussions.

Keynes's essay was more optimistic than Ricardo's chapter "On Machinery." In the same essay, he wrote: "For many ages to come the old Adam will be so strong in us that everybody will need to do some work if he is to be contented. We shall do more things for ourselves than is usual with the rich to-day, only too glad to have small duties and tasks and routines. But beyond this, we shall endeavor to spread the bread then on the butter—to make what work there is still to be done to be as widely shared as possible. Three-hour shifts or a fifteen-hour week may put off the problem for a great while" (1930 [1966], 368–369). He also followed the statement we provide in the text with this line: "But this is only a temporary phase of maladjustment. All this means in the long-run *that mankind is solving its economic problem*" (364, italics in original).

Despite Keynes's stature in the profession, his views on technological unemployment, like those of Ricardo before him, did not have a major impact on the mainstream. Paul Douglas (1930a, 1930b) discussed technological unemployment independently of Keynes, at the same time or even before him. But Douglas, like Gottfried Haberler (1932), argued that the market mechanism would almost automatically restore employment even if machinery displaced some workers from their jobs. Indeed, until recently the economics mainstream did not even pay much attention to the concerns of Ricardo, Keynes, and Leontief.

Finally, the concept of general-purpose technology introduced in this chapter goes back to David (1989), Bresnahan and Trajtenberg (1995), Helpman and Trajtenberg (1998), and David and Wright (2003). Its importance for us stems from the fact that choice over the direction of technology is particularly relevant when technologies are general purpose, as emphasized in Acemoglu and Restrepo (2019b).

Opening epigraphs. Bacon (1620 [2017], 128); Wells (1895 [2005], 49).

"The 340 years that have . . ." is from *Time* (1960), page 2 of the online version. "I can imagine no period . . ." is from Kennedy (1963). "This means unemployment . . ." is from Keynes (1930 [1966], 364).

"Machinery did not lessen the demand for labour" is from Ricardo (1951–1973, 5:30), an edited version of the Hansard record for December, 16, 1819. "It is more incumbent on me to declare my opinion . . ." is from Ricardo (1821 [2001], 282). "If machinery could do all the work . . ." is from Ricardo (1951–1973, 8:399–400, letter dated June 30, 1821).

Bill Gates, "The [digital] technologies . . .," is from an event at Stanford University on January 28, 1998 (no online version currently available). Steve Jobs, "Let's go and invent . . .," is from a 2007 conference (https://allthingsd.com/20070531/d5-gates-jobs-transcript). Labor market developments, including wage inequality by education, are examined in more detail in Chapter 8; see the notes for that chapter for details on our sources and calculations.

The Bandwagon of Progress. "What can we do . . ." is from a TED talk by Erik Brynjolfsson in April 2017 (www.techpolicy.com/Blog/April-2017/Erik-Brynjolfsson-Racing-with-the-Machine-Beats-R.aspx). Automotive industry facts are from McCraw (2009, 14, 17, 23). Auto industry employment in the 1920s is from CQ Researcher (1945). The evolution of tasks in the auto industry is discussed further in chapters 7 and 8; full sources are in the notes for those chapters. The statement on the factory of the future is commonly attributed to Warren Bennis. However, a closer examination (https://quoteinvestigator.com/2022/01/30/future-factory) indicates that "Warren Bennis did employ this joke in 1988 and 1989, but he disclaimed authorship as indicated further below" and that a reasonable assessment is "Bennis deserves credit for helping to popularize the joke."

Why Worker Power Matters. Educational attainment of US workers for 2016 is from the Bureau of Labor Statistics, included in Brundage (2017).

Optimism, with Caveats. The discussion of the heliocentric system and its acceptance is covered in https://galileo.ou.edu/exhibits/revolutions-heavenly-spheres-1543. On Moderna's vaccine development, see www.bostonmagazine.com/health/2020/06/04/moderna-coronavirus-vaccine. On February 24, 2020, Moderna announced it had shipped the first batch of mRNA-1273 forty-two days after sequence identification. For steam engines, see Tunzelmann (1978). On the social credit system in China, see www.wired.co.uk/article/china-social-credit-system-explained. On the 2018 Facebook algorithmic change, see www.wsj.com/articles/facebook-algorithm-change-zuckerberg-11631654215.

Fire, This Time. This interpretation of the evidence from Swartkrans is from Pyne (2019, 25). Sundar Pichai, "AI is probably . . .," is from https://money.cnn.com/2018/01/24/technology/sundar-pichai-google-ai-artificial-intelligence/index.html. Kai-Fu Lee, "AI could be . . .," is from Lee (2021). Demis Hassabis, "[By] deepening our capacity," is from https://theworldin.economist.com/edition/2020/article/17385/demis-hassabis-ais-potential; "Either we need . . ." is from www.techrepublic.com/article/google-deepmind-founder-demis-hassabis-three-truths-about-ai. "The intelligent revolution . . ." is from Li (2020). On Ray Kurzweil's ideas, see Kurzweil (2005). Reid Hoffman, "Could we have a bad . . .," is from www.city-journal.org/html/disrupters-14950.html.

Chapter 2: Canal Vision

This chapter draws on the following histories: Wilson (1939), Mack (1944), DuVal (1947), Beatty (1956), Marlowe (1964), Kinross (1969), Silvestre (1969), McCullough (1977), Karabell (2003), and Bonin (2010). The emphasis of this chapter—that the Panama Canal debacle was rooted in Lesseps's social power and vision, which became amplified because of the success of the Suez Canal—is based on our reading of those sources and the specific items mentioned below.

The debate at the Paris Congress of 1879 was reported by Ammen (1879), Johnston (1879), and Menocal (1879). Lesseps (1880) and (1887 [2011]) provided his own spin on events. The Napoleonic episode is covered by Chandler (1966) and Wilkinson (2020). Saint-Simon's writings are in Manuel (1956). The "spirit of Saint-Simon" in the Panama project is suggested by Siegfried (1940, 239).

Opening epigraphs. Lewis (1964, 7); Ferdinand de Lesseps from DuVal (1947, 58).

Lesseps's statements and actions at the 1879 Congress are from Johnston (1879) and Ammen (1879), neither of whom was particularly sympathetic. Mack (1944, Chapter 25) has details on the work done by various committees and the complaints of American delegates. The *Compte Rendu des Séances* of the Congrès International d'Études du Canal Interocéanique (1879) is the official record of plenary sessions and work by the individual commissions.

Lesseps, "*à l'Américaine,*" is from Johnston (1879, 174), a colorful first-hand account. Unlike Ammen, Menocal, or Lesseps himself, he seems a bit more dispassionate. Mack (1944, 290) reports a more elegant version from the official transcript: "I ask the congress to conduct its proceedings in the American fashion, that is with speed and in a practical manner, yet with scrupulous care. . . ."

We Must Go to the Orient. "The general in chief of the Army of the Orient . . ." is from Karabell (2003, 20). Casualties at the "Battle of the Pyramids" are from Chandler (1966, 226), which says the French suffered "a nominal loss of 29 killed and perhaps 260 wounded."

Capital Utopia. The Saint-Simon quotation is from Taylor (1975). For more discussion, see also Chapter 25, "The Natural Elite," in Manuel (1956). The Enfantin quotation is from Karabell (2003, 205).

Lesseps Finds Vision. Erie Canal details are from Bernstein (2005). Karabell (2003) has the early history of discussion around building the Suez Canal. Lesseps's early efforts are in Wilson (1939), Beatty (1956), Marlowe (1964), Kinross (1969), Silvestre (1969), and Karabell (2003). The "Men of genius" point is highlighted by McCullough (1977, 79).

Little People Buy Small Shares. "The names of those Egyptian sovereigns . . ." is from Lesseps (1887 [2011], 170–175). A slightly different translation is in Karabell (2003, 74): "The Names of the Egyptian sovereigns who erected the Pyramids, those useless monuments of human pride, will be ignored. The name of the prince who will have opened the grand canal through the Suez will be blessed century after century for posterity." Financial details on the share offering are in Beatty (1956, 181–183), which includes this line from the prospectus: "The capital of the Company is limited to 200 million francs apportioned as between 400,000 shares of 500 francs each" (182). Palmerston, "Little men have been induced to buy small shares," is from Beatty (1956, 187). Chapter 10 of Beatty contains more details on this phase of the fund-raising.

One Cannot Say That They Are Exactly Forced Labor. "This forced labour system . . ." was said by Lord Russell, quoted in Kinross (1969, 174). "It is true that without the intervention . . ." is from Beatty (1956, 218). Lesseps was quoting Lord Henry Scott.

Frenchmen of Genius. This section draws directly on Karabell (2003). Early financial results from the Suez Canal are on page 270 of Beatty (1956); pages 271–278 of the same source discuss subsequent political events as Britain moved to increase its sway over Egypt and the canal. The increase in share value and dividend by 1880 is from McCullough (1977, 125).

Panama Dreaming. "I do not hesitate to declare . . ." and "To create a harbour . . ." are both from Lesseps (1880, 14). "[Lesseps] is the great canal digger . . ." is from Johnston (1879, 172).

On whether lives could have been saved with a different approach, Godin de Lépinay made this point effectively at the congress (see, for example, Mack, 1944, 294). Lépinay advocated for a lock canal, centered on an artificial lake created above sea level—very much in line with what the Americans eventually built. In refusing to vote for the sea-level plan, Lépinay predicted that building a canal with locks would save the lives of fifty thousand men; see Congrès International d'Études du Canal Interocéanique (1879, 659). (Lépinay's reasoning was provided in a letter included as an annex to that report from the congress.)

Mack (1944, 295) points out that Lépinay's argument was based in part on the "then prevalent but mistaken theory that tropical fevers were caused by a mysterious toxic emanation from earth freshly excavated and exposed to the air, and that therefore the less ground was disturbed the less illness there would be." Nevertheless, Lépinay was proved right, if for partially the wrong reasons.

Regarding our claim that the French, the British, and other Europeans had developed practical health measures over more than a century of military operations in tropical countries, see Curtin (1998). When European militaries could choose the timing of their tropical campaigns—and avoid a large presence of troops during the rainy season—mortality could be curtailed, at least in some places and for some time. See Curtin (1998, Chapter 3, 73) on the Asante expedition of 1874, with his important caveat: "Whether success was based on skill or luck, it was hard to duplicate."

Waking the Envy of the Happy Gods. "Now that I have gone over . . ." is from McCullough (1977, 118). The cost revisions by Lesseps are discussed in DuVal (1947, 40, 56–57, 64); see also McCullough (1977, 117–118, 125–128) on cost estimates, commissions, and "publicity." "Remember, when you have anything important to accomplish . . ." is from Lesseps (1880, 9).

Death on the Chagres. "Any homage paid . . ." was said by Philippe Bunau-Varilla, quoted in McCullough (1977, 187).

Vision Trap. "The failure of this Congress . . ." is from Johnston (1879, 180).

Chapter 3: Power to Persuade

The material in this chapter is a synthesis of Michael Mann's (1986) treatise on social power, which draws key distinctions among economic, political, military, and ideological power; works in social psychology on influence and persuasion (for example, Cialdini 2006, Turner 1991); and our own past work on institutions and political power (Acemoglu, Johnson, and Robinson 2005a; Acemoglu and Robinson 2006b, 2012, and 2019), which in turn builds on, among others, Brenner (1976), North (1982), and North, Wallis, and Weingast (2009).

The distinctive aspects of our approach in this chapter are our emphasis on the primacy of persuasion power, even when there are coercive opportunities, and our theory that persuasion power is in turn shaped by networks and institutions. In this way, our approach builds on the literature on the political economy of institutions but goes beyond this literature by emphasizing the role of ideas and persuasion power and highlighting the role of institutions in structuring how persuasion power works.

Opening epigraphs. Deutsch (1963, 111); Bernays (1928 [2005], 1).

You Can Shoot Your Emperor If You Dare. "Soldiers of the 5th . . ." is from Chandler (1966, 1011). This section draws on the account in Chandler's Chapter 88.

Wall Street on Top. The discussion of Wall Street's power in this section draws on Johnson and Kwak (2010). For evidence on how power affects behavior and others' perceptions, see Keltner, Gruenfeld, and Anderson (2003). On whether and in what sense big banks were "too big to jail," see www.pbs.org/wgbh/frontline/article/eric-holder-backtracks-remarks-on -too-big-to-jail, which includes a discussion of Eric Holder, the attorney general, walking back earlier statements. See also this interview with Lanny Breuer, assistant attorney general in the Department of Justice's Criminal Division: www.pbs.org/wgbh/frontline/article/lanny-breuer-financial-fraud -has-not-gone-unpunished. For use of "too big to jail" by critics, see https:// financialservices.house.gov/uploadedfiles/07072016_oi_tbtj_sr.pdf.

The Power of Ideas. The details about *Liar's Poker* are from Lewis (1989) and were previously cited in this way by Johnson and Kwak (2010).

It's Not a Fair Marketplace. On memes and their spread, see Dawkins (1976). On imitation in children and social learning, see Tomasello, Carpenter, Call, Behne, and Moll (2005) and Henrich (2016) for general discussion; see also Tomasello (2019) for a more holistic view. See also Shteynberg and Apfelbaum (2013). On overimitation in children, see Gergely, Bekkering, and Király (2002) and Carpenter, Call, and Tomasello (2005). The experiment discussed in the text is from Lyons, Young, and Keil (2007). On lack of overimitation in chimpanzees, see Buttelmann,

Carpenter, Call, and Tomasello (2007) and Tomasello (2019), Chapter 5. On the experiments showing the effects of bystanders' behavior on learning by children, see Chudek, Heller, Birch, and Henrich (2012).

Agenda Setting. Brain consumption of total energy is from Swaminathan (2008).

The Bankers' Agenda. The material in this section again draws on Johnson and Kwak (2010). On the decision not to help home owners, see Hundt (2019). On "lavish bonuses" of more than a million dollars per person, see Story and Dash (2009): "Nine of the financial firms that were among the largest recipients of federal bailout money paid about 5,000 of their traders and bankers bonuses of more than $1 million apiece for 2008, according to a report released Thursday by Andrew M. Cuomo, the New York attorney general."

Ideas and Interests. Blankfein, "God's work," was widely reported, including by Reuters Staff (2009).

When the Rules of the Game Keep You Down. "We have entered upon a struggle . . ." is from Foner (1989, 33). For discussion of the pre–Civil War restrictions on slaves learning how to read and other behavior, see Woodward (1955). Foner (1989, 111) puts it this way: "Before the war, every Southern state except Tennessee had prohibited the instruction of slaves, and while many free blacks had attended school and a number of slaves became literate through their own efforts or the aid of sympathetic masters, over 90 percent of the South's adult black population was illiterate in 1860."

On Black political representation in states in the South and in the federal government after the Civil War, see Woodward (1955, 54). "The South's adoption . . ." is from Woodward (1955, 69). "Simply an armed camp . . ." is from Du Bois (1903, 88). "Of what avail . . ." is from Congressional Globe, 1864 (38th Congress, 1st Session), 2251; part of this quote is in Wiener (1978, 6); the same source discusses landholdings and the agricultural basis of power. Ager, Boustan, and Eriksson (2021) study how White slave owners recovered from the wealth shock of emancipation. On the Dunning school, see Foner (1989). "Whatever blessings . . ." is from the *Atlantic Monthly* (October 1901, 1).

A Matter of Institutions. For our view on institutions, democracy, and economic development, see Acemoglu, Johnson, and Robinson (2005a).

The Power to Persuade Corrupts Absolutely. Lord Acton's statement is from a letter to the archbishop of Canterbury (https://oll.libertyfund.org /title/acton-acton-creighton-correspondence). On the behavior of powerful individuals, see Keltner (2016). The experiments reported in the text are summarized in Piff, Stancato, Côté, Mendoza-Denton, and Keltner (2012).

Choosing Vision and Technology. This section draws on the general sources listed at the start of this section.

What's Democracy Got to Do with It? For a discussion of Condorcet's ideas and their applicability today, see Landemore (2017). For evidence that democracy increases GDP per capita, introduces additional reforms, and invests more in education and health care, see Acemoglu, Naidu, Restrepo, and Robinson (2019). On people's attitudes toward democracy depending on democracy's performance concerning economic growth and redistribution, see Acemoglu, Ajzeman, Aksoy, Fiszbein, and Molina (2021). This paper finds that people are unwilling to delegate power to unaccountable experts, especially when the people have experience with democracy. On decision making and attitudes in diverse groups, see Gaither, Apfelbaum, Birnbaum, Babbitt, and Sommers (2018) and Levine, Apfelbaum, Bernard, Bartelt, Zajac, and Stark (2014).

Vision Is Power; Power Is Vision. On the views of "those who believe in democracy" not wanting to cede political voice in favor of the experts and their priorities, see Acemoglu, Ajzeman, Aksoy, Fiszbein, and Molina (2021). On the relationship between status and overconfidence, see Anderson, Brion, Moore, and Kennedy (2012).

Chapter 4: Cultivating Misery

Our interpretation in this chapter draws on the theoretical ideas in Brenner (1976) and Acemoglu and Wolitzky (2011). See also Naidu and Yuchtman (2013). Although these works emphasize the role of the balance of power between lords and peasants (or employers and employees in agriculture), they do not explore the implications of technological change. We are not aware of other approaches to agricultural technology that have pointed out its immiserizing consequences depending on institutional structure and balance of power.

Opening epigraphs. Bertolt Brecht, from Kuhn and Constantine (2019, 675); Arthur Young (1801), quoted in Gazley (1973, 436–437). The title of this Brecht poem is often translated as "A Worker Reads History."

The list of technological improvements in the Middle Ages is based on Carus-Wilson (1941), White (1964, 1978), Cipolla (1972b), Duby (1972), Thrupp (1972), Gimpel (1976), Fox (1986), Hills (1994), Smil (1994, 2017), Gies and Gies (1994), and Centennial Spotlight (2021).

The discussion of mills and their impact on productivity draws on Gimpel (1976), Smil (1994, 2017), Langdon (1986, 1991), and Reynolds (1983). Smil (2017), 154, estimates that a small water mill with fewer than 10 workers could grind as much flour in a 10-hour day as 250 people working by hand. The same source reports 6,500 places with mills "in eleventh-century England" (Smil 2017, 149); while the Domesday Book reported 5,624 mills in 1085 (Gimpel 1976, 12); the same source provides details on the earliest water mills in his Chapter 1. Total and urban population is discussed in Russell (1972), e.g., Table 1, 36, and there is a very interesting analysis of London in Galloway, Kane, and Murphy (1996). Our core references for the overall economy and living conditions are Dyer (1989, 2002), supplemented by May (1973) and Keene (1998). On the impact of the Norman Conquest, see the same sources plus Welldon (1971) and Kapelle (1979). Medieval Europe is covered more broadly by Pirenne (1937, 1952) and Wickham (2016). Postan (1966) and Barlow (1999) are also informative.

In 1100, 2 million rural residents fed 2.2 million people, while in 1300, the respective numbers were 4 million feeding 5 million. If the age composition of rural areas was about the same, with a working age population of about half the total, this suggests the ratio of fed people to active agricultural workers rose from 2.2 to 2.5, a rise of agricultural productivity, crudely measured, of just under 15 percent.

The building and operation of monasteries, churches, and cathedrals is from Gimpel (1983), Burton (1994), Swanson (1995), and Tellenbach (1993). More economic detail is in Kraus (1979). Details on the clerical population are in Russell (1944). England in the thirteenth century is covered by Harding (1993). Details on the number of religious houses and "date of foundation" are in Knowles (1940, 147). Abbot Suger, "Those who criticize us . . .," is from Gimpel (1983, 14). The cost of building cathedrals in France is from Denning (2012).

On the size of the population in religious orders, Burton (1994, 174) says that "by the thirteenth century the total number of monks, canons, nuns, and members of military orders was in the region of 18,000–20,000, or, at a rough calculation, one in every 150 of the population." Harding (1993, 233) puts the thirteenth-century numbers at 30,000 "secular" clergy in 9,500 parishes, plus 20,000–25,000 monks, nuns, and friars in "530 major monasteries and 250 smaller establishments."

A Society of Orders. Walsingham, "Crowds of them . . .," is from Dobson (1970, 132). Knighton, "No longer restricting . . .," is from Dobson (1970, 136). Walsingham and Knighton should be read with care, for they were clearly biased against the peasants. Becket, "This will certainly not . . .," is from Guy (2012, 177). The society of orders is discussed in Duby (1982). On the Peasants' Revolt of 1381, see also Barker (2014).

A Broken Bandwagon. This section uses the general sources mentioned at the start of the notes for this chapter.

The Synergy Between Coercion and Persuasion. Jocelin of Brakelond, "Hearing this . . .," and the abbot, "I thank you . . .," are from Gimpel (1983, 25); the original text is de Brakelond (1190s [1903]). Gimpel (1983) uses the H. E. Butler translation, available here: https://archive.org/details

/chronicleofjoceoojoceuoft/page/n151/mode/2up, 59–60. Gimpel (1983) provides the details on Saint Albans and its confrontations.

A Malthusian Trap. The famous line "Population, when unchecked . . ." is from Malthus (1798 [2018], 70); this is a highlight of the 1798 edition and a central statement in Chapter 1 but does not appear in the commonly cited and reprinted 1803 edition. Our view of the effects of the Black Death on peasant-lord relations draws on Brenner (1976), Hatcher (1981, 1994), and Hatcher (2008, 180–182, 242, inter alia). See Hatcher's (1981, 37–38) summary of the literature on the relationship between population and wages. The interpretation of how this altered because of changes in the balance of power between lords and peasants is based on Brenner (1976) and Hatcher (1994), especially 14–20. The alarm of the king and his advisers is described in Hatcher (1994, 11). "Because a great part of the people . . ." and "Let no one . . ." are from the Statute of Labourers (1351, first and second paragraph, respectively). Our reading of the Statute of Labourers is consistent with Hatcher (1994, 10–11). Knighton, "[the workmen were] so arrogant and obstinate . . .," is from Hatcher (1994, 11). Gower, "And on the other hand . . .," is from Hatcher (1994, 16); this was written before 1378. The two excerpts from the House of Commons petition of 1376, "as soon as . . ." and "they are taken into service . . .," are from Hatcher (1994, 12). Knighton, "the elation of the inferior . . .," is from Hatcher (1994, 19). Gower, "Servants are now masters . . .," is from Hatcher (1994, 17). Ancient Greece is discussed in Morris (2004) and Ober (2015b), and the Roman Republic is discussed in Allen (2009b). The fall of Rome is the focus of Goldsworthy (2009). Link (2022) presents evidence on early episodes of growth around the world.

Original Agricultural Sin. Early agriculture is from Smil (1994, 2017), along with Childe (1950), Brothwell and Brothwell (1969), Smith (1995), Mithen (2003), Morris (2013, 2015), and Reich (2018). The material in Scott (2017) is informative on some grains. Flannery and Marcus (2012) discuss the emergence of inequality.

The Pain of Grain. The potential advantages of hunter-gatherer life are in Suzman (2017); McCauley (2019) discusses life expectancy. Living standards across two thousand years are reviewed in Koepke and Baten (2005). Recent DNA evidence on European hunter-gatherers is reviewed in Reich (2018). Wright (2014) has a detailed discussion of Çatalhöyük. Göbekli Tepe is discussed by Collins (2014). Cauvin (2007) discusses the emergence of religion more broadly.

Pyramid Scheme. Detailed work records from the pyramids are in Tallet and Lehner (2022). Lehner (1997) provides more detail on what it took to build the pyramids. The pastoral lifestyle and diet in early Egypt are discussed by Wilkinson (2020, 9–12) and Smil (1994, 57). Rice cultivation in the Indus Valley is discussed in Green (2021); see also Agrawal (2007) and Chase (2010).

One Kind of Modernization. Our discussion of enclosures draws on Tawney (1941), Neeson (1993), and Mingay (1997). Recent findings are reported in Heldring, Robinson, and Vollmer (2021a, 2021b). They find somewhat larger productivity benefits from parliamentary enclosures but also substantial inequality increases, consistent with our discussion. "[H]as no claim . . ." is from Malthus (1803 [2018], 417); it does not appear in the 1798 first edition. Young, "everyone but an idiot," is from Young (1771, 4:361); we have modernized the spelling. "The universal benefit . . ." is from Young (1768, 95). "What is it to a poor man . . ." is from Young (1801, 42) and is also quoted in Gazley (1973, 436). Yields for open-field farmers are from Allen (2009a). Broader social development from 1500 is covered in Wrightson (1982, 2017) and Hindle (1999, 2000). The changes in English agriculture are discussed in Overton (1996) and Allen (1992, 2009a), and the rise of the modern European state is in Ertman (1997).

The Savage Gin. "One man and a horse . . ." is from a letter Whitney wrote to his father, September 11, 1793; a digital image of the original is available online: www.teachingushistory.org/ttrove/documents/Whitney Letter.pdf.

On the American South, see Woodward (1955), Wright (1986), and Baptist (2014). Cotton statistics are from Beckert (2014). Judge Johnson, "Individuals, who were . . .," is from Lyman (1868, 158). "[R]egimented and relentless . . ." is from the National Archives online article on "Eli Whitney's Patent for the Cotton Gin," www.archives.gov/education/lessons/cotton-gin-patent. "When the price rises . . ." is from Brown (1854 [2001], 171); part of this quotation is also in Beckert (2014, 110). The development of accounting on slave plantations is in Rosenthal (2018). The cotton gin is discussed in detail by Lakwete (2003). Hammond's speech is from Hammond (1836). On the "positive good of slavery," see Calhoun (1837).

A Technological Harvest of Sorrow. Soviet agriculture and the famine of the 1930s are discussed by Conquest (1986), Ellman (2002), Allen (2003), Davies and Wheatcroft (2006), and Applebaum (2017). We use the numbers from Allen (2003). "Communism is Soviet power . . ." is from Volume 31 of Lenin's *Collected Works* (1920 [1966], 419); the sentence continues, "since industry cannot be developed without electrification." "The successes of our . . ." is from Volume 12 of Stalin's *Works* (1954, 199). Details on the ten thousand Americans with specific skills, including engineers, teachers, metalworkers, pipefitters, and miners, who came to the Soviet Union to help install and apply industrial technology are from Tzouliadis (2008). For background on agricultural policies during the 1920s, see Johnson and Temin (1993).

Chapter 5: A Middling Sort of Revolution

Our interpretation in this chapter draws on several seminal analyses of the origins of the Industrial Revolution. Particularly important are Mantoux (1927), Ashton (1986), Mokyr (1990, 1993, 2002, 2010, and 2016), Allen (2009a), Voth (2004), Kelly, Mokyr, and Ó Gráda (2014 and forthcoming), Crafts (1977, 2011), Freeman (2018), and Koyama and Rubin (2022). We are not aware of other theories that link the British industrial revolution to the aspirations of the middling sort of entrepreneurs and then explain the development of these aspirations, and their success, via the institutional changes that English and then British society underwent starting in the sixteenth century. Mokyr (2016) points to a "culture of growth" that emerged starting in the eighteenth century as a major contributor to the Industrial Revolution, although his focus is more on scientific advances and the more science-based phase of the revolution in the second half of the nineteenth century.

McCloskey (2006) has a related emphasis, focusing on the rise of "bourgeois virtues." Her interpretation is very different from ours, however. In particular, she does not relate the origins of the middling sort of vision to the institutional changes taking place in England (and then Britain) starting in the fifteenth century. She also views the "bourgeois virtues" as unabashedly positive and does not share our emphasis that the emergent vision was attempting to rise within the existing system and thus was not likely to be conducive to a broad-based enrichment or favor the working classes.

Our discussion of the institutional changes in England draws heavily on Acemoglu, Johnson, and Robinson (2005b) and Acemoglu and Robinson (2012).

Opening epigraphs. Defoe (1697 [1887], first line of the Author's Introduction); Charles Babbage (1851 [1968], 103).

The story about workers visiting the Crystal Palace is from Leapman (2001, Chapter 1). Details on what was on display at the Great Exhibition are from the *Official Catalogue of the Great Exhibition of the Works of Industry of All Nations, 1851* (Spicer Brothers, London). For more context, see Auerbach (1999) and Shears (2017). "About 1760 a wave . . ." is from Ashton (1986, 58). Assessments of living standards over the ages are from Morris (2013). Population estimates are from McEvedy and Jones (1978), and growth rates before industrialization are from Maddison (2001, 28, 90, and 265).

Coals from Newcastle. The Stephenson material draws heavily on Rolt (2009). "I say he . . ." is from 98. "[I]f the railway . . ." is from 59.

Science at the Starting Gate. The quotations from Davy, Losh, and the earl of Strathmore are from Rolt (2009, 28–29). "Communications were received . . ." is from Ferneyhough (1980, 45).

Why Britain? Our discussion of early European growth draws on Acemoglu, Johnson, and Robinson (2005b) and Allen (2009a)—see those papers for more on the relevant literatures. Tunzelmann (1978) assesses how developed the British economy would have been in 1800 without Watt's steam engine. Literacy rates in 1500 and 1800 are from Allen (2009a, Table 2.6, 53). Pomeranz (2001) disputes whether geography favored China, arguing that it lacked sufficient coal in suitable places. The high-level equilibrium trap idea is from Elvin (1973). On why Britain was different, see also Brenner (1993) and Brenner and Isett (2002). See also the sources listed at the start of chapters 5 and 6 of this bibliography for more general background and alternative hypotheses.

A Nation of Upstarts. Information on who founded industrial enterprises is from Crouzet (1985). For more on the notion of individualism and when this may have originated, see Macfarlane (1978) and Wickham (2016).

The Unraveling. William Harrison, "We in England divide . . .," is from Wrightson (1982). Thomas Rainsborough, "For really I think . . ." and "I do not find anything . . .," are from Sharp (1998, 103 and 106, respectively). Thomas Turner, "Oh, what a pleasure . . .," is from Muldrew (2017, 290). Turner's diary was published in 1761.

New Does Not Mean Inclusive. Soame Jenyns, "The merchant vies . . .," is from Porter (1982, 73). Philip Stanhope, "The middle class of people . . .," is from Porter (1982, 73). Gregory King, "decreasing the wealth of the kingdom . . ." is from Green (2017, 256). William Harrison, "neither voice nor authoritie . . .," is from Wrightson (1982, 19). According to Wrightson (1982), this group included "day labourers, poor husbandmen, artificers, and servants." This was the lowest of the four groups in Harrison's classification of the tiers in English society.

Chapter 6: Casualties of Progress

In addition to the main elements of our conceptual framework laid out above, this chapter emphasizes the nonwage implications of the balance of power between capital and labor, including for worker autonomy, working conditions, and worker health. In particular, in line with our discussion of worker monitoring and rent shifting, employers may sometimes be able to use new technologies or changing social conditions in order to increase profits by intensifying work duties or imposing more discipline on workers. These issues were first highlighted in the context of the British industrial revolution by Thompson (1966). Although some of Thompson's ideas—such as those concerning the origins of worker organizations and whether the Luddites should be viewed as the beginning of a coherent labor movement—are controversial, the ideas we emphasize in this chapter, which are related to the intensification of factory discipline and workers' reactions to them, are not, and they are confirmed by later scholarship—for example, de Vries (2008), Mokyr (2010), and Voth (2012).

Our discussion of the direction of technology in the second half of the nineteenth century draws on Habakkuk (1962) and especially on his emphasis that US technologies, especially the American System of Manufacturing, were partly motivated by the need to economize on skilled labor, which was scarce in the United States. Our discussion also draws on Rosenberg (1972).

We are not aware of other conceptual frameworks that combine these elements. Nor do we know of other interpretations of the second phase of the Industrial Revolution that emphasize the onset of technologies that are more worker friendly (for example, by creating new tasks), although Mokyr (1990, 2010) and Frey (2019) also argue that technology started generating greater demand for labor from 1850 onward.

The idea that rapid productivity growth from new technologies can contribute to employment growth when it expands the demand for labor in other sectors, already mentioned in Chapter 1, plays an important role in this chapter. We expand it and use it in the context of the systemic effects of railways in this chapter. The theoretical ideas here also borrow from the literature on "backward and forward linkages." Specifically, backward linkages arise when a sector's expansion triggers growth in other industries that supply inputs to it. Forward linkages refer to a sector contributing to growth in other industries that use its products as inputs and take place, for example, because railway growth reduces the cost of transport to other sectors that depend on transport services. Backward and forward linkages were emphasized as an important factor in economic development by Hirschman (1958) and build on the analysis of input-output linkages pioneered by Leontief (1936). Acemoglu and Restrepo (2019b and 2022) illustrate how large productivity increases and sectoral linkages can increase demand for workers, even in the presence of automation.

Early critiques of industrialization and its negative effects were formulated by Gaskell (1833), Carlyle (1829), and Engels (1845 [1892]). Marx also repeated some of these in *Capital*—for example, when he argued that in early factories, "Every organ of sense is injured in an equal degree by artificial elevation of the temperature, by the dust-laden atmosphere, by the deafening noise, not to mention danger to life and limb among the thickly crowded machinery, which, with the regularity of the seasons, issues its list of the killed and wounded in the industrial battle" (Marx 1867 [1887], 286–287).

The question of whether and how much wages and incomes increased has been debated extensively in the economic history literature. The lack of real income growth was initially dubbed "the living standards paradox."

Important contributions to this debate include Williamson (1985), Allen (1992, 2009a), Feinstein (1998), Mokyr (1988, 2002), and Voth (2004). The increase in working hours is discussed in McCormick (1959), de Vries (2008), and Voth (2004). The disruptive effects of factory discipline and the hardships that it imposes are discussed in Thompson (1966), Pollard (1963), and Freeman (2018).

Opening epigraphs. Greeley (1851, 25); Engels (1845 [1892], 48).

Quotations in the introduction to this chapter are from the Royal Commission of Inquiry into Children's Employment (1842 [1997]). We use an annex to the main report, containing details of interviews in Yorkshire. We quote from page 116 (David Pyrah), 135 (William Pickard), 93 (Sarah Gooder), 124 (Fanny Drake), 120 (Mrs. Day), and 116 (Mr. Briggs). We really appreciate and acknowledge the work that went into digitizing the record of these people's experiences by the Coal Mining History Resource Centre, Picks Publishing, and Ian Winstanley. Technical information on coal mining and steam engines is from Smil (2017).

Less Pay for More Work. Data on income and consumption are from Allen (2009a), and hours worked are from Voth (2012, including Table 4.8, 317). The cotton industry historical details are from Beckert (2014). We also draw on de Vries (2008). The history of military drill is from Lockhart (2021). Arkwright's factory and his career are discussed in Freeman (2018). The folk ballad "So, come all you cotton-weavers . . ." is "Hand-Loom v. Power-Loom," by John Grimshaw, published in Harland (1882, 189); it is also quoted in Thompson (1966, 306), though with a typo. "I have had seven boys . . ." is on 186, paragraph 2643, of the Report from Select Committee on Hand-Loom Weavers' Petitions, published July 1, 1835, House of Commons, testimony of John Scott on April 11, 1835. It also appears in Thompson (1966, 307).

The Luddites' Plight. Byron's speech was first published in Dallas (1824): "The rejected workmen . . .," 208, and "I have traversed . . .," 214. "On every side . . ." is from Greeley (1851, 25). "In fact, the division . . ." is from Ure (1835 [1861], 317, italics in original). The Glasgow weaver, "The theorists in political economy . . .," is from Richmond (1825, 1). Part of this statement also appears in Donnelly (1976, 222), where Richmond is identified as a "self-educated Glasgow weaver." On the Statute of Labourers and Master and Servant Act, see Naidu and Yuchtman (2013) as well as Steinfeld (1991). Pelling (1976) discusses the rise of British trade unions more broadly. Our discussion of the Poor Laws draws on Lewis (1952). "[P]rison system to punish poverty" is from Richardson (2012, 14).

The Entrance to Hell Realized. "A steam-engine of 100 horse-power" is from Baines (1835, 244); he cites "Mr. Farey, in his *Treatise on the Steam-Engine.*" "The manner in which . . ." is from Engels (1845 [1892], 74). "[T]he entrance to hell realized!" is Major General Sir Charles James

Napier's journal entry for July 20, 1839. See Napier (1857 [2011], 57) and Freeman (2018, 27). Death rates in Birmingham and other northern cities are from Finer (1952, 213), and the number of toilets is from the same source (215), citing the 1843–1844 Health of Towns Commission. Cartwright and Biddiss (2004, 152–156) discuss tuberculosis and provide annual deaths from this disease for some years. Annual deaths per year are from official British data in "Deaths Registered in England and Wales," 2021, https:// www.ons.gov.uk/peoplepopulationandcommunity/birthsdeathsand marriages/deaths/datasets/deathsregisteredinenglandandwalesseriesdrrefer encetables. The population of Manchester is from Marcus (1974 [2015], 2). See also the discussion in Chapter 6 of Rosen (1993) and in Harrison (2004). British drinking of gin and other health conditions are discussed in Chapter 7 of Cartwright and Biddiss (2004, 143–145, inter alia).

Where the Whigs Went Wrong. "For the history . . ." is from Macaulay (1848, 1:2). "[S]uch is the factory system . . ." is from Ure (1835 [1861], 307). On the Whig interpretation of history, see Butterfield (1965). The Whigs were a political party, but the Whig interpretation of history encompasses anyone who saw the history of Britain, prior to around 1850, through rose-tinted glasses.

Progress and Its Engines. Numbers on stagecoach transportation are from Wolmar (2007, 6). "The rapid introduction of cast-iron . . ." is from Field (1848), and part of it is also in Jefferys (1945 [1970], 15). On railway development more broadly, see Ferneyhough (1975), Buchanan (2001), and Jones (2011).

Gifts from Across the Atlantic. Joseph Whitworth, "The labouring classes are comparatively . . .," is quoted in Habakkuk (1962, 6); Whitworth made this statement in an 1854 report to Parliament. "The inventive genius . . ." is from Levasseur (1897, 9). Eli Whitney, "to substitute correct . . .," is from Habakkuk (1962, 22). British Parliamentary Committee, "The workman whose business . . .," is from Rosenberg (1972, 94). The superintendent at Colt's factory is Gage Stickney; "about 50 per cent" and "first-class labour . . ." are from Hounshell (1984, 21). The development of the sewing machine is discussed in Hounshell (1984, 67–123). "As regards the . . ." is from the *Report of the Committee on the Machinery of the U.S.* (128–129), as cited in Rosenberg (1972, 96). "The only obstacle . . ." is from Buchanan (1841, Appendix B, "Remarks on the Introduction of the Slide Principle in Tools and Machines Employed in the Production of Machinery," by James Nasmyth, 395). Part of this passage also appears in Jefferys (1945 [1970], 12). Nasmyth was an engineer who worked with Henry Maudslay, "the greatest of them [engineers designing new machine tools] all" (Jefferys 1945 [1970], 13). See also James and Skinner (1985) for statistical evidence that American technology in the second half of the nineteenth century was complementary to unskilled labor.

The Age of Countervailing Powers. "Now, though every workshop . . ." is from Thelwall (1796, 24), and part of this statement also appears in Thompson (1966, 185). Reverend J. R. Stephens, "the question of universal suffrage . . .," is from Briggs (1959, 34). This seems to be a paraphrasing of what he was reported to have said, on page 6 of the *Northern Star*, September 29, 1838:

> This question of Universal Suffrage was a knife and fork question after all; this question was a bread and cheese question, notwithstanding all that had been said against it; and if any man asked him what he meant by Universal Suffrage, he would answer, that every working man in the land had a right to have a good coat to his back, a comfortable abode in which to shelter himself and his family, a good dinner upon his table, and no more work than was necessary for keeping him in health, and as much wages for that work as would keep him in plenty, and afford him the enjoyment of all the blessings of life which a reasonable man could desire.

Earl Grey, "I do not support . . .," is from Grey (1830). See Hansard, House of Lords Debate, November 22, 1830, volume 1, cc604–18. There are more catchy versions of what Earl Grey said, including in standard references such as Evans (1996, 282). Those versions may have captured the spirit of the prime minister's sentiments, but their origin seems to have been an article by Henry Hetherington in the *Poor Man's Guardian* (November 19, 1831, 171), which claimed that Grey's statement was "If any persons suppose that this Reform will lead to ulterior measures, they are mistaken; for there is no one more decided against annual parliaments, universal suffrage, and the ballot, than I am. My object is not to favour, but to *put an end 'to such hopes and projects'*" (italics in Hetherington's report).

Our discussion of Disraeli is based on Blake (1966). Disraeli's 1872 Manchester speech was delivered at Free Trade Hall on April 3, 1872 (see Disraeli 1872, 22). Discussion of Chadwick draws on Lewis (1952) and Finer (1952).

Poverty for the Rest. The history of cotton in India is based on Beckert (2014). The general assessment of Lord Dalhousie is from Spear (1965). "[W]ill afford to India . . ." is from Dalhousie (1850, paragraph 47). Dalhousie and Indian railways are discussed in Wolmar (2010, 51–52, inter alia) and Kerr (2007). Winston Churchill, "I am quite satisfied . . .," is from Dalton (1986, 126). A slightly different version appears in Roberts (1991, 56). Churchill apparently made this remark to Lord Halifax in private conversation; Halifax later told Dalton.

Confronting Technology's Bias. Chartists are discussed by Briggs (1959).

Chapter 7: The Contested Path

This chapter provides a reinterpretation of twentieth-century economic growth in the United States and Western Europe based on the main elements of our conceptual framework: the balance between automation technologies and the creation of new tasks, and the institutional foundations of rent sharing.

We emphasize that the direction of early twentieth-century technology was shaped in part by choices that had sought to economize on skilled labor in the nineteenth-century US economy. We are not aware of any other accounts that have a similar theory, although many scholars emphasize the importance of interchangeable parts and the American System of Manufacturing in the early twentieth century—for example, in the context of the introduction of new electrical machinery and especially in Ford's automobile factories.

Opening epigraphs. Remarque (1928 [2013], 142); the President's Advisory Committee on Labor-Management Policy, January 11, 1962, cover letter attached to first formal report to President Kennedy.

On the evolution of military technologies between the Middle Ages and Waterloo, see Lockhart (2021). On the numbers of deaths in World War I and from the Spanish flu pandemic, see Mougel (2011) and Centers for Disease Control and Prevention (2019). "Even in the . . ." is from Zweig (1943, 5). On the scarring effects of the Great Depression, see Malmendier and Nagel (2011). Our discussion of technology choices in the early twentieth century draws heavily on Hounshell (1984). Our emphasis on engineer-managers is based on Jefferys (1945 [1970]) and Noble (1977). The central role we give to electricity and the reorganization of factories that enabled the introduction of advanced machinery and more advanced interchangeable parts draws on Hounshell (1984) and Nye (1992, 1998). Our discussion of the Ford factories also follows these references. Rosenberg (1972) is the basis for our interpretation that American technologies, creating demand for skilled and unskilled labor, spread to Britain and the rest of Europe. Examples of specific technologies that were exported from the US to Britain and Canada come from Hounshell (1984). Our discussion of how collective bargaining and the power of unions influence the direction of technology draws on theoretical ideas in Acemoglu and Pischke (1998, 1999) and Acemoglu (1997, 2002b, 2003b), as well as the historical discussion of Noble (1984). The importance of accuracy in manufacturing is covered in detail in Hounshell (1984, 228). The discussion of the key role of sequencing in the organization of production comes from Nye (1998, 142), Nye (1992, Chapter 5), and Hounshell (1984, Chapter 6).

Electrifying Growth. US GDP in 1870 and 1913 is from Maddison (2001, 261), in 1990 international dollars. For the rising scientific position of the United States, see Gruber and Johnson (2019, Chapter 1). The share of US workers in farming in 1860 is from www.digitalhistory.uh.edu/disp _textbook.cfm?smtID=11&psid=3837. The development of the McCormick reaper is discussed in Hounshell (1984, Chapter 4). Labor requirements for hand production and mechanized production for corn, cotton, potatoes, wheat, and other crops are from the Thirteenth Annual Report of the Commissioner of Labor, Vol. I (1898), 24–25, as reported in "Mechanization of Agriculture as a Factor in Labor Displacement," *Monthly Labor Review*, Vol. 33, No. 4, October 1931, Table 3, 9. Data on labor share in value added for industry and agriculture are from Edward Budd: www.nber.org/system /files/chapters/c2484/c2484.pdf. See Acemoglu and Restrepo (2019b) for interpretation. Patent statistics are from https://www.uspto.gov/web/ offices/ac/ido/oeip/taf/h_counts.htm. "The manufacturers judge . . ." is from Levasseur (1897, 18). Part of this statement is in Nye (1998, 132), where Levasseur is described as visiting "American steel mills, silk factories, and packing houses." From Levasseur (1897), it seems that he traveled widely in the United States, with a keen eye for how labor was used relative to machines. "The term *Factory* . . ." is from Ure (1835 [1861], 13). Importance of new applications using electricity builds directly on Nye (1992, 188–191). Factory power from electricity in 1889 and 1919 is from Nye (1992, Table 5.1, 187). "Incandescent electric light . . ." is from Lent (1895, 84), in the context of residential housing. This statement also appears in Nye (1998, 95). "But the greatest advantage . . ." is from Warner (1904, 97), which was based on an address to the Electrical Engineering Society of the Worcester Polytechnic Institute on November 20, 1903. From context, Warner was a senior executive at Westinghouse, with a broad view of how technology was developing. This passage also appears in Nye (1992, 202), where it is attributed to a "Westinghouse technical circular," but Nye's endnote 40 on page 202 and page 416 point to Warner's article. It seems likely that Warner's opinions reflected the official view at Westinghouse. On the new factory organization made possible by electricity, see Nye (1992, Chapter 5, including 195–196). See also the discussion of lighting and productivity in Nye (1992, 222–223). Columbia Mills is discussed in Nye (1992, 197–198). Westinghouse factories are discussed in Hounshell (1984, 240) and Nye (1992, 170–171, 196, 202, 220). Estimates of productivity gains in foundries that introduced these methods are reported in Hounshell (1984, 240).

New Tasks from New Engineers. The share of white-collar workers in manufacturing, 1860, 1910, and 1940, is from Michaels (2007). Data on education achievement (percentage of people with high school diplomas, etc.) are from Goldin and Katz (2008), 194–195, Figure 6.1, 205. Michaels

(2007) finds that new industries with a more diverse set of occupations were at the forefront of overall employment growth and the expansion of white-collar occupations in US manufacturing during this period. The association between faster productivity growth and employment growth from 1909 to 1914 is documented in Alexopoulos and Cohen (2016), which also shows that this association was stronger in new industries relying on electrical machinery and electronics. Fiszbein, Lafortune, Lewis, and Tessada (2020) confirms the same association and shows that the effects of electrification on employment were more positive when there was less concentration, which is consistent with our point that monopoly power can weaken the productivity bandwagon. The importance of organizing machinery for use by unskilled workers in the United States is discussed in detail in Hounshell (1984, 230) and Nye (1992, 211). Nye (1992, 211) emphasizes the goal of reducing labor turnover, which became more expensive "with more capital committed to machines."

In the Driving Seat. General discussion and description of early Highland Park production and the Model N are in Hounshell (1984, Chapter 6). "[W]e are making 40,000 cylinders . . ." is from Hounshell (1984, 221). "System, system, system!" is from Hounshell (1984, 229). "So thoroughly is . . .," from the *American Machinist*, is in Colvin (1913a, 759). This passage is also quoted in Hounshell (1984, 229); on 228, Colvin is described as a "well-known technical journalist." Hounshell (1984) also makes the important point that Colvin's in-depth observations were made immediately before assembly-line production was adopted by Ford. "The provision of . . ." and "Also high-speed tools . . ." are from Ford (1930, 33); parts are also quoted in Nye (1998, 143). Model T prices are from Hounshell (1984, Table 6.1, 224); conversion to prices today uses the Consumer Price Index calculator in www.measuringworth.com/calculators/us compare for 1908–2021. "Mass production is not merely . . ." was published in Ford (1926, 821). The article is signed with the initials "H.F.," but Henry Ford's authorship is confirmed here: www.britannica.com/topic/Encyclopaedia-Britannica-English-language-reference-work/Thirteenth-edition. Part of this passage also appears in Hounshell (1984, 217). Turnover in the Highland Park plant is discussed in Hounshell (1984, 257–259) and Nye (1992, 210). "The chain system . . ." is from Hounshell (1984, 259). The systems approach to increasing wages, reorganizing factories, and reducing turnover is discussed in Nye (1992, 215–216). "The keynote of the whole work is simplicity . . ." is from Colvin (1913b, 442); Colvin was writing about the assembling department and the machining department. This statement is also quoted in Hounshell (1984, 236). Recruitment at Ford during the 1960s is discussed in Murnane and Levy (1996). "If we had a vacancy . . ." is by Art Johnson, a human resource director at Ford Motor

Company; see Murnane and Levy (1996, 19). "Productivity *creates* . . ." is from Alexander (1929, 43, italics in original); also quoted in Noble (1977, 52–53).

An Incomplete New Vision. Magnus Alexander, "[W]hereas *laissez-faire* . . .," is from Alexander (1929, 47); a partial version appears in Noble (1977, 53). In the original, "laissez faire" appears in quotation marks. John R. Commons is discussed in Nye (1998, 147–148).

Nordic Choices. The German discussion and numbers are from Evans (2005). Our discussion of the Scandinavian case is based on Berman (2006, Chapter 5), Baldwin (1990), and Gourevitch (1986). Branting, "In a backward land . . .," is from Berman (2006, 157). "The party does not aim . . ." is from Berman (2006, 172). For the idea that industry-level wage setting can increase investment, see Moene and Wallerstein (1997), and for union-imposed wage compression encouraging investment, see Acemoglu (2002b).

New Deal Aspirations. Our discussion of the New Deal builds on Katznelson (2013) and Fraser and Gerstle (1989). "A strong government . . ." is from Tugwell (1933). "The interests of society . . ." is from Cooke (1929, 2). Part of this passage also appears in Fraser and Gerstle (1989, 60–61). "Certainly anyone . . ." is from Fraser and Gerstle (1989, 75–76). On aircraft carriers, see Dunnigan and Nofi (1995, 364), which shows eleven carrier launchings in 1945. This is not an aberration: there were eight such launchings in 1944 and twelve in 1943. In addition, the US built smaller escort carriers—the same source shows twenty-five such launchings in 1943, thirty-five in 1944, and nine in 1945. The six operational aircraft carriers on December 7, 1941, were *Enterprise, Lexington*, and *Saratoga* in the Pacific, and *Yorktown, Ranger*, and *Wasp* in the Atlantic. On difficulties with supplies for the military during the early US involvement in World War II, see Atkinson (2002); "It appears . . ." is on page 50, and "The American Army . . ." is on 415. Atkinson (2002, 414) also quotes a British report opining that the American "genius lay in creating resources rather than in using them economically."

Glorious Years. "Great Compression" is from Goldin and Margo (1992). Numbers on the income share of the top 1 percent are our calculations from the World Income Database, https://wid.world. In all cases we report pretax income numbers for individuals over age twenty. Data on mean and median real wage growth by different groups are our calculations from various sources, as described in greater detail in the bibliographic notes at the start of Chapter 8. TFP numbers are also our calculations; details and alternative estimates are presented in the notes to the next chapter.

Clash over Automation and Wages. On Jacquard's loom, see Essinger (2004). Our discussion in the section draws on Noble (1977, 1984); see Noble (1984, 84, inter alia) for how the general approach—programmable machine tool automation—became numerical control. "[T]he threat and promise . . ." and "clean, spacious, and . . ." are from an unsigned editorial comment in *Fortune* (November 1, 1946, 160) and are quoted in Leaver and Brown (1946, 165). These also appear in Noble (1984, on 67 and 68, respectively). The air force and navy approach to automation is discussed in Noble (1984, 84–85). At his press conference on February 14, 1962, President Kennedy was asked, "Mr. President, our Labor Department estimates that approximately 1.8 million persons holding jobs are replaced every year by machines. How urgent do you view this problem—automation." His response, "I regard it as . . .," is from www.jfklibrary.org/archives/other-resources/john-f-kennedy-press-conferences/news-conference-24. The discussion and numbers for Bell Company switchboard operators are from Feigenbaum and Gross (2022). Lin (2011) provides the first empirical study of new tasks in the US labor market, and the numbers we report on the growth of professional, administrative, and clerical occupations are from Autor, Chin, Salomons, and Seegmiller (2022). Harold Ickes, "You are on your way . . .," is from Brinkley (1989, 123). "[T]he most concentrated period . . ." refers to the first six months of 1946 and is from the Bureau of Labor Statistics, "Work Stoppages Caused by Labor-Management Disputes in 1946" (1947, *Bulletin* no. 918, 9). The UAW-GM arbitration and the discussion of skilling/deskilling caused by machinery are from Noble (1984, 253, 255). The UAW statement, "We offer our cooperation . . .," is from Noble (1984, 253), which also discusses the UAW's general approach. This resolution, which was issued at its 1955 convention, began with "The UAW-CIO welcomes automation, technological progress. . . ." The arbitrator's statement, "This is not a case . . .," is from Noble (1984, 254). "[H]as to acquire . . ." is from Earl Via, a numerical-control maintenance technician, in Noble (1984, 256). "[T]he increased effort . . ." is from the United Electrical, Radio, and Machine Workers (UE), in Noble (1984, 257). From context, both statements were made in the 1970s. The recent study by Boustan, Choi, and Clingingsmith (2022) provides evidence that numerically controlled machinery displaced workers from some manual tasks but also created new tasks, especially for those who were union members. Harry Bridges, "Those guys who . . .," is from Levinson (2006, 109–110). "We believe that . . ." is from Levinson (2006, 110). "Every longshoreman . . ." is from Levinson (2006, 112). "The days of sweating . . ." is from Levinson (2006, 117). The discussion of the rates of displacement caused by automation and job creation resulting from new tasks, as well as the numbers we use, are from Acemoglu and Restrepo (2019b). The effects of automation and new tasks on the demand for skills and inequalities are from Acemoglu and Restrepo (2020b and 2022).

Abolition of Want. General discussion, population numbers, displacement, and the situation in Europe are from Judt (2006). Beveridge (1942) is the source for "a revolutionary moment . . ." (6) and "Abolition of want . . ." (7). Discussion of the reception of the report and the Labour Party's attitude is in Baldwin (1990).

Social Progress and Its Limits. Details of growth in ancient Greece are from Ober (2015b). Growth rates in ancient Rome are from Morris (2004). See also Allen (2009b). For health statistics and related discussion, see Deaton (2013). Education statistics are from Organisation for Economic Co-operation and Development (https://data.oecd.org/education.htm) and Goldin and Katz (2008). Preindustrial and early-industrial growth rates are for total GDP; see Maddison (2001, 28, 126, inter alia). Life expectancy at birth in 1900 is from Maddison (2001, 30). Life expectancy in 1970 is from the World Bank's Development Indicators (online database).

Chapter 8: Digital Damage

The conceptual framework of this chapter is as we outlined in Chapter 1 and used in chapters 6 and 7. The emphasis is how, within this framework, the two supports for shared prosperity both became unwound in the United States after 1980. In particular, we stress technologies becoming more focused on automation, building on Acemoglu and Restrepo (2019b), and a decline in the countervailing powers of labor (see, for example, Phillips-Fein 2010, Andersen 2021, and Gerstle 2022). See also Perlstein (2009), Burgin (2015), and Appelbaum (2019). Inspired by Noble's (1984) discussion, we also argue that the decline of labor's bargaining power contributed to technology moving more in an automation direction.

The empirical patterns documented in this chapter draw heavily on Acemoglu and Autor (2011) and Autor (2019). In most cases they have been replicated and extended for this book based on the same data sources and with the superb research assistance of Carlos Molina. The evidence on the role of automation in the decline in the labor share, slow growth in median wages, and surge in inequality comes from Acemoglu and Restrepo (2022).

Our interpretation of the ethos and approaches of early computer enthusiasts and hackers, and the idea that their focus was not on top-down automation, are inspired by the discussion in Levy (2010) and Isaacson (2014). Noble (1984) and Zuboff (1988) provide the basis of our view of modern automation in factories and offices, and workers' reactions to it.

Our discussion of disappointing productivity benefits from digital technologies is based on Gordon (2016) as well as the theoretical ideas discussed in Acemoglu and Restrepo (2019b).

Opening epigraphs. Any internet search will confirm that the Ted Nelson statement is widely attributed to him, without a confirmed source; and Leontief (1983), 405.

Lee Felsenstein quoting *Revolt in 2100*, "Secrecy is the keystone . . .," is from Levy (2010, 131). Ted Nelson, "THE PUBLIC DOES NOT . . ." and "THIS BOOK . . .," are from Levy (2010, 144). Grace Hopper is discussed at length in Isaacson (2014, Chapter 3).

A Reversal. US inequality trends are explored and discussed in Goldin and Margo (1992), Katz and Murphy (1992), Piketty and Saez (2003), Goldin and Katz (2008), and Autor and Dorn (2013). Our approach builds on Acemoglu and Autor (2011), Autor (2019), and Acemoglu and Restrepo (2022), which also provide related numbers. Here we give additional details of the methods and data sources. For most of the numbers on labor market inequality, employment, and wage trends, we combine the US Census of Population data for 1940, 1950, 1970, 1980, 1990, and 2000 with annual data from the March Current Population Survey (March CPS) and the American Community Survey (ACS). All these data are sourced from the IPUMS repository. Occupational classifications are harmonized across decades using the classification scheme developed by Dorn (2009). When the yearly income is top-coded as defined by the survey instrument, we impute it as 1.5 times the value of the top-code amount (which varies across years and even states in most recent years). Only a small fraction of observations is affected by top-coding. In 2019, for example, less than 0.5 percent of observations are top-coded. To deal with misreporting in the lowest part of the income distribution, we impose a minimum hourly wage equal to the first percentile of the hourly wage distribution. We compute hourly wage by dividing yearly income by self-reported number of hours in a year, unless these exceed the maximum number of hours (3570 = 70 hours per week for 51 weeks per year). For top-coded observations, we use annual hours of 1750 in the denominator (35 hours per week for 50 weeks per year). We define weekly and yearly wage as the product of hourly wage and the number of hours worked per week and per year, respectively (after the adjustment to the upper and lower bound of the hourly wage distribution).

In terms of educational classifications, we follow those described in detail in Acemoglu and Autor (2011) and in Autor (2019). Throughout, all numbers are composition-adjusted mean or median log wage for full-time, full-year workers ages 16 to 64 in the indicated group (e.g., all workers or high school graduates, etc.). For the composition adjustment, we sort the data into gender-education-experience groups of two genders, five education categories (high school dropout, high school graduate, some college, college graduate, and post-college degree), and four potential experience categories (0–9, 10–19, 20–29, and 30–39 years). Educational categories are harmonized following the procedures in Autor, Katz, and Kearney (2008). Mean log wages for broader groups in each year represent weighted averages of the relevant (composition-adjusted) cell means using a fixed set of weights, equal to the mean share of total hours worked by each group over 1963–2005. Median log wages are computed similarly. All earnings numbers are converted to real earnings by being deflated using the chain-weighted (implicit) price deflator for personal consumption expenditures.

Labor-force participation for prime-age workers for the US is computed from the same data, and for other countries we use data from the Organisation for Economic Co-operation and Development (OECD), https://data.oecd.org/emp/labour-force-participation-rate.htm.

The Pew Research Center report is by Schumacher and Moncus (2021). The numbers for Black-White wage differentials are computed from the same sources as above. For related discussion and analysis, see Daly, Hobijn, and Pedtke (2017). Numbers on the evolution of aggregate capital and labor shares of national income across countries are from Karabarbounis and Neiman (2014).

What Happened? Changes in the US auto industry are discussed in Murnane and Levy (1996) and Krzywdzinski (2021). The numbers on blue-collar jobs are based on our calculations from the same sources as above. On the China shock, the standard reference is Autor, Dorn, and Hanson (2013). The estimates of job losses in the United States caused by merchandise imports from China are from Acemoglu, Autor, Dorn, Hanson, and Price (2016). The list of areas affected by these imports is from these studies. The evidence on the effects of industrial robots on employment and wages is from Acemoglu and Restrepo (2020a). See also Graetz and Michaels (2018). The list of areas most affected by the introduction of robots is from this study as well. Our discussion of good jobs builds on Harrison and Bluestone (1990, including Chapter 5) and Acemoglu (1999, 2001). Acemoglu and Restrepo (2022) estimate the relative contribution of industrial automation (including robots, dedicated equipment, and specialized software), offshoring, and merchandise imports from China. Their estimates suggest that between 50 and 70 percent of the changes in wage inequality among five hundred demographic groups (defined by education, age, gender, ethnicity, and domestic versus foreign-born status) is explained by automation. Offshoring and merchandise imports from China have smaller impacts. Part of the reason for this is a result of which sorts of industries are affected by Chinese imports as opposed to by automation, as discussed in Acemoglu and Restrepo (2020a). "Deaths of despair" is used by Case and Deaton (2020) to describe deaths from alcoholism (liver disease), drug overdose, and suicides. They discuss in detail the potential effects of negative economic shocks on deaths of despair. A statistical analysis of the effects of import shocks from China on marriage, out-of-wedlock childbirth, teenage pregnancy, and other social problems is reported in Autor, Dorn, and Hanson (2019).

For more-general discussions of the effects of globalization on US labor markets, see Autor, Dorn, and Hanson (2013); for the effects of increasing market power of firms, see Philippon (2019); for the role of the financial industry, see Philippon and Reshef (2012); and for a broader discussion of the consequences of ideological shifts, see Sandel (2020).

The Liberal Establishment and Its Discontents. A particular version of consumer protection history is provided by Digital History (2021). Opposition to the New Deal from various business organizations and leading companies is discussed in detail in Phillips-Fein (2010). On M. Stanton Evans, see Evans (1965) and Phillips-Fein (2010). "[T]he chief point about . . ." is from Evans (1965, 18). On the U.S. system of welfare, see Hacker (2002).

What Is Good for General Motors. "What was good for . . ." is from Charles Wilson's confirmation hearing, Committee on Armed Services, United States Senate, January 15, 1953 (hearing transcript, 26). Senator Henrickson asked whether Wilson could, hypothetically, make a decision that was "extremely adverse to the interests of your stock and General Motors Corp." if that was in the interest of the US government. Wilson's full reply is recorded as follows:

> Yes, sir; I could. I cannot conceive of one because for years I thought what was good for our country was good for General Motors, and vice versa. The difference does not exist.
>
> Our company is too big. It goes with the welfare of the country. Our contribution to the Nation is quite considerable.

On Buckley, see Judis (1988) and Schneider (2003). "[I]n its maturity, literate America . . ." and "Since ideas . . ." are from Buckley (1955). The discussion of the Business Roundtable and the Chamber of Commerce's changing attitudes is from Phillips-Fein (2010, Chapter 9). "[B]usiness has very serious . . ." is in Phillips-Fein (2010, 192). "The way we earn . . .," "free enterprise system . . .," and "free enterprise concentrates . . ." are from Phillips-Fein (2010, 193). George H. W. Bush, "Less than fifty . . .," is from Phillips-Fein (2010, 185). On Hayek, see Phillips-Fein (2010, Chapter 2) and Appelbaum (2019). Background on pro-market views at the University of Chicago and Stanford's Hoover Institution can be found in Appelbaum (2019).

On the Side of Angels and Shareholders. "A Friedman Doctrine" is the title of Friedman (1970). Background and context for Friedman are in Appelbaum (2019, Chapter 1). For what we call the Jensen amendment, see Jensen and Meckling (1976) and Jensen (1986). "[T]he Business Roundtable believes . . ." is from Phillips-Fein (2010, 194). On the Enron scandal, see McLean and Elkind (2003). On wage policies and consequences of CEOs with business degrees, see Acemoglu, He, and LeMaire (2022), which is also the source for all the other related numbers on this topic. See also the general discussion in Marens (2011).

Big Is Beautiful. "People of the same trade . . ." is from Smith (1776 [1999], 232). On the Arrow replacement effect, see Arrow (1962). "We may have democracy . . ." is from Lonergan (1941, 42). Lonergan claimed that Brandeis said this to a "younger friend." Lonergan's tribute was originally published in *Labor*, the "Organ of the 15 Recognized Standard Railroad Labor Organizations," shortly after Brandeis's death.

On the innovativeness of smaller, younger firms, see Acemoglu, Akcigit, Alp, Bloom, and Kerr (2018). Specifically, this paper shows that conditioning on the sample of innovative firms, small-young firms are much more innovative than large-old firms (where large firms are those with more than two hundred employees, small firms are those with fewer than two hundred employees, and young firms have existed fewer than nine years). For example, the R&D-to-sales ratio is about twice for small-young firms as for large-old firms. The probability of patenting is also higher for small-young firms than for large-old firms. Robert Bork is discussed in Appelbaum (2019). On the Manne Economics Institute for Federal Judges and its effects on rulings, see Ash, Chen, and Naidu (2022). On current Supreme Court justices' relationships with the Federalist Society, see Feldman (2021), although some of the details are disputed.

A Lost Cause. See the general discussion in Phillips-Fein (2010). On the Taft-Hartley Act, see Phillips-Fein (2010, 31–33). General statistics on work stoppages, including the Annual Historical Table from 1947, are available from the US Bureau of Labor Statistics, www.bls.gov/wsp.

A Grim Reengineering. The term *reengineering the corporation* was coined and advocated in Hammer and Champy (1993). See also Davenport (1992) for related ideas. "Much of the old . . ." is from Hammer and Champy (1993, 74). On the IBM word-processing machine, see Haigh (2006).

"Office automation is simply . . ." is from Hammer and Sirbu (1980, 38). The Xerox vice president quoted as saying "We may, in fact . . ." is from Spinrad (1982, 812). "[T]he automation of all phases" is from Menzies (1981, xv). "We don't know . . ." is from Zuboff (1988, 3). See Autor, Levy, and Murnane (2002) on automation of check processing in a large bank. The numbers on the fraction of American women working in clerical jobs and its evolution are based on our calculations using the same sources as above.

Lee Felsenstein, "The industrial approach is grim . . ." and "the user's ability . . .," are from Levy (2010, 201). Bob Marsh, "We wanted to make . . .," is from Levy (2010, 203). "As the majority . . ." is from Bill Gates's letter, available here: https://lettersofnote.com/2009/10/08/most-of-you-steal -your-software. This letter is also quoted and discussed in Levy (2010, 193).

The evolution of US robot adoption is discussed in Acemoglu and Restrepo (2020a). Evidence that demographic factors have triggered rapid robot adoption in Germany, Japan, and South Korea and that differing demographic factors have caused a relatively slower adoption in the US is in Acemoglu and Restrepo (2021). Numbers on the evolution of blue-collar occupations are computed by us from the same sources as above.

Once Again, a Matter of Choice. The effects of industrial robots in Germany are estimated in Dauth, Findeisen, Suedekum, and Woessner (2021). They follow the same methodology as Acemoglu and Restrepo (2020a). They also estimate negative effects on blue-collar jobs and wages, but not on overall jobs, for there appears to be an increase in white-collar jobs. The differential evolution of white-collar jobs in German and Japanese manufacturing and their different approach to technology, including "Industry 4.0" and "Digital Factory" initiatives, are discussed in Krzywdzinski (2021) and Krzywdzinski and Gerber (2020). The comparison of auto sales and trends in employment and blue-collar occupations across automobile manufacturers in these three countries is from Krzywdzinski (2021). The German apprenticeship system is discussed in Acemoglu and Pischke (1998) and Thelen (1991), and worker voice via work councils that place worker representatives on corporate boards is discussed in Thelen (1991) and Jäger, Schoefer, and Heining (2021). The latter paper finds that this type of participation gives a voice to workers in technology choices. Effective taxes on equipment, software, and other capital, as well as on labor, are estimated in Acemoglu, Manera, and Restrepo (2020), and the numbers we report are from their paper. On the evolution of US federal support for research, see Gruber and Johnson (2019).

Digital Utopia. "Show me a problem . . ." is from Gates (2021, 14). Zuckerberg's early motto, "Move fast and break things," is reported in Blodget (2009). A detailed discussion of the attitudes we summarize is in Ferenstein (2017), which also reports the statements "very few are . . ." and "I've become an expert. . . ."

Not in the Productivity Statistics. On innovation slowing down, see Gordon (2016) and Gruber and Johnson (2019). Bloom, Jones, Van Reenen, and Webb (2020) show that more spending is going into R&D to produce the same rate of improvement across a number of sectors. On the number of patents and productivity growth trends, see also Acemoglu, Autor, and Patterson (forthcoming). "You can see . . ." is from Solow (1987).

Total factor productivity (TFP) estimates are computed using the standard formulae with a Cobb-Douglas production function, with weights for labor and capital of, respectively, 0.7 and 0.3, as in Gordon (2016). Thus, TFP growth is computed as

GDP growth minus 0.7*labor input growth minus 0.3*capital input growth.

Labor input growth is adjusted for an index of quality, which takes into account evolution of educational composition of the workforce, using Goldin and Katz's (2008) estimates. Data for GDP are from the Bureau of Economic Analysis National Income and Product Accounts tables. We have also computed TFP estimates using different data sources and alternative methodologies—for example, following the methodology in Fernald (2014), Bergeaud, Cette, and Lecat (2016), and Feenstra, Inklaar, and Timmer (2015)—with very similar results. For instance, in the periods 1948–1960, 1961–1980, 1981–2000, and 2001–2019, the average annual TFP growth estimates from Gordon (2016) are 2, 1, 0.7, and 0.6 percent, respectively. The same numbers using Fernald's data and methodology (2014) are, respectively, 2.2, 1.5, 0.8, and 0.8. From Bergeaud, Cette, and Lecat (2016), they are 2.4, 1.5, 1.3, and 0.9. Finally, from Feenstra, Inklaar, and Timmer (2015), they are 1.3, 0.7, 0.6, and 0.6.

"We're in the . . ." is from Irwin (2016). On Varian's arguments regarding mismeasurement, see Varian (2016) and Pethokoukis (2017a). Hatzius, "We think it . . .," is from Pethokoukis (2016). See also Pethokoukis (2017b).

Evidence that manufacturing industries investing more in digital technologies are not showing faster productivity growth or any evidence of more mismeasurement is from Acemoglu, Autor, Dorn, Hanson, and Price (2014). Robert Gordon's views are in Gordon (2016). For Tyler Cowen's views, see Cowen (2010).

The discussion of Japanese robot adoption and later attempts to introduce flexibility is provided in Krzywdzinski (2021). On the Fremont plant before and after Toyota's investments and comparisons to other US car manufacturers, see Shimada and MacDuffie (1986) and MacDuffie and Krafcik (1992).

On followers diverging from industry leaders, see Andrews, Criscuolo, and Gal (2016). On the costs of unbalanced investment in R&D across sectors, see Acemoglu, Autor, and Patterson (forthcoming). On automation at Tesla, see Boudette (2018) and Büchel and Floreano (2018). Musk, "Yes, excessive automation . . .," is from this tweet: https://twitter .com/elonmusk/status/984882630947753984 (@elonmusk, April 13, 2018). Čapek, "Only years of practice . . .," is from Čapek (1929 [2004]).

Toward Dystopia. Zuboff (1988) has an early and prescient discussion.

Chapter 9: Artificial Struggle

Our interpretation in this chapter has three key building blocks. The first draws on our overall framework and especially our discussion of so-so automation. Specifically, we argue that artificial intelligence is likely to generate more limited productivity benefits than many of its enthusiasts hope because it is expanding into tasks in which machine capabilities are still quite limited and because human productivity builds on tacit knowledge, accumulated expertise, and social intelligence. This interpretation is inspired by Larson's (2021) account of human reasoning that is currently out of the reach of AI, Mercier and Sperber's (2017) discussion of the social nature of human intelligence, and evidence of flexible adaptation by human groups (e.g., Henrich, 2016), as well as Pearl's (2021) discussion of the limits of machine learning and Chomsky's views on the shortcomings of AI-based language models (e.g., as shown in this panel discussion: http://languagelog.ldc.upenn.edu/myl/PinkerChomskyMIT.html). General discussions of AI technologies, machine learning methods, and deep learning/neural networks are provided in Russell and Norvig (2009), Neapolitan and Jiang (2018), and Wooldridge (2020). For the focus of AI technologies on prediction, see Agrawal, Gans, and Goldfarb (2018).

Second, we emphasize, again in line with our overall conceptual framework, that the malleability of technology, especially within this broad area, enables many different trajectories of development. Moreover, even if AI-based automation turns out to be so-so, it may still proceed rapidly. This may be because of market incentives, such as the profitability of automation, worker monitoring, and other rent-shifting activities, or because of the specific visions of powerful actors in the tech industry.

The third is the emphasis that rather than machine intelligence we should think about "machine usefulness." We are not aware of other works that have made this point, but our ideas here heavily draw on Wiener (1954) and Licklider (1960). An excellent account of Engelbart's life and work, with an explicit discussion of two visions of how computers can be used, is the highly readable book by Markoff (2015).

We should note that these ideas are still far from the mainstream in the area, which tends to be much more optimistic about the benefits of AI and even the possibility of artificial general intelligence. See, for example, Bostrom (2017), Christian (2020), Stuart Russell (2019), and Ford (2021) on advances in artificial intelligence, and Kurzweil (2005) and Diamandis and Kotler (2014) on the economic abundance that this would create.

Our discussion of routine and nonroutine tasks builds on Autor, Levy, and Murnane's (2003) seminal paper and Autor's (2014) discussion of limits to automation. Our interpretation that current AI still mostly focuses on routine tasks is based on the evidence in Acemoglu, Autor, Hazell, and Restrepo (2022). Frey and Osborne's famous (2013) study also supports the notion that AI is primarily about automation; they estimate that close to 50 percent of US jobs can be automated by AI within the next several decades. On the difficulties of using machine learning to improve on human decision making, see Kleinberg, Lakkaraju, Leskovec, Ludwig, and Mullainathan (2018).

Finally, our emphasis that current AI is being used for extensive worker monitoring is influenced by Zuboff (1988) on the use of digital technologies in offices and by her more recent work, Zuboff (2019), by Pasquale (2015), as well as by O'Neil (2016). The interpretation of worker monitoring as a way of shifting rents or payments away from labor toward capital, and the negative social applications of this, draws on Acemoglu and Newman (2002).

Opening epigraphs. Poe (1836 [1975], 421); Wiener (1964, 43).

The *Economist*: "Since the dawn . . ." and "popular perceptions . . ." are from the first section, "A Bright Future for the World of Work," in Williams (2021). "In fact, by lowering costs . . ." is from the fifth section, "Robots Threaten Jobs Less Than Fearmongers Claim." "A Bright Future" is the title of the first section. McKinsey, "For many members . . .," is from Luchtenberg (2022) and is the written introduction to a *McKinsey Talks Operations* podcast. This quote appears on the McKinsey website under capabilities/operations/our-insights; see the reference for Luchtenberg (2022) for the full web address. The McKinsey Global Institute has produced reports that explicitly recognize the possibility of job losses from AI. See for example, Manyika et al. (2017). "In the coming 12 years . . ." and "[o]f course, there will be . . ." are in Anderson and Rainie (2018). "[T]he challenge is . . .," "improving lives . . .," "creative capitalism," and "to take on . . ." are from Gates (2008). On various definitions of AI, see the leading textbook, Russell and Norvig (2009), which provides several different definitions.

From the Field of AI Dreams. On Jacquard's loom, see Essinger (2004). On robotic process automation, see AIIM (2022) and Roose (2021). On RPAs' mixed results, see Trefler (2018). On the classification of routine tasks, see Autor, Levy, and Murnane (2003) and Acemoglu and Autor (2011). The prediction that AI can perform close to 50 percent of jobs is in Frey and Osborne (2013). Further discussion can be found in Susskind (2020). Kai-Fu Lee, "And like most technologies, . . .," is from his introduction to Lee and Qiufan (2021, xiv). The evidence that the rollout of AI concentrates on firms and establishments that have jobs that can be replaced by artificial intelligence and the negative effects of this activity on job postings in these establishments are in Acemoglu, Autor, Hazell, and Restrepo (2022). On the aggregate job consequences of industrial robots, see Acemoglu and Restrepo (2020a).

The Imitation Fallacy. Background on Turing can be found in Isaacson (2014, Chapter 2) and Dyson (2012). "You cannot make . . ." is from Turing (1951 [2004], 105). "I do not wish . . ." is from Turing (1950, 447).

Boom and Mostly Bust. Background on the digesting duck and the Mechanical Turk can be found in Wood (2002) and Levitt (2000). On the Dartmouth conference, see Isaacson (2014) and Markoff (2015). Minsky, "In from three to eight years . . .," is reported in Heaven (2020). "If you work in AI . . ." is from Romero (2021). Hassabis, "[S]olving intelligence . . .," is from Simonite (2016). "Someone who is . . ." and "Five great programmers . . ." are from Taylor (2011).

The Underappreciated Human. The concept of "so-so technologies" is from Acemoglu and Restrepo (2019b). On cassava and other adaptations in the Yucatán, see Henrich (2016, 97–99). On naked streets, see McKone (2010). On the theory of the mind, see Baron-Cohen, Leslie, and Frith (1985), Tomasello (1995), and Sapolsky (2017). On the growing demand for social skills, see Deming (2017). On the relationship between IQ and success in technical and nontechnical fields, see Strenze (2007). "[P]eople should stop . . ." is from Hinton (2016, at the 0:29 mark). To be fair, Hinton does go on to say, "It might be ten years." On how this prediction has fared, see Smith and Funk (2021), which says, "Yet, the number of radiologists working in the US has gone up, not down, increasing by about 7% between 2015 and 2019. Indeed, there is now a shortage of radiologists that is predicted to increase over the next decade."

On the diagnosis of diabetic retinopathy and the combination of AI algorithms and specialists, see Raghu, Blumer, Corrado, Kleinberg, Obermeyer, and Mullainathan (2019).

On the wishes of Google's chief of self-driving cars, see Fried (2015). For Elon Musk's comments on self-driving cars, see Hawkins (2021).

General AI Illusion. On superintelligence, see Bostrom (2017). On AlphaZero, see https://www.deepmind.com/blog/alphazero-shedding-new-light-on-chess-shogi-and-go. For an interesting critique of the current AI approach to intelligence, which also emphasizes the social and situational aspects of intelligence, see Larson (2021). See also Tomasello (2019) for an excellent general discussion, although he does not use the terms *social intelligence* and *situational intelligence.* For more discussion of the social and situational aspects of intelligence, see Mercier and Sperber (2017) and Chollet (2017, 2019). On social intelligence, see Riggio (2014) and Henrich (2016). On the shortcomings of GPT-3, see Marcus and Davis (2020). Overfitting is discussed in many standard references, including Russell and Norvig (2009). A more general discussion is provided in Everitt and Skrondal (2010). Our definition of overfitting is a little more general and encompasses ideas that are sometimes discussed under the heading of "misalignment" to capture the inability of models to be identified from irrelevant dimensions of the sample and thus fail to generalize in the appropriate way. For more references on this, see Gilbert, Dean, Lambert,

Zick, and Snoswell (2022), Pan, Bhatia, and Steinhardt (2022), and Ilyas, Santurkar, Tsipras, Engstrom, Tran, and Mądry (2019). "The marketing power . . ." is from Romero (2021).

The Modern Panopticon. "The ETS . . ." is from Zuboff (1988, 263). "One of the things we hear . . ." is from Lecher (2019). "They basically can see . . ." is from Greene (2021). The OSHA numbers are from Greene and Alcantara (2021). A general discussion of flexible scheduling, zero-hour contracts, and clopening is provided by O'Neil (2016). "There is no career . . ." is from Ndzi (2019).

A Road Not Taken. "The best material model . . ." is from Rosenblueth and Wiener (1945, 320). "Let us remember . . ." is from Wiener (1954, 162). "It is necessary to realize . . ." and "when a machine . . ." are from Wiener (1960, 1357). "We can be humble . . ." is from Wiener (1949). The story behind Wiener's op ed, and why none of it appeared in print for more than six decades, is explained in Markoff (2013). The story of Apple/Macintosh and background on J. C. R. Licklider can be found in Isaacson (2014). Licklider's statements are directly from his paper, Licklider (1960). More information on human-centered design can be obtained from Norman (2013) and especially Shneiderman (2022). For more on the contrast between the two visions of machine intelligence, see Markoff (2015).

Machine Usefulness in Action. The material in this section builds on Acemoglu (2021). Kai-Fu Lee, "Robots and AI will . . .," is from Lee (2021). "Today, what people call learning . . ." is from Asimov (1989, 267). Gains from personalized, adaptive teaching are discussed in Bloom (1984), Banerjee, Cole, Duflo, and Linden (2007), and Muralidharan, Singh, and Ganimian (2019). See also the discussion and additional references in Acemoglu (2021). For more details on the origins of the World Wide Web, see Isaacson (2014). The discussion of the consequences of cell phones in the fishing industry in Kerala is based on Jensen (2006). On M-Pesa, see Jack and Suri (2011). Other examples of the use of digital technologies to build new platforms are provided in Acemoglu, Jordan, and Weyl (2021). Estimates of AI spending are for 2016, from McKinsey Global Institute (2017).

Mother of All Inappropriate Technologies. On Frances Stewart's ideas, see Stewart (1977). More-modern discussions of inappropriate technology are provided in Basu and Weil (1998) and Acemoglu and Zilibotti (2001). The discussion of resistance of new crop varieties to different pests and pathogens and examples of innovations targeted at the US and West European agriculture, which are then inappropriate for the conditions in Africa, are from Moscona and Sastry (2022). The agricultural examples are also from Moscona and Sastry (2022). On the Green Revolution, see Evanson and Gollin (2003), and on Borlaug, see Hesser (2019). The within-country and in-between-country inequality implications of inappropriate technologies are discussed in Acemoglu and Zilibotti (2001).

Rebirth of the Two-Tiered Society. This section uses the general sources listed at the start of this chapter's bibliographic note.

Chapter 10: Democracy Breaks

The high-level idea of this chapter—that the current use of AI is mostly about data collection, which brings control over individuals as consumers, citizens, and workers—builds on and extends Pasquale (2015), O'Neil (2016), Lanier (2018), Zuboff (2019), and Crawford (2021). Sunstein (2001) provided an early analysis of the pernicious effects of digital echo chambers; see also Cinelli et al. (2021). The idea that this type of data collection distorts how social media platforms work is also explored in Acemoglu, Ozdaglar, and Siderius (2022) and Acemoglu (forthcoming). To the best of our knowledge, the parallel that we draw between the approaches of the Chinese government and the leading technology companies in the US—and how both of these approaches are fueled by access to abundant data—is new. Our discussion of surveillance and censorship in China is influenced by McGregor (2010) for the early phase and Dickson (2021) for the more recent period. We have been particularly inspired by various works of David Yang and coauthors, which we cite below, as well as extensive discussions with David.

Opening epigraphs. Chris Cox from Frenkel and Kang (2021, 224); Arendt (1978).

On the growth of AI spending in China, see Beraja, Yang, and Yuchtman (2020). We use the translation from the State Council's official planning document, "is founded on laws, regulations . . ." available here: https://chinacopyrightandmedia.wordpress.com/2014/06/14/plan ning-outline-for-the-construction-of-a-social-credit-system-2014-2020. Supreme People's Court, "Defaulters [on court orders] . . .," is from https:// english.court.gov.cn/2019-07/11/c_766610.htm, an official Chinese government website via *China Daily*. Protests surrounding President Joseph Estrada's impeachment are described in Shirky (2011). Wael Ghonim, "I want to meet . . .," is from an NPR interview on January 17, 2012: www.npr .org/2012/01/17/145326759/revolution-2-0-social-medias-role-in-removing -mubarak-from-power. The Twitter cofounder Biz Stone, "[S]ome Tweets may . . .," is in https://blog.twitter.com/en_us/a/2011/the-tweets-must -flow. Secretary of State Hillary Rodham Clinton's thoughts on the internet and freedom are in Clinton (2010).

A Politically Weaponized System of Censorship. On developments in China after Mao's death, see MacFarquhar and Schoenhals (2008), and on censorship in the 2000s, see McGregor (2010). Details on the

Tiananmen Square massacre and the "seven demands" can be found in Zhang, Nathan, Link, and Schell (2002). The "major research effort" on censorship and limited freedoms in the early 2010s is in King, Pan, and Roberts (2013). "Another team of researchers" is in Qin, Strömberg, and Wu (2017), which provides evidence of limited collective action using social media. The 2017 "New Generation AI Development Plan" can be found at www.newamerica.org/cybersecurity-initiative/digichina/blog/full-trans lation-chinas-new-generation-artificial-intelligence-development-plan-2017. Xiao Qiang, "China has . . .," is from Zhong, Mozur, Krolik, and Kao (2020).

A Braver New World. On media censorship, including for corruption cases, see Xu and Albert (2017). On foreign media stories being censored, specifically regarding allegations of corruption in the Namibian office of a company run by the son of a high Chinese official, see McGregor (2010), 148. This case involved Hu Haifeng, son of Hu Jintao, then China's top leader.

Curriculum reform and its implications are studied in Cantoni, Chen, Yang, Yuchtman, and Zhang (2017). The experimental study of the implications of the Great Firewall, and more context on its implications, is presented in Chen and Yang (2019). "What Orwell feared . . ." is from Postman (1985, xxi). "[U]nder a scientific . . ." is from Huxley (1958, 37).

From Prometheus to Pegasus. On VK (*VKontakte*)'s spread and role in protests, see Enikolopov, Makarin, and Petrova (2020). On the NSO Group, see Bergman and Mazzetti (2022). The Pegasus story has been confirmed in widespread reporting by media sources that include the *Washington Post*, National Public Radio, the *New York Times*, the *Guardian*, and *Foreign Policy*: www.washingtonpost.com/investigations/interactive/2021 /nso-spyware-pegasus-cellphones; www.washingtonpost.com/world/2021/07 /19/india-nso-pegasus; www.npr.org/2021/02/25/971215788/biden-administra tion-poised-to-release-report-on-killing-of-jamal-khashoggi; www.nytimes.com /2021/07/17/world/middleeast/israel-saudi-khashoggi-hacking-nso.html; www.theguardian.com/world/2021/jul/18/nso-spyware-used-to-target-family- of-jamal-khashoggi-leaked-data-shows-saudis-pegasus; and https://foreignpolicy .com/2021/07/21/india-israel-nso-pegasus-spyware-hack-modi-bjp-democracy -watergate.

For the Saudi claims about a "rogue operation," see www.reuters.com /article/us-saudi-khashoggi/saudi-arabia-calls-khashoggi-killing-grave -mistake-says-prince-not-aware-idUSKCN1MV0HI.

The NSO response to Forbidden Stories appeared here: www.the guardian.com/news/2021/jul/18/response-from-nso-and-governments, beginning with "NSO Group firmly denies false claims made in your report." NSO specifically rejected any involvement in the killing of Khashoggi: "As NSO has previously stated, our technology was not associated in any way with the heinous murder of Jamal Khashoggi." More broadly, NSO sums up its policy regarding how its technology is used this way: NSO "does not operate the systems that it sells to vetted government customers, and does not have access to the data of its customers' targets[,] yet [its customers] are obligated to provide us with such information under investigations. NSO does not operate its technology, does not collect, nor possesses, nor has any access to any kind of data of its customers."

Snowden, "I, sitting at my desk . . .," is from Sorkin (2013). Clearview's CEO, "Our belief . . .," is in Hill (2020), which also discusses Clearview AI more broadly.

Surveillance and the Direction of Technology. "Technology favors tyranny" and "digital dictatorship" are from Harari (2018). Evidence on how AI tools are being used by local governments in China and how data sharing encourages more AI monitoring is from Beraja, Yang, and Yuchtman (2020). This paper also provides evidence of the effect of these activities on the size of the police force. Evidence of the effectiveness of AI deployment against protests is from Beraja, Kao, Yang, and Yuchtman (2021), which is also the source on the exporting of monitoring technologies to other authoritarian governments. On the role of Huawei in the export of surveillance technologies to other authoritarian nations, see also Feldstein (2019), from which we also take the estimate that this company has exported these technologies to more than fifty countries.

Social Media and Paper Clips. The paper-clip parable is from Bostrom (2017). The discussion of Facebook's entry and policies in Myanmar draws on Frenkel and Kang (2021). Thein Sein on "Rohingya Terrorists" crossing borders is from Human Rights Watch (2013), www.hrw.org /report/2013/04/22/all-you-can-do-pray/crimes-against-humanity-and -ethnic-cleansing-rohingya-muslims. "I accept the term extremist . . ." is from a CBS *60 Minutes* interview with Ashin Wirathu; the transcript is available here: www.cbsnews.com/news/new-burma-aung-san-suu-kyi-60 -minutes. Facebook's response to government demands in 2019—by the labeling of ethnic organizations as "dangerous" and banning them from the platform—is discussed in Frenkel and Kang (2021). Banning the four groups is discussed in Jon Russell (2019). "Think Before You Share" is in Chapter 9 of Frenkel and Kang (2021). The point about anti-Muslim comments propagated via Facebook in Sri Lanka and "There's incitements . . ." are from Taub and Fisher (2018). T. Raja Singh's comments on Facebook are from Purnell and Horwitz (2020).

Misinformation Machine. Statistics on social media use and sources of news are from Levy (2021), Allcott, Gentzkow, and Yu (2019), and Allcott and Gentzkow (2017). "[F]alsehood diffused . . ." is from Vosoughi, Roy, and Aral (2018). See Guess, Nyhan, and Reifler (2020) on the 2015–2016 election. Pariser's 2010 TED talk is here: www.youtube.com/watch?v=B8of WFx525s. The discussion of the doctored video of House Speaker Nancy

Pelosi is from Frenkel and Kang (2021). Nick Clegg, "Our job . . .," is from Timberg, Romm, and Harwell (2019). The discussion of the Oath Keepers is from Frenkel and Kang (2021). YouTube radicalization and "I fell down the alt-right rabbit hole" are from Roose (2019). Robert Evans's statement on "15 out of 75 fascist activists . . ." is from Evans (2018). Minmin Chen, "We can really . . .," is from Ditum (2019). Evidence on anti-Muslim posts and violence following Trump's tweets is from Müller and Schwarz (2021). For more on Twitter, see Halberstam and Knight (2016). The material on Reddit builds on Marantz (2020).

The Ad Bargain. Material in this section builds on Isaacson (2014) and Markoff (2015). "In this paper . . ." is from the abstract in Brin and Page (1998). Page, "amazingly . . .," is from Isaacson (2014, 458).

The Socially Bankrupt Web. Material in this section draws on Frenkel and Kang (2021), which is also the source for Sheryl Sandberg, "[W]hat we believe we've done is . . ." (2021, 61). Look-alike audiences and "a way your ads can reach . . ." are from the Meta Business Help Center, www.facebook.com/business/help/164749007013531?id=401668390442328. The mental health effects of Facebook's expansion are from Braghieri, Levy, and Makarin (2022) and O'Neil (2022). On social media use and outrage, see Rathje, Van Bavel, and van der Linden (2021) and O'Neil (2022). On the effects of algorithms on such emotional responses, see Stella, Ferrara, and De Domenico (2018). See also general discussions in Brady, Wills, Jost, Tucker, and Van Bavel (2017), Tirole (2021), and Brown, Bisbee, Lai, Bonneau, Nagler, and Tucker (2022). On the Facebook "ambitious research project" and its happiness and other activity implications, see Allcott, Gentzkow, and Song (2021) and Allcott, Braghieri, Eichmeyer, and Gentzkow (2020). "Fuck it, ship it" is from Frenkel and Kang (2021). "This is about giving people, including some of the most reprehensible people on earth, the biggest platform in history to reach a third of the planet" is from Cohen (2019).

The Antidemocratic Turn. On Habermas's theory of the public sphere, see Habermas ([1962] 1991). "Most of the fears . . ." and "perhaps it's only . . ." are from Vassallo (2021); the author is a general partner at Foundation Capital. Mark Zuckerberg's statement to *Time* magazine, "Whenever any technology . . .," is in Grossman (2014). The editorial statement concerning the large Facebook study in the *Proceedings of the National Academy of Sciences* is in Verna (2014). Google's strategy in establishing Google Books and Google Maps is discussed in Zuboff (2019). On ImageNet, see www.image-net.org. Fei-Fei Li, "In the age of the Internet . . .," is from Markoff (2012). On the *New York Times* reporting on Clearview AI, see "The Secretive Company That Might End Privacy as We Know It," by Kashmir Hill, https://www.nytimes.com/2020/01/18/technology/

clearview-privacy-facial-recognition.html, including this assessment, "The system—whose backbone is a database of more than three billion images that Clearview claims to have scraped from Facebook, YouTube, Venmo and millions of other websites—goes far beyond anything ever constructed by the United States government or Silicon Valley giants." For more on the thinking behind Clearview and Peter Thiel's early involvement, see Chafkin (2021, 296–297, inter alia).

"[L]aws have to . . ." are the words of David Scalzo, an investor in Clearview AI; see Hill (2020).

Radio Days. Background on Father Coughlin can be found in Brinkley (1983). The effects of Coughlin's radio speeches are explored in Wang (2021). Joseph Goebbels said, "[O]ur way of taking . . ." in August 1933; see Tworek (2019). The effects of radio propaganda on support for Nazis are documented in Adena, Enikolopov, Petrova, Santarosa, and Zhuravskaya (2015), and see also Satyanath, Voigtländer, and Voth (2017). For the German constitution, freedom of speech, and *Volksverhetzung*, see www .gesetze-im-internet.de/englisch_gg/englisch_gg.html.

Digital Choices. Limited improvements in Reddit and YouTube against hate speech are discussed in www.nytimes.com/2019/06/05/busi ness/youtube-remove-extremist-videos.html and https://variety.com/2020/ digital/news/reddit-bans-hate-speech-groups-removes-2000-subreddits -donald-trump-1234692898, but also see https://time.com/6121915/reddit -international-hate-speech. Wikipedia's arbitration procedures and bureaucratic structure are described in https://en.wikipedia.org/wiki/Wikipedia: Administration. On Facebook facilitating exports by small businesses, see Fergusson and Molina (forthcoming).

Democracy Undermined When We Most Need It. "For, after all . . ." is from Orwell (1949, 92).

Chapter 11: Redirecting Technology

The importance of redirecting technology and some of the tax-subsidy schemes that might help in this effort are discussed in Acemoglu (2021). To the best of our knowledge, the emphasis that any redirection of technology needs to build on a change in narrative—about how we should use technology and who should control it—and new countervailing powers is new.

Opening epigraphs. The People's Computer Company is from www .digibarn.com/collections/newsletters/peoples-computer/peoples-1972 -oct/index.html; Brandeis is from Baron (1996), which gives the origin as "Arbitration Proceedings, N.Y., Cloak Industry, October 13, 1913."

An earlier discussion of the Progressive movement is in Acemoglu and Johnson (2017). For background on the Progressive movement, see

McGerr (2003). "There are two . . ." is widely attributed to Mark Hanna—for example, by Safire (2008, 237). On Ida Tarbell, see Tarbell (1904). On "Mother" Jones and the march of the mill children, see McFarland (1971). For the work of the Pujo Committee, the breakup of Standard Oil, and early antitrust thinking, see Johnson and Kwak (2010).

Redirecting Technological Change. The role of policy to redirect technological choices in energy is discussed in Acemoglu (2021). Data on green or renewable patents across countries are reported in Acemoglu, Aghion, Barrage, and Hemous (forthcoming). Data on renewable costs and evolution over time are from www.irena.org/publications/2021/Jun/Renewable -Power-Costs-in-2020, assessing the "levelized cost of electricity" generated from various sources. "In 50 years . . ." is from McKibben (2013).

Remaking Countervailing Powers. On the economic and broader implications of growing concentration of power in the hands of Big Tech, see Foer (2017). For blue-collar production workers as a share of the US labor force, see https://bluecollarjobs.us/2017/04/10/highest-to-lowest-share-of -blue-collar-jobs-by-state. Starbucks unionization is discussed in Eavis (2022). On Hong Kong protests, see Cantoni, Yang, Yuchtman, and Zhang (2019). On the GM sit-down strike, see Fine (1969). On Botswana's tribal assemblies (*kgotla*), see Acemoglu, Johnson, and Robinson (2003). On New_Public and for Ursula Le Guin, "what we can learn to do," see Chan (2021). The expression, "what we can learn to do," is from Le Guin (2004); a more complete statement is "That's the neat thing about technologies. They're what we can learn to do." On Audrey Tang's efforts and the presidential hackathon, see Tang (2019). On Taiwan's COVID response involving civil society and private companies, see Lanier and Weyl (2020). "With the advent of . . ." is from Justice Anthony Kennedy, writing in January 2010 for the majority of the Supreme Court in its 5-to-4 *Citizens United* ruling, which allowed unlimited corporate contributions to political campaigns. See *Citizens United v. Federal Election Commission*, 558 U.S. 310 (2010), https://www.supremecourt.gov/opinions/boundvol umes/558bv.pdf, beginning on p. 310.

Policies for Redirecting Technology. On tax reform, see Acemoglu, Manera, and Restrepo (2020). On training, see Becker (1993) and Acemoglu and Pischke (1999). On the development of antibiotics and their use in World War II, see Gruber and Johnson (2019). On the negative effects of GDPR regulation on small firms, see Prasad (2020). On the problems of data markets when individuals reveal information on their social network, see Acemoglu, Makhdoumi, Malekian, and Ozdaglar, forthcoming. On data ownership, see Lanier (2018, 2019) and Posner and Weyl (2019). Zuckerberg, "I just believe strongly . . .," is reported in McCarthy (2020). On removing asymmetries of taxation between capital and labor and the implications for automation, see Acemoglu, Manera, and Restrepo (2020). The digital advertising tax is proposed by Romer (2021). On Section 230, see Waldman (2021). South Korea and Finland's industrial policies are discussed in, respectively, Lane (2022) and Mitrunen (2019).

Other Useful Policies. On wealth taxes, see *Boston Review* (2020). On social mobility across countries, see Corak (2013) and Chetty, Hendren, Kline, and Saez (2014). The estimates about income differences between families eliminated within a generation in Denmark and in the United States are based on Figure 1 in Corak (2013). On current state and federal minimum wages, see www.dol.gov/agencies/whd/minimum-wage/state. On the effects of the minimum wage, see Card and Krueger (2015). On higher minimum wages encouraging more worker-friendly investments, see Acemoglu and Pischke (1999). On the potential impact of the pandemic on automation, see Chernoff and Warman (2021).

The Future Path of Technology Remains to Be Written. The discussion of HIV activism and responses draws on Shilts (2007) and Specter (2021).

References

Acemoglu, Daron. 1997. "Training and Innovation in an Imperfect Labor Market." *Review of Economic Studies* 64, no. 2: 445–464.

Acemoglu, Daron. 1998. "Why Do New Technologies Complement Skills? Directed Technical Change and Wage Inequality." *Quarterly Journal of Economics* 113, no. 4: 1055–1089.

Acemoglu, Daron. 1999. "Changes in Unemployment and Wage Inequality: An Alternative Theory and Some Evidence." *American Economic Review* 89, no. 5: 1259–1278.

Acemoglu, Daron. 2001. "Good Jobs vs. Bad Jobs." *Journal of Labor Economics* 19, no. 1: 1–21.

Acemoglu, Daron. 2002a. "Directed Technical Change." *Review of Economic Studies* 69, no. 4: 781–810.

Acemoglu, Daron. 2002b. "Technical Change, Inequality, and the Labor Market." *Journal of Economic Literature* 40, no. 1: 7–72.

Acemoglu, Daron. 2003a. "Labor- and Capital-Augmenting Technical Change." *Journal of European Economic Association* 1, no. 1: 1–37.

Acemoglu, Daron. 2003b. "Patterns of Skill Premia." *Review of Economic Studies* 70, no. 2: 199–230.

Acemoglu, Daron. 2009. *Introduction to Modern Economic Growth*. Princeton, NJ: Princeton University Press.

Acemoglu, Daron. 2010. "When Does Labor Scarcity Encourage Innovation?" *Journal of Political Economy* 118, no. 6: 1037–1078.

Acemoglu, Daron. 2021. "AI's Future Doesn't Have to Be Dystopian." *Boston Review*, May 20, 2021. https://www.bostonreview.net/forum/ais-future-doesnt-have-to-be-dystopian/.

Acemoglu, Daron. Forthcoming. "Harms of AI." In *The Handbook of AI Governance*, edited by Justin Bullock, Yu-Che Chen, Johannes Himmelreich, Valerie M. Hudson, Anton Korinek, Matthew Young, and Baobao Zhang. New York: Oxford University Press.

Acemoglu, Daron, Philippe Aghion, Lint Barrage, and David Hemous. Forthcoming. "Climate Change, Director Innovation, and the Energy Transition: The Long-Run Consequences of the Shale Gas Revolution."

Acemoglu, Daron, Philippe Aghion, Leonardo Bursztyn, and David Hemous. 2012. "The Environment and Directed Technical Change." *American Economic Review* 102, no. 1: 131–166.

Acemoglu, Daron, Nicolás Ajzeman, Cevat Giray Aksoy, Martin Fiszbein, and Carlos Molina. 2021. "(Successful) Democracies Breed Their Own Support." NBER Working Paper no. 29167. DOI:10.3386/w29167.

Acemoglu, Daron, Ufuk Akcigit, Harun Alp, Nicholas Bloom, and William Kerr. 2018. "Innovation, Reallocation, and Growth." *American Economic Review* 108, no. 11: 3450–3491.

Acemoglu, Daron, and David H. Autor. 2011. "Skills, Tasks and Technologies: Implications for Employment and Earnings." *Handbook of Labor Economics* 4:1043–1171.

Acemoglu, Daron, David H. Autor, David Dorn, Gordon H. Hanson, and Brendan Price. 2014. "Return of the Solow Paradox? IT, Productivity, and Employment in US Manufacturing." *American Economic Review* 104, no. 5: 394–399.

Acemoglu, Daron, David H. Autor, David Dorn, Gordon H. Hanson, and Brendan Price. 2016. "Import Competition and the Great U.S. Employment Sag of the 2000s." *Journal of Labor Economics* 34:S141–S198.

Acemoglu, Daron, David H. Autor, Jonathon Hazell, and Pascual Restrepo. 2022. "AI and Jobs: Evidence from Online Vacancies." *Journal of Labor Economics* 40 (S1): S293–S340.

Acemoglu, Daron, David H. Autor, and Christina H. Patterson. Forthcoming. "Bottlenecks: Sectoral Imbalances in the U.S. Productivity Slowdown." Prepared for the NBER Macroeconomics Annual, 2023.

Acemoglu, Daron, Alex Xi He, and Daniel LeMaire. 2022. "Eclipse of Rent-Sharing: The Effects of Managers Business Education on Wages and the Labor Share in the US and Denmark." NBER Working Paper no. 29874. DOI:10.3386/w29874.

Acemoglu, Daron, and Simon Johnson. 2005. "Unbundling Institutions." *Journal of Political Economy* 113:949–995.

Acemoglu, Daron, and Simon Johnson. 2017. "It's Time to Found a New Republic." *Foreign Policy*, August 15. https://foreignpolicy.com/2017/08/15/its-time-to-found-a-new-republic.

Acemoglu, Daron, Simon Johnson, and James A. Robinson. 2003. "An African Success Story: Botswana." In *In Search of Prosperity: Analytical Narratives on Economic Growth*, edited by Dani Rodrik, 80–119. Princeton, NJ: Princeton University Press.

Acemoglu, Daron, Simon Johnson, and James A. Robinson. 2005a. "Institutions as Fundamental Determinants of Long-Run Growth." In *Handbook of Economic Growth*, edited by Philippe Aghion and Steven Durlauf, 1A:385–472. Amsterdam: North-Holland.

Acemoglu, Daron, Simon Johnson, and James A. Robinson. 2005b. "The Rise of Europe: Atlantic Trade, Institutional Change and Economic Growth." *American Economic Review* 95:546–579.

Acemoglu, Daron, Michael Jordan, and Glen Weyl. 2021. "The Turing Test Is Bad for Business." *Wired*, www.wired.com/story/artificial-intelligence-turing-test-economics-business.

Acemoglu, Daron, Claire Lelarge, and Pascual Restrepo. 2020. "Competing with Robots: Firm-Level Evidence from France." *American Economic Review Papers and Proceedings* 110:383–388.

Acemoglu, Daron, and Joshua Linn. 2004. "Market Size in Innovation: Theory and Evidence from the Pharmaceutical Industry." *Quarterly Journal of Economics* 119:1049–1090.

Acemoglu, Daron, Ali Makhdoumi, Azarakhsh Malekian, and Asu Ozdaglar. Forthcoming. "Too Much Data: Prices and Inefficiencies in Data Markets." *American Economic Journal*.

Acemoglu, Daron, Andrea Manera, and Pascual Restrepo. 2020. "Does the US Tax Code Favor Automation?" *Brookings Papers on Economic Activity*, no. 1, 231–285.

Acemoglu, Daron, Suresh Naidu, Pascual Restrepo, and James A. Robinson. 2019. "Democracy Does Cause Growth." *Journal of Political Economy* 127, no. 1: 47–100.

Acemoglu, Daron, and Andrew F. Newman. 2002. "The Labor Market and Corporate Structure." *European Economic Review* 46, no. 10: 1733–1756.

Acemoglu, Daron, Asu Ozdaglar, and James Siderius. 2022. "A Model of Online Misinformation." NBER Working Paper no. 28884. DOI:10.3386/w28884.

Acemoglu, Daron, and Jörn-Steffen Pischke. 1998. "Why Do Firms Train? Theory and Evidence." *Quarterly Journal of Economics* 113, no. 1: 79–119.

Acemoglu, Daron, and Jörn-Steffen Pischke. 1999. "The Structure of Wages and Investment in General Training." *Journal of Political Economy* 107, no. 3: 539–572.

Acemoglu, Daron, and Pascual Restrepo. 2018. "The Race Between Machine and Man: Implications of Technology for Growth, Factor Shares and Employment." *American Economic Review* 108, no. 6: 1488–1542.

Acemoglu, Daron, and Pascual Restrepo. 2019a. "Artificial Intelligence, Automation, and Work." In *The Economics of Artificial Intelligence: An*

Agenda, edited by Ajay Agarwal, Joshua S. Gans, and Avi Goldfarb, 197–236. Chicago: University of Chicago Press.

Acemoglu, Daron, and Pascual Restrepo. 2019b. "Automation and New Tasks: How Technology Changes Labor Demand." *Journal of Economic Perspectives* 33, no. 2: 330.

Acemoglu, Daron, and Pascual Restrepo. 2020a. "Robots and Jobs: Evidence from U.S. Labor Markets." *Journal of Political Economy* 128, no. 6: 2188–2244.

Acemoglu, Daron, and Pascual Restrepo. 2020b. "Unpacking Skill Bias: Automation and New Tasks." *American Economic Review, Papers and Proceedings* 110:356–361.

Acemoglu, Daron, and Pascual Restrepo. 2020c. "The Wrong Kind of AI." *Cambridge Journal of Regions, Economy, and Society* 13:25–35.

Acemoglu, Daron, and Pascual Restrepo. 2021. "Demographics and Automation." *Review of Economic Studies* 89, no. 1: 1–44.

Acemoglu, Daron, and Pascual Restrepo. 2022. "Tasks, Automation and the Rise in US Wage Inequality." *Econometrica* 90, no. 5: 1973–2016.

Acemoglu, Daron, and James A. Robinson. 2006a. "Economic Backwardness in Political Perspective." *American Political Science Review* 100, no. 1: 15–31.

Acemoglu, Daron, and James A. Robinson. 2006b. *Economic Origins of Dictatorship and Democracy*. New York: Cambridge University Press.

Acemoglu, Daron, and James A. Robinson. 2012. *Why Nations Fail: The Origins of Power, Prosperity, and Poverty*. New York: Crown.

Acemoglu, Daron, and James A. Robinson. 2019. *The Narrow Corridor: States, Societies, and the Fate of Liberty*. New York: Penguin.

Acemoglu, Daron, and Alexander Wolitzky. 2011. "The Economics of Labor Coercion." *Econometrica* 79, no. 2: 555–600.

Acemoglu, Daron, and Fabrizio Zilibotti. 2001. "Productivity Differences." *Quarterly Journal of Economics* 116, no. 2: 563–606.

Adena, Maja, Ruben Enikolopov, Maria Petrova, Veronica Santarosa, and Ekaterina Zhuravskaya. 2015. "Radio and the Rise of the Nazis in Prewar Germany." *Quarterly Journal of Economics* 130, no. 4: 1885–1939.

Ager, Philipp, Leah Boustan, and Katherine Eriksson. 2021. "The Intergenerational Effects of a Large Wealth Shock: White Southerners After the Civil War." *American Economic Review* 111, no. 11: 3767–3794.

Agrawal, Ajay, Joshua S. Gans, and Avi Goldfarb. 2018. *Prediction Machines: The Simple Economics of Artificial Intelligence*. Cambridge, MA: Harvard Business Review Press.

Agrawal, D. P. 2007. *The Indus Civilization: An Interdisciplinary Perspective.* New Delhi: Aryan.

AIIM (Association for Intelligent Information Management). 2022. "What Is Robotic Process Automation?" www.aiim.org/what-is-robotic-process-automation.

Alexander, Magnus W. 1929. "The Economic Evolution of the United States: Its Background and Significance." Address presented at the World Engineering Congress, Tokyo, Japan, November 1929. National Industrial Conference Board, New York.

Alexopoulos, Michelle, and Jon Cohen. 2016. "The Medium Is the Measure: Technical Change and Employment, 1909–1949." *Review of Economics and Statistics* 98, no. 4: 792–810.

Allcott, Hunt, Luca Braghieri, Sarah Eichmeyer, and Matthew Gentzkow. 2020. "The Welfare Effects of Social Media." *American Economic Review* 110, no. 3: 629–676.

Allcott, Hunt, and Matthew Gentzkow. 2017. "Social Media and Fake News in the 2016 Election." *Journal of Economic Perspectives* 31:211–236.

Allcott, Hunt, Matthew Gentzkow, and Lena Song. 2021. "Digital Addiction." NBER Working Paper no. 28936. DOI:10.3386/w28936.

Allcott, Hunt, Matthew Gentzkow, and Chuan Yu. 2019. "Trends in the Diffusion of Misinformation on Social Media." *Research and Politics* 6, no. 2: 1–8.

Allen, Robert C. 1992. *Enclosure and the Yeoman: The Agricultural Development of the South Midlands, 1450–1850.* Oxford: Clarendon.

Allen, Robert C. 2003. *Farm to Factory: A Reinterpretation of the Soviet Industrial Revolution.* Princeton, NJ: Princeton University Press.

Allen, Robert C. 2009a. *The British Industrial Revolution in Global Perspective.* New York: Cambridge University Press.

Allen, Robert C. 2009b. "How Prosperous Were the Romans? Evidence from Diocletian's Price Edict (301 AD)." In *Quantifying the Roman Economy: Methods and Problems*, edited by Alan Bowman and Andrew Wilson, 327–345. Oxford: Oxford University Press.

Ammen, Daniel. 1879. "The Proposed Interoceanic Ship Canal Across Nicaragua." In "Appendix A, Proceedings in the General Session of the Canal Congress in Paris, May 23, and in the 4th Commission." *Journal of the American Geographical Society of New York* 11 (May 26): 153–160.

Andersen, Kurt. 2021. *Evil Geniuses: The Unmaking of America, a Recent History.* New York: Random House.

Anderson, Cameron, Sebastien Brion, Don A. Moore, and Jessica A. Kennedy. 2012. "A Status-Enhancement Account of Overconfidence." *Journal of Personality and Social Psychology* 103, no. 4: 718–735.

Anderson, Janna, and Lee Rainie. 2018. "Improvements Ahead: How Humans and AI Might Evolve Together in the Next Decade." Pew Research Center, December 10. www.pewresearch.org/internet /2018/12/10/improvements-ahead-how-humans-and-ai-might-evolve -together-in-the-next-decade.

Andrews, Dan, Chiara Criscuolo, and Peter N. Gal. 2016. "The Best vs. the Rest: The Global Productivity Slowdown, Divergence across Firms and the Role of Public Policy." OECD Working Paper no. 5, www .oecd-ilibrary.org/economics/the-best-versus-the-rest_63629cc9-en.

Appelbaum, Binyamin. 2019. *Economists' Hour: False Prophets, Free Markets, and the Fracture of Society*. New York: Little, Brown.

Applebaum, Anne. 2017. *Red Famine: Stalin's War on Ukraine*. New York: Doubleday.

Arendt, Hannah. 1978. "Totalitarianism: Interview with Roger Errera." *New York Review of Books*, www.nybooks.com/articles/1978/10/26 /hannah-arendt-from-an-interview.

Arrow, Kenneth J. 1962. "The Economic Implications of Learning by Doing." *Review of Economic Studies* 29:155–173.

Ash, Elliott, Daniel L. Chen, and Suresh Naidu. 2022. "Ideas Have Consequences: The Impact of Law and Economics on American Justice." NBER Working Paper no. 29788. DOI:10.3386/w29788.

Ashton, T. S. 1986. *The Industrial Revolution 1760–1830*. Oxford: Oxford University Press.

Asimov, Isaac. 1989. "Interview with Bill Moyers." In *Bill Moyers: A World of Ideas*, edited by Betty Sue Flowers, 265–278. New York: Doubleday.

Atkinson, Anthony B., and Joseph E. Stiglitz. 1969. "A New View of Technological Change." *Economic Journal* 79, no. 315: 573–578.

Atkinson, Rick. 2002. *An Army at Dawn: The War in North Africa, 1942–1943*. New York: Henry Holt.

Auerbach, Jeffrey A. 1999. *The Great Exhibition of 1851: A Nation on Display*. New Haven, CT: Yale University Press.

Autor, David H. 2014. "Skills, Education and the Rise of Earnings Inequality Among the Other 99 Percent." *Science* 344, no. 6186: 843–851.

Autor, David H. 2019. "Work of the Past, Work of the Future." *American Economic Review, Papers and Proceedings* 109:1–32.

Autor, David H., Caroline Chin, Anna Salomons, and Bryan Seegmiller. 2022. "New Frontiers: The Origins and Content of New Work, 1940–2018." NBER Working Paper no. 30389. DOI:10.3386/ w30389.

Autor, David H., and David Dorn. 2013. "The Growth of Low-Skill Service Jobs and the Polarization of the U.S. Labor Market." *American Economic Review* 103, no. 5: 1553–1597.

Autor, David H., David Dorn, and Gordon H. Hanson. 2013. "The China Syndrome: Local Labor Market Effects of Import Competition in the United States." *American Economic Review* 103:2121–2168.

Autor, David H., David Dorn, and Gordon Hanson. 2019. "When Work Disappears: How Adverse Labor Market Shocks Affect Fertility, Marriage, and Children's Living Circumstances." *American Economic Review: Insights* 1, no. 2: 161–178.

Autor, David H., Lawrence Katz, and Melissa Kearney. 2008. "Trends in U.S. Wage Inequality: Revising the Revisionists." *Review of Economics and Statistics* 90, no. 2: 300–323.

Autor, David H., Frank Levy, and Richard J. Murnane. 2002. "Upstairs, Downstairs: Computers and Skills on Two Floors of a Large Bank." *Industrial Labor Relations Review* 55, no. 3: 432–447.

Autor, David H., Frank Levy, and Richard J. Murnane. 2003. "The Skill Content of Recent Technological Change: An Empirical Exploration." *Quarterly Journal of Economics* 118, no. 4: 1279–1333.

Babbage, Charles. 1851 [1968]. *The Exposition of 1851; Or, Views of the Industry, the Science, and the Government, of England*, 2nd ed. Abingdon: Routledge.

Bacon, Francis. 1620 [2017]. *The New Organon: Or True Directions Concerning the Interpretation of Nature*. Translated by Jonathan Bennett. www.earlymoderntexts.com/assets/pdfs/bacon1620.pdf.

Baines, Edward. 1835. *History of the Cotton Manufacture in Great Britain*. London: Fisher, Fisher, and Jackson.

Baldwin, Peter. 1990. *The Politics of Social Solidarity: Class Bases of the European Welfare State 1875–1975*. Cambridge: Cambridge University Press.

Banerjee, Abhijit V., Shawn Cole, Esther Duflo, and Leigh Linden. 2007. "Remedying Education: Evidence from Two Randomized Experiments in India." *Quarterly Journal of Economics* 122, no. 3: 1235–1264.

Baptist, Edward E. 2014. *The Half Has Never Been Told: Slavery and the Making of American Capitalism*. New York: Basic Books.

Barker, Juliet. 2014. *1381: The Year of the Peasants' Revolt*. Cambridge, MA: Harvard University Press.

Barlow, Frank. 1999. *The Feudal Kingdom of England, 1042–1216*, 5th ed. London: Routledge.

Baron, Joseph L. 1996. *A Treasury of Jewish Quotations*, rev. ed. Lanham, MD: Jason Aronson.

Baron-Cohen, Simon, Alan M. Leslie, and Uta Frith. 1985. "Does the Autistic Child Have a 'Theory of Mind'?" *Cognition* 21, no. 1: 37–46.

Barro, Robert, and Xavier Sala-i-Martin. 2004. *Economic Growth*. Cambridge, MA: MIT Press.

Basu, Susanto, and David N. Weil. 1998. "Appropriate Technology and Growth." *Quarterly Journal of Economics* 113, no. 4: 1025–1054.

Beatty, Charles. 1956. *De Lesseps of Suez: The Man and His Times*. New York: Harper.

Becker, Gary S. 1993. *Human Capital*, 3rd ed. Chicago: University of Chicago Press.

Beckert, Sven. 2014. *Empire of Cotton: A Global History*. New York: Vintage.

Bentham, Jeremy. 1791. *Panopticon, or The Inspection House*. Dublin: Thomas Payne.

Beraja, Martin, Andrew Kao, David Y. Yang, and Noam Yuchtman. 2021. "AI-tocracy." NBER Working Paper no. 29466. DOI:10.3386/w29466.

Beraja, Martin, David Y. Yang, and Noam Yuchtman. 2020. "Data-Intensive Innovation and the State: Evidence from AI Firms in China." NBER Working Paper no. 27723. DOI:10.3386/w27723. Forthcoming in *Review of Economic Studies*.

Berg, Maxine. 1980. *The Machinery Question in the Making of Political Economy 1815–1848*. Cambridge: Cambridge University Press.

Bergeaud, Antonin, Gilbert Cette, and Remy Lecat. 2016. "Productivity Trends in Advanced Countries Between 1890 and 2012." *Review of Income and Wealth* 62, no. 3: 420–444.

Bergman, Ronen, and Mark Mazzetti. 2022. "The Battle for the World's Most Powerful Cyberweapon." *New York Times Magazine*, January 28 (updated January 31).

Berman, Sheri. 2006. *The Primacy of Politics: Social Democracy in the Making of Europe's 20th Century*. New York: Cambridge University Press.

Bernays, Edward L. 1928 [2005]. *Propaganda*. Brooklyn: Ig Publishing.

Bernstein, Peter L. 2005. *Wedding of the Waters: The Erie Canal and the Making of a Great Nation*. New York: W.W. Norton.

Besley, Timothy, and Torsten Persson. 2011. *The Pillars of Prosperity*. Princeton, NJ: Princeton University Press.

Beveridge, William H. 1942. "Social Insurance and Allied Services." Presented to Parliament, November 1942. http://pombo.free.fr/beveridge42.pdf.

Blake, Robert. 1966. *Disraeli*. London: Faber and Faber.

Blodget, Henry. 2009. "Mark Zuckerberg on Innovation." *Business Insider*, October 1. www.businessinsider.com/mark-zuckerberg-innovation-2009-10.

Bloom, Benjamin. 1984. "The Two Sigma Problem: The Search for Methods of Proof Instruction as Effective as One-To-One Tutoring." *Educational Researcher* 13, no. 6: 4–16.

Bloom, Nicholas, Charles I. Jones, John Van Reenen, and Michael Webb. 2020. "Are Ideas Getting Harder to Find?" *American Economic Review* 110, no. 4: 1104–1144.

Bonin, Hubert. 2010. *History of the Suez Canal Company, 1858–2008: Between Controversy and Utility.* Geneva: Librarie Droz.

Boston Review. 2020. "Taxing the Superrich." Forum, March 17, https://bostonreview.net/forum/gabriel-zucman-taxing-superrich.

Bostrom, Nick. 2017. *Superintelligence.* New York: Dunod.

Boudette, Neal. 2018. "Inside Tesla's Audacious Push to Reinvent the Way Cars Are Made." *New York Times*, June 30. www.nytimes.com/2018/06/30/business/tesla-factory-musk.html.

Boustan, Leah Platt, Jiwon Choi, and David Clingingsmith. 2022. "Automation After the Assembly Line: Computerized Machine Tools, Employment and Productivity in the United States." NBER Working Paper no. 30400, October.

Brady, William J., Julian A. Wills, John T. Jost, Joshua A. Tucker, and Jay J. Van Bavel. 2017. "Emotion Shapes the Diffusion of Moralized Content in Social Networks." *Proceedings of the National Academy of Sciences* 114, no. 28: 7313–7318.

Braghieri, Luca, Ro'ee Levy, and Alexey Makarin. 2022. "Social Media and Mental Health." SSRN working paper. https://papers.ssrn.com/sol3/papers.cfm?abstract_id=3919760.

Brenner, Robert. 1976. "Agrarian Class Structure and Economic Development in Preindustrial Europe." *Past and Present* 70:30–75.

Brenner, Robert. 1993. *Merchants and Revolution.* Princeton, NJ: Princeton University Press.

Brenner, Robert, and Christopher Isett. 2002. "England's Divergence from China's Yangzi Delta: Property Relations, Microeconomics, and Patterns of Development." *Journal of Asian Studies* 61, no. 2: 609–662.

Bresnahan, Timothy F., and Manuel Trajtenberg. 1995. "General-Purpose Technologies: Engines of Growth?" *Journal of Econometrics* 65, no. 1: 83–108.

Briggs, Asa. 1959. *Chartist Studies.* London: Macmillan.

Brin, Sergey, and Lawrence Page. 1998. "The Anatomy of a Large-Scale Hypertextual Web Search Engine." *Computer Networks and ISDN Systems* 30:107–117.

Brinkley, Alan. 1983. *Voices of Protests: Huey Long, Father Coughlin, and the Great Depression.* New York: Vintage.

Brinkley, Alan. 1989. "The New Deal and the Idea of the State." In *The Rise and Fall of the New Deal Order, 1930–1980*, edited by Steve Fraser and Gary Gerstle, 85–121. Princeton, NJ: Princeton University Press.

Broodbank, Cyprian. 2013. *The Making of the Middle Sea: A History of the Mediterranean from the Beginning to the Emergence of the Classical World*. Oxford: Oxford University Press.

Brothwell, Don, and Patricia Brothwell. 1969. *Food in Antiquity: A Survey of the Diet of Early Peoples*. Baltimore: Johns Hopkins University Press.

Brown, John. 1854 [2001]. *Slave Life in Georgia: A Narrative of the Life, Sufferings, and Escape of John Brown, a Fugitive Slave, Now in England*. Edited by Louis Alexis Chamerovzow. https://docsouth.unc.edu/neh/jbrown/jbrown.html.

Brown, Megan A., James Bisbee, Angela Lai, Richard Bonneau, Jonathan Nagler, and Joshua A. Tucker. 2022. "Echo Chambers, Rabbit Holes, and Algorithmic Bias: How YouTube Recommends Content to Real Users." May 25. https://ssrn.com/abstract=4114905.

Brundage, Vernon Jr. 2017. "Profile of the Labor Force by Educational Attainment." US Bureau of Labor Statistics, Spotlight on Statistics, www.bls.gov/spotlight/2017/educational-attainment-of-the-labor-force.

Brynjolfsson, Erik, and Andrew McAfee. 2014. *The Second Machine Age: Work, Progress, and Prosperity in a Time of Brilliant Technologies*. New York: W.W. Norton.

Buchanan, Angus. 2001. *Brunel: The Life and Times of Isambard Kingdom Brunel*. London: Bloomsbury.

Buchanan, Robertson. 1841. *Practical Essays on Millwork and Other Machinery*, 3rd ed. London: John Weale.

Büchel, Bettina, and Dario Floreano. 2018. "Tesla's Problem: Overestimating Automation, Underestimating Humans." *Conversation*, May 2. https://theconversation.com/teslas-problem-overestimating-automation-underestimating-humans-95388.

Buckley, William F. Jr. 1955. "Our Mission Statement." *National Review*, November 19. www.nationalreview.com/1955/11/our-mission-statement-william-f-buckley-jr.

Burgin, Angus. 2015. *The Great Persuasion: Reinventing Free-Markets Since the Great Depression*. Cambridge, MA: Harvard University Press.

Burke, Edmund. 1795. *Thoughts and Details on Scarcity*. London: F. and C. Rivington.

Burton, Janet. 1994. *Monastic and Religious Orders in Britain, 1000–1300*, Cambridge Medieval Textbooks. Cambridge: Cambridge University Press.

Buttelmann, David, Malinda Carpenter, Josep Call, and Michael Tomasello. 2007. "Enculturated Chimpanzees Imitate Rationally." *Developmental Science* 10, no. 4: F31–F38.

Butterfield, Herbert. 1965. *The Whig Interpretation of History*. New York: W.W. Norton.

Calhoun, John C. 1837. "The Positive Good of Slavery." Speech before the US Senate, February 6.

Cantoni, Davide, Yuyu Chen, David Y. Yang, Noam Yuchtman, and Y. Jane Zhang. 2017. "Curriculum and Ideology." *Journal of Political Economy* 125, no. 1: 338–392.

Cantoni, Davide, David Y. Yang, Noam Yuchtman, and Y. Jane Zhang. 2019. "Protests as Strategic Games: Experimental Evidence from Hong Kong's Antiauthoritarian Movement." *Quarterly Journal of Economics* 134, no. 2: 1021–1077.

Čapek, Karel. 1920 [2001]. *R.U.R. (Rossum's Universal Robots)*. Translated by Paul Selvir and Nigel Playfair. New York: Dover.

Čapek, Karel. 1929 [2004]. *The Gardener's Year*. London: Bloomsbury.

Card, David, and Alan Krueger. 2015. *Myth and Measurement: The New Economics of the Minimum Wage*, 20th anniversary ed. Princeton, NJ: Princeton University Press.

Carlyle, Thomas. 1829. "Signs of the Times." *Edinburgh Review* 49:490–506.

Carpenter, Malinda, Josep Call, and Michael Tomasello. 2005. "Twelve- and 18-Month-Olds Copy Actions in Terms of Goals." *Developmental Science* 8, no. 1: F13–F20.

Cartwright, Frederick F., and Michael Biddiss. 2004. *Disease & History*, 2nd ed. Phoenix Mill: Sutton.

Carus-Wilson, E. M. 1941. "An Industrial Revolution of the Thirteenth Century." *Economic History Review* 11, no. 1: 39–60.

Case, Anne, and Angus Deaton. 2020. *Deaths of Despair and the Future of Capitalism*. Princeton, NJ: Princeton University Press.

Cauvin, Jacques. 2007. *The Birth of the Gods and the Origins of Agriculture*. Cambridge: Cambridge University Press.

Centennial Spotlight. 2021. *The Complete Guide to the Medieval Times*. Miami: Centennial Media.

Centers for Disease Control and Prevention. 2019. "1918 Pandemic (H1N1 Virus)." www.cdc.gov/flu/pandemic-resources/1918-pandemic -h1n1.html.

Chafkin, Max. 2021. *The Contrarian: Peter Thiel and Silicon Valley's Pursuit of Power*. New York: Penguin Press.

Chan, Wilfred. 2021. "A First Look at Our New Magazine." *New_ Public*, September 12. https://newpublic.substack.com/p/-a-first-look -at-our-new-magazine?s=r.

Chandler, David G. 1966. *The Campaigns of Napoleon*. New York: Scribner.

Chase, Brad. 2010. "Social Change at the Harappan Settlement of Gola Dhoro: A Reading from Animal Bones." *Antiquity* 84:528–543.

Chen, Yuyu, and David Y. Yang. 2019. "The Impact of Media Censorship: *1984* or *Brave New World?*" *American Economic Review* 109, no. 6: 2294–2332.

Chernoff, Alex, and Casey Warman. 2021. "COVID-19 and Implications for Automation." Bank of Canada, Staff Working Paper 2021–25, May 31. www.bankofcanada.ca/wp-content/uploads/2021/05/swp 2021-25.pdf.

Chetty, Raj, Nathaniel Hendren, Patrick Kline, and Emmanuel Saez. 2014. "Where Is the Land of Opportunity? The Geography of Intergenerational Mobility in the United States." *Quarterly Journal of Economics* 129, no. 4 (November): 1553–1623.

Childe, Gordon. 1950. "The Urban Revolution." *Town Planning Review* 21, no. 1 (April): 3–17.

Chollet, François. 2017. "The Implausibility of Intelligence Explosion." *Medium*, November 27, https://medium.com/@francois.chollet/the -impossibility-of-intelligence-explosion-5be4a9eda6ec.

Chollet, François. 2019. "On the Measure of Intelligence." Working paper, https://arxiv.org/pdf/1911.01547.pdf?ref=https://githubhelp.com.

Christian, Brian. 2020. *The Alignment Problem: Machine Learning and Human Values*. New York: W.W. Norton.

Chudek, Maciej, Sarah Heller, Susan Birch, and Joseph Henrich. 2012. "Prestige-Biased Cultural Learning: Bystander's Differential Attention to Potential Models Influences Children's Learning." *Evolution and Human Behavior* 33, no. 1: 46–56.

Cialdini, Robert B. 2006. *Influence: The Psychology of Persuasion*, rev. ed. New York: Harper Business.

Cinelli, Matteo, Gianmarco De Francisci Morales, Alessandro Galeazzi, Walter Quattrociocchi, and Michele Starnini. 2021. "The Echo Chamber Effect on Social Media." *Proceedings of the National Academy of Sciences* 118, no. 9. www.pnas.org/doi/10.1073 /pnas.2023301118.

Cipolla, Carlo M., ed. 1972a. *The Fontana Economic History of Europe: The Middle Ages*. London: Collins/Fontana.

Cipolla, Carlo M. 1972b. "The Origins." In *The Fontana Economic History of Europe: The Middle Ages*, edited by Cipolla, 11–24. London: Collins/Fontana.

Clinton, Hillary Rodham. 2010. "Remarks on Internet Freedom." *Newseum*, January 10. https://2009-2017.state.gov/secretary/2009 2013clinton/rm/2010/01/135519.htm.

Cohen, Sacha Baron. 2019. "Keynote Address." ADL's 2019 Never Is Now Summit on Anti-Semitism and Hate, November 21. www.adl.org/ news/article/sacha-baron-cohens-keynote-address-at-adls-2019-never -is-now-summit-on-anti-semitism.

Collins, Andrew. 2014. *Göbekli Tepe: Genesis of the Gods, The Temple of the Watchers and the Discovery of Eden*. Rochester: Bear.

Colvin, Fred H. 1913a. "Building an Automobile Every 40 Seconds." *American Machinist* 38, no. 19 (May 8): 757–762.

Colvin, Fred H. 1913b. "Special Machines for Auto Small Parts." *American Machinist* 39, no. 11 (September 11): 439–443.

Congrès International d'Études du Canal Interocéanique. 1879. *Compte Rendu des Séances*. Du 15 au 29 Mai. Paris: Émile Martinet.

Conquest, Robert. 1986. *The Harvest of Sorrow: Soviet Collectivization and the Terror Famine*. Oxford: Oxford University Press.

Cooke, Morris Llewellyn. 1929. "Some Observations on Workers' Organizations." Presidential Address Before the Fifteenth Annual Meeting of the Taylor Society, December 6, 1928. *Bulletin of the Taylor Society* 14, no. 1 (February): 2–10.

Corak, Miles. 2013. "Income Inequality, Equality of Opportunity, and Intergenerational Mobility." *Journal of Economic Perspectives* 27, no. 3 (Summer): 79–102.

Cowen, Tyler. 2010. *The Great Stagnation*. New York: Dutton.

CQ Researcher. 1945. "Automobiles in the Postwar Economy." https://library.cqpress.com/cqresearcher/document.php?id=cqresrre1945082100.

Crafts, Nicholas F. R. 1977. "Industrial Revolution in England and France: Some Thoughts on the Question, Why Was England First?" *Economic History Review* 30, no. 3: 429–441.

Crafts, Nicholas F. R. 2011. "Explaining the First Industrial Revolution: Two Views." *European Economic History Review* 15, no. 1: 153–168.

Crawford, Kate. 2021. *Atlas of AI: Power, Politics, and the Planetary Cost of Artificial Intelligence*. New Haven, CT: Yale University Press.

Crouzet, François. 1985. *The First Industrialists: The Problem of Origins*. New York: Cambridge University Press.

Curtin, Philip D. 1998. *Disease and Empire: The Health of European Troops in the Conquest of Africa*. Cambridge: Cambridge University Press.

Dalhousie, Lord. 1850. "Minute by Dalhousie on Introduction of Railways in India." In *Our Indian Railway*, edited by Roopa Srinivasan, Manish Tiwari, and Sandeep Silas, Chapter 2. Delhi: Foundation Books, 2006.

Dallas, R. C. 1824. *Recollections of the Life of Lord Byron, from the Year 1808 to the End of 1814*. London: Charles Knight.

Dalton, Hugh. 1986. *The Second World War Diary of Hugh Dalton, 1940–45*. Edited by Ben Pimlott. London: Jonathan Cape.

Daly, Mary C., Bart Hobijn, and Joseph H. Pedtke. 2017. "Disappointing Facts About the Black-White Wage Gap." *FRBSF Economic Letter*, Federal Reserve Bank of San Francisco, September 5.

Dauth, Wolfgang, Sebastian Findeisen, Jens Suedekum, and Nicole Woessner. 2021. "The Adjustment of Labor Markets to Robots." *Journal of the European Economic Association* 19, no. 6: 3104–3153.

Davenport, Thomas H. 1992. *Process Innovation: Reengineering Work Through Information Technology*. Cambridge, MA: Harvard Business Review Press.

David, Paul A. 1989. "Computer and Dynamo: The Modern Productivity Paradox in a Not-Too-Distant Mirror." www.gwern.net/docs/economics/automation/1989-david.pdf.

David, Paul A., and Gavin Wright. 2003. "General Purpose Technologies and Surges in Productivity: Historical Reflections on the Future of the ICT Revolution." In *The Economic Future in Historical Perspective*, edited by Paul A. David and Mark Thomas, 135–166. Oxford: Oxford University Press.

Davies, R. W., and Stephen G. Wheatcroft. 2006. "Stalin and the Soviet Famine of 1932–33: A Reply to Ellman." *Europe-Asia Studies* 58, no. 4 (June): 625–633.

Dawkins, Richard. 1976. *The Selfish Gene*. Oxford: Oxford University Press.

De Brakelond, Jocelin. 1190s [1903]. *The Chronicle of Jocelin of Brakelond: A Picture of Monastic Life in the Days of Abbot Samson*. London: De La More.

De Vries, Jan. 2008. *The Industrious Revolution: Consumer Behavior and the Household Economy, 1650 to the Present*. Cambridge: Cambridge University Press.

Deaton, Angus. 2013. *The Great Escape: Health, Wealth, and the Origins of Inequality*. Princeton, NJ: Princeton University Press.

Defoe, Daniel. 1697 [1887]. *An Essay on Projects*. London: Cassell.

Deming, David J. 2017. "The Growing Importance of Social Skills in the Labor Market." *Quarterly Journal of Economics* 132, no. 4: 1593–1640.

Denning, Amy. 2012. "How Much Did the Gothic Churches Cost? An Estimate of Ecclesiastical Building Costs in the Paris Basin Between 1100–1250." Bachelor's thesis, Florida Atlantic University. www.medievalists.net/2019/04/how-much-did-the-gothic-churches-cost-an-estimate-of-ecclesiastical-building-costs-in-the-paris-basin-between-1100-1250.

Deutsch, Karl. 1963. *The Nerves of Government: Models of Political Communication and Control*. New York: Free Press.

Diamandis, Peter H., and Steven Kotler. 2014. *Abundance: The Future Is Better Than You Think*, rev. ed. New York: Free Press.

Diamond, Peter. 1982. "Wage Determination and Efficiency in Search Equilibrium." *Review of Economic Studies* 49, no. 2: 217–227.

Dickson, Bruce J. 2021. *The Party and the People: Chinese Politics in the 21st Century*. Princeton, NJ: Princeton University Press.

Digital History. 2021. "Ralph Nader and the Consumer Movement." www.digitalhistory.uh.edu/disp_textbook.cfm?smtid=2&psid=3351.

Disraeli, Benjamin. 1872. "Speech of the Right Hon. B. Disraeli, M.P." Free Trade Hall, Manchester, April 3.

Ditum, Sarah. 2019. "How YouTube's Algorithms to Keep Us Watching Are Helping to Radicalise Viewers." *New Statesman*, July 31. www.newstatesman.com/science-tech/2019/07/how-youtube-s-algorithms-keep-us-watching-are-helping-radicalise.

Dobson, R. B. 1970. *The Peasants' Revolt of 1381*. London: Macmillan.

Donnelly, F. K. 1976. "Ideology and Early English Working-Class History: Edward Thompson and His Critics." *Social History* 1, no. 2: 219–238.

Dorn, David. 2009. *Essays on Inequality, Spatial Interaction, and the Demand for Skills*. PhD diss., University of St. Gallen.

Douglas, Paul H. 1930a. "Technological Unemployment." *American Federationist* 37, no. 8 (August): 923–950.

Douglas, Paul H. 1930b. "Technological Unemployment: Measurement of Elasticity of Demand as a Basis of Prediction of Labor Displacement." *Bulletin of Taylor Society* 15, no. 6: 254–270.

Drandakis, E. M., and Edmund Phelps. 1966. "A Model of Induced Invention, Growth and Distribution." *Economic Journal* 76:823–840.

Du Bois, W. E. B. 1903. *The Souls of Black Folk*. New York: AC McClurg.

Duby, Georges. 1972. "Medieval Agriculture." In *The Fontana Economic History of Europe: The Middle Ages*, edited by Carlo M. Cipolla, 175–220. London: Collins/Fontana.

Duby, Georges. 1982. *The Three Orders: Feudal Society Imagined*. Chicago: University of Chicago Press.

Dunnigan, James F., and Albert A. Nofi. 1995. *Victory at Sea: World War II in the Pacific*. New York: William Morrow.

DuVal, Miles P. Jr. 1947. *And the Mountains Will Move*. Stanford, CA: Stanford University Press.

Dyer, Christopher. 1989. *Standards of Living in the Later Middle Ages: Social Change in England c. 1200–1520*, Cambridge Medieval Textbooks. Cambridge: Cambridge University Press.

Dyer, Christopher. 2002. *Making a Living in the Middle Ages: The People of Britain, 850–1520*. New Haven, CT: Yale University Press.

Dyson, George. 2012. *Turing's Cathedral: The Origins of the Digital World*. New York: Pantheon.

Eavis, Peter. 2022. "A Starbucks Store in Seattle, the Company's Hometown, Votes to Unionize." *New York Times*, March 22.

Ellman, Michael. 2002. "Soviet Repression Statistics: Some Comments." *Europe-Asia Studies* 54, no. 7: 1151–1172.

Elvin, Mark. 1973. *The Pattern of the Chinese Past*. Stanford, CA: Stanford University Press.

Engels, Friedrich. 1845 [1892]. *The Condition of the Working-Class in England in 1844 with a Preface Written in 1892*. Translated by Florence Kelley Wischnewetzky. London: George Allen & Unwin.

Enikolopov, Ruben, Alexey Makarin, and Maria Petrova. 2020. "Social Media and Protest Participation: Evidence from Russia." *Econometrica* 88, no. 4: 1479–1514.

Ertman, Thomas. 1997. *Birth of the Leviathan: Building States and Regimes in Medieval and Early Modern Europe*. New York: Cambridge University Press.

Essinger, Jesse. 2004. *Jacquard's Web: How a Hand-Loom Led to the Birth of the Information Age*. Oxford: Oxford University Press.

Evans, Eric J. 1996. *The Forging of the Modern State: Early Industrial Britain, 1783–1870*, 2nd ed. New York: Longman.

Evans, M. Stanton. 1965. *The Liberal Establishment: Who Runs America . . . and How*. New York: Devin-Adair.

Evans, Richard J. 2005. *The Coming of the Third Reich*. New York: Penguin.

Evans, Robert. 2018. "From Memes to Infowars: How 75 Fascist Activists Were 'Red-Pilled.'" bell¿ngcat. www.bellingcat.com/news/americas/2018/10/11/memes-infowars-75-fascist-activists-red-pilled.

Evanson, Robert E., and Douglas Gollin. 2003. "Assessing the Impact of the Green Revolution, 1960 to 2000." *Science* 300, no. 5620: 758–762.

Everitt, B. S., and Anders Skrondal. 2010. *Cambridge Dictionary of Statistics*. Cambridge: Cambridge University Press.

Feenstra, Robert C., Robert Inklaar, and Marcel P. Timmer. 2015. "The Next Generation of the Penn World Table." *American Economic Review* 105, no. 10: 3150–3182. www.ggdc.net/pwt.

Feigenbaum, James, and Daniel P. Gross. 2022. "Answering the Call of Automation: How the Labor Market Adjusted to the Mechanization of Telephone Operation." NBER Working Paper no. w28061, revised April 30. DOI:10.3386/w28061.

Feinstein, Charles H. 1998. "Pessimism Perpetuated: Real Wages and the Standard of Living in Britain During and After the Industrial Revolution." *Journal of Economic History* 58, no. 3: 625–658.

Feldman, Noah. 2021. *Takeover: How a Conservative Student Club Captured the Supreme Court*. Audiobook. www.pushkin.fm/audiobooks/takeover-how-a-conservative-student-club-captured-the-supreme-court.

Feldstein, Steven. 2019. "The Global Expansion of AI Surveillance." Carnegie Endowment for International Peace working paper. https://carnegieendowment.org/2019/09/17/global-expansion-of-ai-surveillance-pub-79847.

Ferenstein, Gregory. 2017. "The Disrupters: Silicon Valley Elites' Vision of the Future." *City Journal*, Winter 2017. www.city-journal.org/html/disrupters-14950.html.

Fergusson, Leopoldo, and Carlos Molina. Forthcoming. 2022. "Facebook and International Trade."

Fernald, John. 2014. "A Quarterly, Utilization-Adjusted Series on Total Factor Productivity." Federal Reserve Bank of San Francisco Working Paper 2012–19. https://doi.org/10.24148/wp2012-19.

Ferneyhough, Frank. 1975. *The History of Railways in Britain*. Reading: Osprey.

Ferneyhough, Frank. 1980. *Liverpool & Manchester Railway, 1830–1980*. London: Hale.

Field, Joshua. 1848. "Presidential Address." *Proceedings of the Institute of Civil Engineers*, February 1. www.icevirtuallibrary.com/doi/epdf/10.1680/imotp.1848.24213.

Fine, Sidney. 1969. *Sit-Down: The General Motors Strike of 1936–1937*. Michigan: University of Michigan Press.

Finer, S. E. 1952. *The Life and Times of Sir Edwin Chadwick*. London: Routledge.

Finkelstein, Amy. 2004. "Static and Dynamic Effects of Health Policy: Evidence from the Vaccine Industry." *Quarterly Journal of Economics* 119:527–564.

Fiszbein, Martin, Jeanne Lafortune, Ethan G. Lewis, and José Tessada. 2020. "New Technologies, Productivity, and Jobs: The (Heterogeneous) Effects of Electrification on US Manufacturing." NBER Working Paper no. 28076. DOI:10.3386/w28076.

Flannery, Kent, and Joyce Marcus. 2012. *The Creation of Inequality: How Our Prehistoric Ancestors Set the Stage for Monarchy, Slavery, and Empire*. Cambridge, MA: Harvard University Press.

Foer, Franklin. 2017. *World Without Mind: The Existential Threat of Big Tech*. New York: Penguin.

Foner, Eric. 1989. *Reconstruction: America's Unfinished Revolution, 1863–1877*, 2014 Anniversary Edition. New York: Harper Perennial.

Ford, Henry. 1926. "Mass Production." In *Encyclopedia Britannica*, edited by J. L. Garvin, 13th ed., supplementary volume 2: 821–823.

Ford, Henry, in collaboration with Samuel Crowther. 1930. *Edison as I Know Him*. New York: Cosmopolitan.

Ford, Martin. 2021. *Rule of the Robots: How Artificial Intelligence Will Transform Everything*. New York: Basic Books.

Fox, H. S. A. 1986. "The Alleged Transformation from Two-Field to Three-Field Systems in Medieval England." *Economic History Review* 39, no. 4 (November): 526–548.

Fraser, Steve, and Gary Gerstle. 1989. *The Rise and Fall of the New Deal Order, 1930–1980*. Princeton, NJ: Princeton University Press.

Freeman, Joshua B. 2018. *Behemoth: A History of the Factory and the Making of the Modern World*. New York: W.W. Norton.

Frenkel, Sheera, and Cecelia Kang. 2021. *An Ugly Truth: Inside Facebook's Battle for Domination*. New York: HarperCollins.

Frey, Carl Benedikt. 2019. *The Technology Trap: Capital, Labor, and Power in the Age of Automation*. Princeton, NJ: Princeton University Press.

Frey, Carl Benedikt, and Michael A. Osborne. 2013. "The Future of Employment: How Susceptible Are Jobs to Computerisation?" Mimeo. Oxford: Oxford Martin School.

Fried, Ina. 2015. "Google Self-Driving Car Chief Wants Tech on the Market Within Five Years." *Vox*, March 17. www.vox.com/2015/3/17/11560406/google-self-driving-car-chief-wants-tech-on-the-market-within-five.

Friedman, Milton. 1970. "A Friedman Doctrine—the Social Responsibility of Business Is to Increase Its Profits." *New York Times*, September 13. www.nytimes.com/1970/09/13/archives/a-friedman-doctrine-the-social-responsibility-of-business-is-to.html.

Gaither, Sarah E., Evan P. Apfelbaum, Hannah J. Birnbaum, Laura G. Babbitt, and Samuel R. Sommers. 2018. "Mere Membership in Racially Diverse Groups Reduces Conformity." *Social Psychological and Personality Science* 9, no. 4: 402–410.

Galbraith, John Kenneth. 1952. *American Capitalism: The Concept of Countervailing Power*. New York: Houghton Mifflin.

Galloway, James A., Derek Kane, and Margaret Murphy. 1996. "Fuelling the City: Production and Distribution of Firewood and Fuel in London's Region, 1290–1400." *Economic History Review* 49 (n.s.), no. 3 (August): 447–472.

Gancia, Gino, and Fabrizio Zilibotti. 2009. "Technological Change and the Wealth of Nations." *Annual Review of Economics* 1:93–120.

Gaskell, P. 1833. *The Manufacturing Population of England: Its Moral, Social, and Physical Conditions, and the Changes Which Have Arisen from the Use of Steam Machinery, with an Examination of Infant Labor*. London: Baldwin and Cradock.

Gates, Bill. 2008. "Prepared Remarks." 2008 World Economic Forum, January 24. www.gatesfoundation.org/ideas/speeches/2008/01/bill-gates-2008-world-economic-forum.

Gates, Bill. 2021. *How to Avoid a Climate Disaster: The Solutions We Have and the Breakthroughs We Need*. New York: Alfred A. Knopf.

Gazley, John G. 1973. *The Life of Arthur Young*. Philadelphia: American Philosophical Society.

Geertz, Clifford. 1963. *Peddlers and Princes*. Chicago: University of Chicago Press.

Gergely, György, Harold Bekkering, and Ildikó Király. 2002. "Rational Imitation in Preverbal Infants." *Nature* 415, no. 6873: 755.

Gerstle, Gary. 2022. *The Rise and Fall of the Neoliberal Order: America and the World in the Free Market Era.* New York: Oxford University Press.

Gies, Frances, and Joseph Gies. 1994. *Cathedral, Forge, and Waterwheel: Technology and Invention in the Middle Ages.* New York: HarperCollins.

Gilbert, Thomas Krendl, Sarah Dean, Nathan Lambert, Tom Zick, and Aaron Snoswell. 2022. "Reward Reports for Reinforcement Learning." https://arxiv.org/abs/2204.10817.

Gimpel, Jean. 1976. *The Medieval Machine: The Industrial Revolution of the Middle Ages.* New York: Penguin.

Gimpel, Jean. 1983. *The Cathedral Builders.* New York: Grove.

Goldin, Claudia, and Lawrence F. Katz. 2008. *The Race Between Education and Technology.* Cambridge, MA: Harvard University Press.

Goldin, Claudia, and Robert A. Margo. 1992. "The Great Compression: The Wage Structure in the United States at Midcentury." *Quarterly Journal of Economics* 107, no. 1: 1–34.

Goldsworthy, Adrian. 2009. *How Rome Fell: Death of a Superpower.* New Haven, CT: Yale University Press.

Gordon, Robert. 2016. *The Rise and Fall of American Growth.* Princeton, NJ: Princeton University Press.

Gourevitch, Peter. 1986. *Politics in Hard Times: Comparative Responses to International Economic Crises.* Ithaca, NY: Cornell University Press.

Graetz, Georg, and Guy Michaels. 2018. "Robots at Work." *Review of Economics and Statistics* 100, no. 5: 753–768.

Greeley, Horace. 1851. *The Crystal Palace and Its Lessons: A Lecture.* New York: Dewitt and Davenport.

Green, Adam S. 2021. "Killing the Priest-King: Addressing Egalitarianism in the Indus Civilization." *Journal of Archaeological Research* 29:153–202.

Green, Adrian. 2017. "Consumption and Material Culture." In *A Social History of England, 1500–1750,* edited by Keith Wrightson, 242–266. Cambridge: Cambridge University Press.

Greene, Jay. 2021. "Amazon's Employee Surveillance Fuels Unionization Efforts: 'It's Not Prison, It's Work.'" *Washington Post,* December 2. www.washingtonpost.com/technology/2021/12/02/amazon-workplace-monitoring-unions.

Greene, Jay, and Chris Alcantara. 2021. "Amazon Warehouse Workers Suffer Serious Injuries at Higher Rates Than Other Firms." *Washington Post,* June 1. www.washingtonpost.com/technology/2021/06/01/amazon-osha-injury-rate.

Grey, Earl. 1830. Speech in House of Lords Debate. *Hansard*, November 22, 1830, volume 1, cc604–18.

Grossman, Lev. 2014. "Inside Facebook's Plan to Wire the World." *Time*, December 15. https://time.com/facebook-world-plan.

Gruber, Jonathan, and Simon Johnson. 2019. *Jump-Starting America: How Breakthrough Science Can Revive Economic Growth and the American Dream*. New York: PublicAffairs.

Guess, Andrew M., Brendan Nyhan, and Jason Reifler. 2020. "Exposure to Untrustworthy Websites in the 2016 US Election." *Nature Human Behaviour* 4, no. 5: 472–480.

Guy, John. 2012. *Thomas Becket: Warrior, Priest, Rebel*. New York: Random House.

Habakkuk, H. J. 1962. *American and British Technology in the Nineteenth Century: The Search for Labour-Saving Inventions*. Cambridge: Cambridge University Press.

Haberler, Gottfried. 1932. "Some Remarks on Professor Hansen's View on Technological Unemployment." *Quarterly Journal of Economics* 46, no. 3: 558–562.

Habermas, Jürgen. [1962] 1991. *The Structural Transformation of the Public Sphere*. Cambridge, MA: MIT Press.

Hacker, Jacob S. 2002. *The Divided Welfare State: The Battle over Public and Private Social Benefits in the United States*. New York: Cambridge University Press.

Haigh, Thomas. 2006. "Remembering the Office of the Future: The Origins of Word Processing and Office Automation." *IEEE Annals of the History of Computing* 28, no. 4: 6–31.

Halberstam, Yosh, and Brian Knight. 2016. "Homophily, Group Size, and the Diffusion of Political Information in Social Networks: Evidence from Twitter." *Journal of Public Economics* 143, no. 1: 73–88.

Hammer, Michael, and James Champy. 1993. *Reengineering the Corporation: A Manifesto for Business Revolution*. New York: HarperBusiness Essentials.

Hammer, Michael, and Marvin Sirbu. 1980. "What Is Office Automation?" 1980 Automation Conference, March 3–5, Georgia World Congress Center.

Hammond, James Henry. 1836. "Remarks of Mr. Hammond of South Carolina on the Question of Receiving Petitions for the Abolition of Slavery in the District of Columbia." Delivered in the House of Representatives, February 1.

Hanlon, W. Walker. 2015. "Necessity Is the Mother of Invention: Input Supplies and Directed Technical Change." *Econometrica* 83, no. 1: 67–100.

Harari, Yuval Noah. 2018. "Why Technology Favors Tyranny." *Atlantic*, October. www.theatlantic.com/magazine/archive/2018/10/yuval-noah-harari-technology-tyranny/568330.

Harding, Alan. 1993. *England in the Thirteenth Century*. Cambridge Medieval Textbooks. Cambridge: Cambridge University Press.

Harland, John. 1882. *Ballads and Songs of Lancashire, Ancient and Modern*, 3rd ed. Manchester: John Heywood.

Harrison, Bennett, and Barry Bluestone. 1990. *The Great U-Turn: Corporate Restructuring and the Polarizing of America*. New York: Basic Books.

Harrison, Mark. 2004. *Disease and the Modern World*. Cambridge, UK: Polity.

Hatcher, John. 1981. "English Serfdom and Villeinage: Towards a Reassessment." *Past and Present* 90:3–39.

Hatcher, John. 1994. "England in the Aftermath of the Black Death." *Past and Present* 144:3–35.

Hatcher, John. 2008. *The Black Death: A Personal History*. Philadelphia: Da Capo.

Hawkins, Andrew J. 2021. "Elon Musk Just Now Realizing That Self-Driving Cars Are a 'Hard Problem.'" *Verge*, July 5. www.theverge.com/2021/7/5/22563751/tesla-elon-musk-full-self-driving-admission-autopilot-crash.

Heaven, Will Douglas. 2020. "Artificial General Intelligence: Are We Close, and Does It Even Make Sense to Try?" *MIT Technology Review*, October 15. www.technologyreview.com/2020/10/15/1010461/artificial-general-intelligence-robots-ai-agi-deepmind-google-openai.

Heldring, Leander, James Robinson, and Sebastian Vollmer. 2021a. "The Economic Effects of the English Parliamentary Enclosures." NBER Working Paper no. 29772. DOI:10.3386/w29772.

Heldring, Leander, James Robinson, and Sebastian Vollmer. 2021b. "The Long-Run Impact of the Dissolution of the English Monasteries." *Quarterly Journal of Economics* 136, no. 4: 2093–2145.

Helpman, Elhanan, and Manuel Trajtenberg. 1998. "Diffusion of General-Purpose Technologies." In *General-Purpose Technologies and Economic Growth*, edited by Helpman, 85–120. Cambridge, MA: MIT Press.

Henrich, Joseph. 2016. *The Secret of Our Success: How Culture Is Driving Human Evolution, Domesticating Our Species, and Making Us Smarter*. Princeton, NJ: Princeton University Press.

Hesser, Leon. 2019. *The Man Who Fed the World*. Princeton, NJ: Righter's Mill.

Hicks, John. 1932. *The Theory of Wages*. London: Macmillan.

Hill, Kashmir. 2020. "The Secretive Company That Might End Privacy as We Know It." *New York Times*, January 18 (updated November 2,

2021). www.nytimes.com/2020/01/18/technology/clearview-privacy -facial-recognition.html.

Hills, Richard L. 1994. *Power from Wind: A History of Windmill Technology*. Cambridge: Cambridge University Press.

Hindle, Steve. 1999. "Hierarchy and Community in the Elizabethan Parish: The Swallowfield Articles of 1596." *Historical Journal* 42, no. 3: 835–851.

Hindle, Steve. 2000. *The State and Social Change in Early Modern England, 1550–1640*. New York: Palgrave Macmillan.

Hinton, Geoff. 2016. "On Radiology." Creative Destruction Lab: Machine Learning and the Market for Intelligence, November 24. www .youtube.com/watch?v=2HMPRXstSvQ.

Hirschman, Albert O. 1958. *The Strategy of Economic Development*. New Haven, CT: Yale University Press.

Hochschild, Adam. 1999. *King Leopold's Ghost: A History of Greed, Terror, and Heroism in Colonial Africa*. Boston: Mariner.

Hollander, Samuel. 2019. "Ricardo on Machinery." *Journal of Economic Perspectives* 33, no. 2: 229–242.

Hounshell, David A. 1984. *From the American System to Mass Production, 1800–1932: The Development of Manufacturing Technology in the United States*. Baltimore: Johns Hopkins University Press.

Human Rights Watch. 2013. "'All You Can Do Is Pray': Crimes Against Humanity and Ethnic Cleansing of Rohingya Muslims in Burma's Arakan State," April. www.hrw.org/report/2013/04/22/all-you-can -do-pray/crimes-against-humanity-and-ethnic-cleansing-rohingya -muslims.

Hundt, Reed. 2019. *A Crisis Wasted: Barack Obama's Defining Decisions*. New York: Rosetta.

Huxley, Aldous. 1958. *Brave New World Revisited*. www.huxley.net/bnw -revisited.

Ilyas, Andrew, Shibani Santurkar, Dimitris Tsipras, Logan Engstrom, Brandon Tran, and Aleksander Mądry. 2019. "Adversarial Examples Are Not Bugs, They Are Features." *Gradient Science*, May 6. https:// gradientscience.org/adv.

Irwin, Neil. 2016. "What Was the Greatest Era for Innovation? A Brief Guided Tour." *New York Times*, May 13. www.nytimes.com /2016/05/15/upshot/what-was-the-greatest-era-for-american-inno vation-a-brief-guided-tour.html.

Isaacson, Walter. 2014. *The Innovators: How a Group of Hackers, Geniuses, and Geeks Created the Digital Revolution*. New York: Simon & Schuster.

Jack, William, and Tavneet Suri. 2011. "Mobile Money: The Economics of M-PESA." NBER Working Paper no. 16721. DOI:10.3386/w16721.

Jäger, Simon, Benjamin Schoefer, and Jörg Heining. 2021. "Labor in the Boardroom." *Quarterly Journal of Economics* 136, no. 2: 669–725.

James, John A., and Jonathan S. Skinner. 1985. "The Resolution of the Labor-Scarcity Paradox." *Journal of Economic History* 45:513–540.

Jefferys, James B. 1945 [1970]. *The Story of the Engineers, 1800–1945.* New York: Johnson Reprint.

Jensen, Michael C. 1986. "Agency Costs of Free Cash Flow, Corporate Finance, and Takeovers." *American Economic Review* 76, no. 2: 323–329.

Jensen, Michael C., and William H. Meckling. 1976. "Theory of the Firm: Managerial Behavior, Agency Costs and Ownership Structure." *Journal of Financial Economics* 3, no. 4: 305–360.

Jensen, Robert. 2006. "The Digital Provide: Information (Technology), Market Performance, and Welfare in the Indian Fisheries Sector." *Quarterly Journal of Economics* 122, no. 3: 879–924.

Johnson, Simon, and James Kwak. 2010. *13 Bankers: The Wall Street Takeover and the Next Financial Meltdown.* New York: Pantheon.

Johnson, Simon, and Peter Temin. 1993. "The Macroeconomics of NEP." *Economic History Review* 46, no. 4: 750–767.

Johnston, W. E. 1879. "Report." Part of "The Interoceanic Ship Canal Meeting at Chickering Hall." *Journal of the American Geographical Society of New York* 11:172–180.

Jones, Charles I. 1998. *Introduction to Economic Growth.* New York: Norton.

Jones, Robin. 2011. *Isambard Kingdom Brunel.* Barnsley: Pen and Sword.

Judis, John B. 1988. *William F. Buckley: Patron Saint of Conservatives.* New York: Simon & Schuster.

Judt, Tony. 2006. *Postwar: A History of Europe Since 1945.* New York: Penguin.

Kapelle, William E. 1979. *The Norman Conquest of the North: The Region and Its Transformation, 1000–1135.* Chapel Hill: University of North Carolina Press.

Karabarbounis, Loukas, and Brent Neiman. 2014. "The Global Decline of the Labor Share." *Quarterly Journal of Economics* 129, no. 1: 61–103.

Karabell, Zachary. 2003. *Parting the Desert.* New York: Knopf Doubleday.

Katz, Lawrence F., and Kevin M. Murphy. 1992. "Changes in Relative Wages, 1963–1987: Supply and Demand Factors." *Quarterly Journal of Economics* 107, no. 1: 35–78.

Katznelson, Ira. 2013. *Fear Itself: The New Deal and the Origins of Our Time.* New York: W.W. Norton.

Keene, Derek. 1998. "Feeding Medieval European Cities, 600–1500." Institute of Historical Research, University of London: School of Advanced Study. https://core.ac.uk/download/pdf/9548918.pdf.

Kelly, Morgan, Joel Mokyr, and Cormac Ó Gráda. 2014. "Precocious Albion: A New Interpretation of the British Industrial Revolution." *Annual Review of Economics* 6, no. 1: 363–389.

Kelly, Morgan, Joel Mokyr, and Cormac Ó Gráda. Forthcoming. "The Mechanics of the Industrial Revolution." *Journal of Political Economy*, https://papers.ssrn.com/sol3/papers.cfm?abstract_id=3628205.

Keltner, Dacher. 2016. *The Power Paradox: How We Gain and Lose Influence*. New York: Penguin.

Keltner, Dacher, Deborah H. Gruenfeld, and Cameron Anderson. 2003. "Power, Approach, and Inhibition." *Psychological Review* 110, no. 2: 265–284.

Kennedy, Charles. 1964. "Induced Bias in Innovation and the Theory of Distribution." *Economic Journal* 74:541–547.

Kennedy, John F. 1963. "Address at the Anniversary Convocation of the National Academy of Sciences," October 22. www.presidency.ucsb.edu/documents/address-the-anniversary-convocation-the-national-academy-sciences.

Kerr, Ian. 2007. *Engines of Change: The Railroads That Made India*. Santa Barbara, CA: Praeger.

Keynes, John Maynard. 1930 [1966]. "Economic Possibilities for Our Grandchildren." In Keynes, *Essays in Persuasion*. New York: W.W. Norton.

Kiley, Michael T. 1999. "The Supply of Skilled Labor and Skill-Biased Technological Progress." *Economic Journal* 109, no. 458: 708–724.

King, Gary, Jennifer Pan, and Margaret Roberts. 2013. "How Censorship in China Allows Government Criticism but Silences Collective Expression." *American Political Science Review* 107, no. 2: 326–343.

Kinross, Lord. 1969. *Between Two Seas: The Creation of the Suez Canal*. New York: William Morrow.

Kleinberg, Jon, Himabindu Lakkaraju, Jure Leskovec, Jens Ludwig, and Sendhil Mullainathan. 2018. "Human Decisions and Machine Predictions." *Quarterly Journal of Economics* 133, no. 1: 237–293.

Knowles, Dom David. 1940. *The Religious Houses of Medieval England*. London: Sheed & Ward.

Koepke, Nikola, and Joerg Baten. 2005. "The Biological Standard of Living in Europe during the Last Two Millennia." *European Review of Economic History* 9:61–95.

Koyama, Mark, and Jared Rubin. 2022. *How the World Became Rich: The Historical Origins of Economic Growth*. New York: Polity.

Kraus, Henry. 1979. *Gold Was the Mortar: The Economics of Cathedral Building*, Routledge Library Editions: The Medieval World, vol. 30. London: Routledge.

Krusell, Per, and José-Víctor Ríos-Rull. 1996. "Vested Interests in a Theory of Stagnation and Growth." *Review of Economic Studies* 63:301–330.

Krzywdzinski, Martin. 2021. "Automation, Digitalization, and Changes in Occupational Structure in the Automobile Industry in Germany, Japan, and the United States: A Brief History from the Early 1990s Until 2018." *Industrial and Corporate Change* 30, no. 3: 499–535.

Krzywdzinski, Martin, and Christine Gerber. 2020. "Varieties of Platform Work: Platforms and Social Inequality in Germany and the United States." Weizenbaum Series, number 7, May. DOI:10.34669/wi.ws/7.

Kuhn, Tom, and David Constantine, trans. and ed. 2019. *The Collected Poems of Bertolt Brecht.* New York: Liveright/Norton.

Kurzweil, Ray. 2005. *The Singularity Is Near: When Humans Transcend Biology.* New York: Penguin.

Lakwete, Angela. 2003. *Inventing the Cotton Gin: Machine and Myth in Antebellum America.* Baltimore: Johns Hopkins University Press.

Landemore, Helene. 2017. *Democratic Reason: Politics, Collective Intelligence, and the Rule of the Many.* Princeton, NJ: Princeton University Press.

Lane, Nathan. 2022. "Manufacturing Revolutions: Industrial Policy and Industrialization in South Korea." University of Oxford working paper, http://nathanlane.info/assets/papers/ManufacturingRevolutions_Lane_Live.pdf.

Langdon, John. 1986. *Horses, Oxen, and Technological Innovation: The Use of Draft Animals in English Farming from 1066 to 1500.* Cambridge: Cambridge University Press.

Langdon, John. 1991. "Water-Mills and Windmills in the West Midlands, 1086–1500." *Economic History Review* 44, no. 3: 424–444.

Lanier, Jaron. 2018. *Ten Arguments for Deleting Your Social Media Accounts Right Now.* New York: Hoffmann.

Lanier, Jaron. 2019. "Jaron Lanier Fixes the Internet." *New York Times*, September 23. www.nytimes.com/interactive/2019/09/23/opinion/data-privacy-jaron-lanier.html.

Lanier, Jaron, and E. Glen Weyl. 2020. "How Civic Technology Can Help Stop a Pandemic. Taiwan's Initial Success Is a Model for the Rest of the World." *Foreign Affairs*, March 20. www.foreignaffairs.com/articles/asia/2020-03-20/how-civic-technology-can-help-stop-pandemic.

Larson, Erik J. 2021. *The Myths of Artificial Intelligence: Why Computers Can't Think the Way We Do.* Cambridge, MA: Harvard University Press.

Le Guin, Ursula. 2004. "A Rant About 'Technology.'" www.ursulakleguinarchive.com/Note-Technology.html.

Leapman, Michael. 2001. *The World for a Shilling: How the Great Exhibition of 1851 Shaped a Nation*. London: Headline.

Leaver, E. W., and J. J. Brown. 1946. "Machines Without Men." *Fortune*, November 1.

Lecher, Colin. 2019. "How Amazon Automatically Tracks and Fires Warehouse Workers for 'Productivity.'" *Verge*, April 25. www.theverge.com/2019/4/25/18516004/amazon-warehouse-fulfillment-centers-productivity-firing-terminations.

Lee, Kai-Fu. 2021. "How AI Will Completely Change the Way We Live in the Next 20 Years." *Time*, September 14. https://time.com/6097625/kai-fu-lee-book-ai-2041.

Lee, Kai-Fu, and Chen Qiufan. 2021. *AI 2041: Ten Visions for Our Future*. New York: Currency.

Lehner, Mark. 1997. *The Complete Pyramids*. London: Thames & Hudson.

Lenin, Vladimir I. 1920 [1966]. *Collected Works*, vol. 31. Moscow: Progress.

Lent, Frank. 1895. *Suburban Architecture, Containing Hints, Suggestions, and Bits of Practical Advice for the Building of Inexpensive Country Houses*, 2nd ed. New York: W.T. Comstock.

Leontief, Wassily W. 1936. "Quantitative Input and Output Relations in the Economic Systems of the United States." *Review of Economic Statistics* 18, no. 3: 105–125.

Leontief, Wassily. 1983. "Technological Advance, Economic Growth, and the Distribution of Income." *Population and Development Review* 9, no. 3: 403–410.

Lesseps, Ferdinand de. 1880. "The Interoceanic Canal." *North American Review* 130, no. 278 (January): 1–15.

Lesseps, Ferdinand de. 1887 [2011]. *Recollections of Forty Years*, vol. 2. Translated by C. B. Pitman. Cambridge: Cambridge University Press.

Levasseur, E. 1897. "The Concentration of Industry, and Machinery in the United States." *Annals of the American Academy of Political and Social Science* 9 (March): 6–25.

Levine, Sheen S., Evan P. Apfelbaum, Mark Bernard, Valerie L. Bartelt, Edward J. Zajac, and David Stark. 2014. "Ethnic Diversity Deflates Price Bubbles." *Proceedings of the National Academy of Sciences* 111, no. 4: 18524–18529.

Levinson, Marc. 2006. *The Box: How the Shipping Container Made the World Smaller and the World Economy Bigger*. Princeton, NJ: Princeton University Press.

Levitt, Gerald M. 2000. *The Turk, Chess Automaton*. Jefferson, NC: McFarland.

Levy, Ro'ee. 2021. "Social Media, News Consumption, and Polarization: Evidence from a Field Experiment." *American Economic Review* 111, no. 3: 831–870.

Levy, Steven. 2010. *Hackers: Heroes of the Computer Revolution*, 25th anniversary ed. New York: O'Reilly.

Lewis, C. S. 1964. *Poems*. New York: Harcourt Brace.

Lewis, Michael. 1989. *Liar's Poker: Rising Through the Wreckage of Wall Street*. New York: W.W. Norton.

Lewis, R. A. 1952. *Edwin Chadwick and the Public Health Movement 1832–1854*. London: Longmans.

Li, Robin. 2020. *Artificial Intelligence Revolution: How AI Will Change Our Society, Economy, and Culture*. New York: Skyhorse. Kindle.

Licklider, J. C. R. 1960. "Man-Computer Symbiosis." *IRE Transactions on Human Factors in Electronics*, HFE-1: 4–11. https://groups.csail.mit.edu/medg/people/psz/Licklider.html.

Lin, Jeffrey. 2011. "Technological Adaptation, Cities, and New Work." *Review of Economics and Statistics* 93, no. 2: 554–574.

Link, Andreas. 2022. "Beasts of Burden, Trade, and Hierarchy: The Long Shadow of Domestication." University of Nuremberg working paper.

Lockhart, Paul. 2021. *Firepower: How Weapons Shaped Warfare*. New York: Basic Books.

Lonergan, Raymond. 1941. "A Steadfast Friend of Labor." In *Mr. Justice Brandeis, Great American*, edited by Irving Dillard, 42–45. Saint Louis: Modern View.

Lucas, Robert E. 1988. "On the Mechanics of Economic Development." *Journal of Monetary Economics* 22:3–42.

Luchtenberg, Daphne. 2022. "The Fourth Industrial Revolution Will Be People Powered." McKinsey: podcast, January 7. www.mckinsey.com/business-functions/operations/our-insights/the-fourth-industrial-revolution-will-be-people-powered.

Lyman, Joseph B. 1868. *Cotton Culture*. New York: Orange Judd.

Lyons, Derek E., Andrew G. Young, and Frank C. Keil. 2007. "The Hidden Structure of Overimitation." *Proceedings of the National Academy of Sciences* 104, no. 50: 19751–19756.

Macaulay, Thomas Babbington. 1848. *Macaulay's History of England, from the Accession of James II*, vol. 1. London: J.M. Dent.

MacDuffie, John Paul, and John Krafcik. 1992. "Integrating Technology and Human Resources for High-Performance Manufacturing: Evidence from the International Auto Industry." In *Transforming Organizations*, edited by Thomas A. Kochan and Michael Useem, 209–225. Oxford: Oxford University Press.

Macfarlane, Alan. 1978. *The Origins of English Individualism*. Oxford: Basil Blackwell.

MacFarquhar, Roderick, and Michael Schoenhals. 2008. *Mao's Last Revolution*. Cambridge, MA: Harvard University Press.

Mack, Gerstle. 1944. *The Land Divided: A History of the Panama Canal and Other Isthmian Canal Projects*. New York: Alfred A. Knopf.

Maddison, Angus. 2001. *The World Economy: A Millennial Perspective*. Paris: OECD Development Centre.

Malmendier, Ulrike, and Stefan Nagel. 2011. "Depression Babies: Do Macroeconomic Experiences Affect Risk Taking?" *Quarterly Journal of Economics* 126, no. 1: 373–416.

Malthus, Thomas. 1798 [2018]. *An Essay on the Principle of Population*. Edited by Joyce E. Chaplin. New York: W.W. Norton.

Malthus, Thomas. 1803 [2018]. *An Essay on the Principle of Population*. Edited by Shannon C. Stimson. New Haven, CT: Yale University Press.

Mankiw, N. Gregory. 2018. *Principles of Economics*, 8th ed. New York: Cengage.

Mann, Michael. 1986. *The Sources of Social Power*. Vol. 1, *A History of Power from the Beginning to AD 1760*. Cambridge: Cambridge University Press.

Mantoux, Paul. 1927. *The Industrial Revolution in the Eighteenth Century: An Outline of the Beginning of the Factory System in England*. Translated by Marjorie Vernon. London: Jonathan Cape.

Manuel, Frank E. 1956. *The New World of Henri Saint-Simon*. Cambridge, MA: Harvard University Press.

Manyika, James, Susan Lund, Michael Chui, Jacques Bughin, Jonathan Woetzel, Parul Batra, Ryan Ko, and Saurabh Sanghvi. 2017. "Jobs Lost, Jobs Gained: Workforce Transitions in a Time of Automation," McKinsey Global Institute, December. https://www.mckinsey.com/~/media/BAB489A30B724BECB5DEDC41E9BB9FAC.ashx.

Marantz, Andrew. 2020. *Antisocial: Online Extremists, Techno-Utopians and the Hijacking of the American Conversation*. New York: Penguin.

Marcus, Gary, and Ernest Davis. 2020. "GPT-3, Bloviator: OpenAI's Language Generator Has No Idea What It's Talking About." *MIT Technology Review*, August 22.

Marcus, Steven. 1974 [2015]. *Engels, Manchester, and the Working Class*. Routledge: London.

Marens, Richard. 2011. "We Don't Need You Anymore: Corporate Social Responsibilities, Executive Class Interests, and Solving Mizruchi and Hirschman's Paradox." https://heinonline.org/HOL/Page?handle=hein.journals/sealr35&id=1215&collection=journals&index.

Markoff, John. 2012. "Seeking a Better Way to Find Web Images." *New York Times*, November 19. www.nytimes.com/2012/11/20/science/for-web-images-creating-new-technology-to-seek-and-find.html.

Markoff, John. 2013. "In 1949, He Imagined an Age of Robots." *New York Times*, May 20. www.nytimes.com/2013/05/21/science/mit-scholars-1949-essay-on-machine-age-is-found.html.

Markoff, John. 2015. *Machines of Loving Grace: The Quest for Common Ground Between Humans and Robots.* New York: HarperCollins.

Marlowe, John. 1964. *The Making of the Suez Canal.* London: Cresset.

Marx, Karl. 1867 [1887]. *Capital: A Critique of Political Economy.* Moscow: Progress. www.marxists.org/archive/marx/works/download/pdf/Capital-Volume-I.pdf.

May, Alfred N. 1973. "An Index of Thirteenth-Century Peasant Impoverishment? Manor Court Fines." *Economic History Review* 26, no. 3: 389–402.

McCarthy, Tom. 2020. "Zuckerberg Says Facebook Won't Be 'Arbiters of Truth' After Trump Threat." *Guardian*, May 28. www.theguardian.com/technology/2020/may/28/zuckerberg-facebook-police-online-speech-trump.

McCauley, Brea. 2019. "Life Expectancy in Hunter-Gatherers." *Encyclopedia of Evolutionary Psychological Science*, January 1: 4552–4554.

McCloskey, Deirdre N. 2006. *The Bourgeois Virtues: Ethics for an Age of Commerce.* Chicago: University of Chicago Press.

McCormick, Brian. 1959. "Hours of Work in British Industry." *ILR Review* 12, no. 3 (April): 423–433.

McCraw, Thomas K. 2009. *American Business Since 1920: How It Worked*, 2nd ed. Chichester: Wiley Blackwell.

McCullough, David. 1977. *The Path Between the Seas: The Creation of the Panama Canal, 1870–1914.* New York: Simon & Schuster.

McEvedy, Colin, and Richard Jones. 1978. *Atlas of World Population History.* London: Penguin.

McFarland, C. K. 1971. "Crusade for Child Labourers: 'Mother' Jones and the March of the Mill Children." *Pennsylvania History: A Journal of Mid-Atlantic Studies* 38, no. 3 (July): 283–296.

McGerr, Michael. 2003. *A Fierce Discontent: The Rise and Fall of the Progressive Movement in America.* Oxford: Oxford University Press.

McGregor, Richard. 2010. *The Party: The Secret World of China's Communist Rulers.* New York: Harper.

McKibben, Bill. 2013. "The Fossil Fuel Resistance." *Rolling Stone*, April 11. www.rollingstone.com/politics/politics-news/the-fossil-fuel-resistance-89916.

McKinsey Global Institute. 2017. "Artificial Intelligence: The Next Digital Frontier." Discussion paper, June.

McKone, Jonna. 2010. "'Naked Streets' Without Traffic Lights Improve Flow and Safety." TheCityFix, October 18. https://thecityfix.com/blog/naked-streets-without-traffic-lights-improve-flow-and-safety.

McLean, Bethany, and Peter Elkind. 2003. *The Smartest Guys in the Room: The Amazing Rise and Scandalous Fall of Enron.* New York: Penguin.

Menocal, A. G. 1879. "Intrigues at the Paris Canal Congress." *North American Review* 129, no. 274 (September): 288–293.

Menzies, Heather. 1981. "Women and the Chip: Case Studies of the Effects of Informatics on Employment in Canada." Montreal: Institute for Research on Public Policy.

Mercier, Hugo, and Dan Sperber. 2017. *The Enigma of Reason*. Cambridge, MA: Harvard University Press.

Michaels, Guy. 2007. "The Division of Labour, Coordination, and the Demand for Information Processing." CEPR Discussion Paper no. DP6358. https://cepr.org/publications/dp6358.

Mill, John Stuart. 1848. *Principles of Political Economy*. Edited by W. G. Ashley. London: Longmans, Green.

Mingay, G. E. 1997. *Parliamentary Enclosure in England: An Introduction to Its Causes, Incidence, and Impact 1750–1850*. London: Routledge.

Mithen, Steven. 2003. *After the Ice: A Global Human History 20,000–5,000 BC*. Cambridge, MA: Harvard University Press.

Mitrunen, Matti. 2019. "War Reparations, Structural Change, and Intergenerational Mobility." Institute for International Economic Studies, Stockholm University, working paper, January 2.

Moene, Karl-Ove, and Michael Wallerstein. 1997. "Pay Inequality." *Journal of Labor Economics* 15, no. 3 (July 1997): 403–430.

Mokyr, Joel. 1988. "Is There Still Life in the Pessimistic Case? Consumption During the Industrial Revolution, 1790–1850." *Journal of Economic History* 48, no. 1: 69–92.

Mokyr, Joel. 1990. *The Lever of Riches, Technological Creativity and Economic Progress*. New York: Oxford University Press.

Mokyr, Joel. 1993. "Introduction." In *The British Industrial Revolution: An Economic Perspective*, edited by Mokyr, 1–131. Boulder, CO: Westview.

Mokyr, Joel. 2002. *The Gifts of Athena: Historical Origins of the Knowledge Economy*. Princeton, NJ: Princeton University Press.

Mokyr, Joel. 2010. *Enlightened Economy: An Economic History of Britain, 1700–1850*. New Haven, CT: Yale University Press.

Mokyr, Joel. 2016. *A Culture of Growth: The Origins of the Modern Economy*. Princeton, NJ: Princeton University Press.

Mokyr, Joel, Chris Vickers, and Nicolas L. Ziebarth. 2015. "The History of Technological Anxiety in the Future of Economic Growth: Is This Time Different?" *Journal of Economic Perspectives* 29, no. 3: 31–50.

Morris, Ian. 2004. "Economic Growth in Ancient Greece." *Journal of Institutional and Theoretical Economics* 160:709–742.

Morris, Ian. 2013. *The Measure of Civilization: How Social Development Decides the Fate of Nations*. Princeton, NJ: Princeton University Press.

Morris, Ian. 2015. *Foragers, Farmers, and Fossil Fuels: How Human Values Evolve*. Princeton, NJ, Princeton University Press.

Mortensen, Dale. 1982. "Property Rights and Efficiency in Mating, Racing and Related Games." *American Economic Review* 72:968–979.

Mortimer, Thomas. 1772. *The Elements of Commerce, Politics and Finances*. London: Hooper.

Moscona, Jacob, and Karthik Sastry. 2022. "Inappropriate Technology: Evidence from Global Agriculture." April 19. Available at SSRN 3886019.

Mougel, Nadège. 2011. "World War I Casualties." REPERES, module 1–0, explanatory notes. http://www.centre-robert-schuman.org/userfiles/files /REPERES%20–%20module%201-1-1%20-%20explanatory%20 notes%20–%20World%20War%20I%20casualties%20–%20EN.pdf.

Muldrew, Craig. 2017. "The 'Middling Sort': An Emergent Cultural Identity." In *A Social History of England, 1500–1750*, edited by Keith Wrightson, 290–309. Cambridge: Cambridge University Press.

Müller, Karsten, and Carlo Schwarz. 2021. "Fanning the Flames of Hate: Social Media and Hate Crime." *Journal of the European Economic Association* 19, no. 4: 2131–2167.

Muralidharan, Karthik, Abhijeet Singh, and Alejandro J. Ganimian. 2019. "Disrupting Education? Experimental Evidence on Technology-Aided Instruction in India." *American Economic Review* 109, no. 4: 1426–1460.

Murnane, Richard J., and Frank Levy. 1996. *Teaching the New Basic Skills: Principles for Educating Children to Thrive in the Changing Economy*. New York: Free Press.

Naidu, Suresh, and Noam Yuchtman. 2013. "Coercive Contract Enforcement: Law and the Labor Market in Nineteenth Century Industrial Britain." *American Economic Review* 103, no. 1: 107–144.

Napier, William. 1857 [2011]. *The Life and Opinions of General Sir Charles James Napier, G.C.B.*, vol. 2. Cambridge: Cambridge University Press.

Ndzi, Ernestine Gheyoh. 2019. "Zero-Hours Contracts Have a Devastating Impact on Career Progression—Labour Is Right to Ban Them." *Conversation*, September 24. https://theconversation.com/zero-hours -contracts-have-a-devastating-impact-on-career-progression-labour -is-right-to-ban-them-123066.

Neapolitan, Richard E., and Xia Jiang. 2018. *Artificial Intelligence: With an Introduction to Machine Learning*, 2nd ed. London: Chapman and Hall/CRC.

Neeson, J. M. 1993. *Commoners, Common Right, Enclosure and Social Change in England, 1700–1820*. Cambridge: Cambridge University Press.

Noble, David. 1977. *America by Design: Science, Technology, and the Rise of Corporate Capitalism*. New York: Alfred A. Knopf.

Noble, David. 1984. *Forces of Production: A Social History of Industrial Automation*. New York: Alfred A. Knopf.

Norman, Douglas. 2013. *The Design of Everyday Things*. New York: Basic Books.

North, Douglass C. 1982. *Structure and Change in Economic History*. New York: W.W. Norton.

North, Douglass C., and Robert Paul Thomas. 1973. *The Rise of the Western World: A New Economic History*. Cambridge: Cambridge University Press.

North, Douglass C., John Wallis, and Barry R. Weingast. 2009. *Violence and Social Orders: A Conceptual Framework for Interpreting Recorded Human History*. New York: Cambridge University Press.

Nye, David E. 1992. *Electrifying America: Social Meanings of a New Technology*. Cambridge, MA: MIT Press.

Nye, David E. 1998. *Consuming Power: A Social History of American Energies*. Cambridge, MA: MIT Press.

Ober, Josiah. 2015a. "Classical Athens." In *Fiscal Regimes and Political Economy of Early States*, edited by Walter Scheidel and Andrew Monson, 492–522. Cambridge: Cambridge University Press.

Ober, Josiah. 2015b. *The Rise and Fall of Classical Greece*. New York: Penguin.

O'Neil, Cathy. 2016. *Weapons of Math Destruction: How Big Data Increases Inequality and Threatens Democracy*. New York: Penguin.

O'Neil, Cathy. 2022. *The Shame Machine: Who Profits in the New Age of Humiliation*. New York: Crown.

Orwell, George. 1949. *Nineteen Eighty-Four*. London: Secker and Warburg.

Overton, Mark. 1996. *Agricultural Revolution in England: The Transformation of the Agrarian Economy, 1500–1850*. Cambridge: Cambridge University Press.

Pan, Alexander, Kush Bhatia, and Jacob Steinhardt. 2022. "The Effects of Reward Misspecification: Mapping and Mitigating Misaligned Models." arxiv.org. https://arxiv.org/abs/2201.03544.

Pasquale, Frank. 2015. *The Black Box Society: The Secret Algorithms That Control Money and Information*. Cambridge, MA: Harvard University Press.

Pearl, Judea. 2021. "Radical Empiricism and Machine Learning Research." *Journal of Causal Inference* 9, no. 1 (May 24): 78–82.

Pelling, Henry. 1976. *A History of British Trade Unionism*, 3rd ed. London: Penguin.

Perlstein, Rick. 2009. *Before the Storm: Barry Goldwater and the Unmaking of the American Consensus*. New York: Bold Type Books.

Pethokoukis, James. 2016. "The Productivity Paradox: Why the US Economy Might Be a Lot Stronger Than the Government Is Saying." AEI Blog, May 20. www.aei.org/technology-and-innovation/the-productivity-paradox-us-economy-might-be-a-lot-stronger.

Pethokoukis, James. 2017a. "Google Economist Hal Varian Tries to Explain America's Productivity Paradox, and How Workers Should Deal with Automation," May 5. www.aei.org/economics/google-economist-hal-varian-tries-to-explain-americas-productivity-paradox-and-how-workers-should-deal-with-automation.

Pethokoukis, James. 2017b. "If Not Mismeasurement, Why Is Productivity Growth So Slow?" AEI Blog, February 14. https://www.aei.org/economics/if-not-mismeasurement-why-is-productivity-growth-so-slow/.

Philippon, Thomas. 2019. *The Great Reversal: How America Gave Up on Free Markets*. Cambridge, MA: Harvard University Press.

Philippon, Thomas, and Ariell Reshef. 2012. "Wages in Human Capital in the U.S. Finance Industry: 1909–2006." *Quarterly Journal of Economics* 127:1551–1609.

Phillips-Fein, Kim. 2010. *Invisible Hands: The Businessmen's Crusade Against the New Deal*. New York: W.W. Norton.

Piff, Paul K., Daniel M. Stancato, Stéphane Côté, Rodolfo Mendoza-Denton, and Dacher Keltner. 2012. "Higher Social Class Predicts Increased Unethical Behavior." *Proceedings of the National Academy of Sciences* 109, no. 11: 4086–4091.

Piketty, Thomas, and Emmanuel Saez. 2003. "Income Inequality in the United States, 1913–1998." *Quarterly Journal of Economics* 118, no. 1: 1–41.

Pirenne, Henri. 1937. *Economic and Social History of Medieval Europe*. New York: Harcourt Brace.

Pirenne, Henri. 1952. *Medieval Cities: Their Origins and the Revival of Trade*. Princeton, NJ: Princeton University Press.

Pissarides, Christopher. 1985. "Short-Run Equilibrium Dynamics of Unemployment, Vacancies, and Real Wages." *American Economic Review* 75, no. 4: 676–690.

Pissarides, Christopher. 2000. *Equilibrium Unemployment Theory*, 2nd ed. Cambridge, MA: MIT Press.

Poe, Edgar Allan. 1836 [1975]. "Maelzel's Chess Player." In *The Complete Tales and Poems of Edgar Allan Poe*. New York: Vintage.

Pollard, Sidney. 1963. "Factory Discipline in the Industrial Revolution." *Economic History Review* 16, no. 2: 254–271.

Pomeranz, Kenneth. 2001. *The Great Divergence: China, Europe and the Making of the Modern World Economy*. Princeton, NJ: Princeton University Press.

Popp, David. 2002. "Induced Innovation and Energy Prices." *American Economic Review* 92:160–180.

Porter, Roy. 1982. *English Society in the Eighteenth Century*, Penguin Social History of Britain. London: Penguin.

Posner, Eric A., and E. Glen Weyl. 2019. *Radical Markets*. Princeton, NJ: Princeton University Press.

Postan, M. M. 1966. "Medieval Agrarian Society in Its Prime: England." In *The Cambridge Economic History of Europe*, edited by Postan, 548–632. London: Cambridge University Press.

Postman, Neil. 1985. *Amusing Ourselves to Death: Public Discourse in the Age of Show Business*. New York: Penguin.

Prasad, Aryamala. 2020. "Two Years Later: A Look at the Unintended Consequences of GDPR." Regulatory Studies Center, George Washington University, September 2. https://regulatorystudies.columbian.gwu.edu/unintended-consequences-gdpr.

Purnell, Newley, and Jeff Horwitz. 2020. "Facebook's Hate-Speech Rules Collide with Indian Politics." *Wall Street Journal*, August 14.

Pyne, Stephen J. 2019. *Fire: A Brief History*, 2nd ed. Seattle: University of Washington Press.

Qin, Bei, David Strömberg, and Yanhui Wu. 2017. "Why Does China Allow Freer Social Media? Protests vs. Surveillance and Propaganda." *Journal of Economic Perspectives* 31, no. 1: 117–140.

Raghu, Maithra, Katy Blumer, Greg Corrado, Jon Kleinberg, Ziad Obermeyer, and Sendhil Mullainathan. 2019. "The Algorithmic Automation Problem: Prediction, Trash, and Human Effort." arxiv.org. https://arxiv.org/abs/1903.12220.

Rathje, Steve, Jay J. Van Bavel, and Sander van der Linden. 2021. "Out-Group Animosity Drives Engagement on Social Media." *Proceedings of the National Academy of Sciences* 118, no. 26: e2024292118.

Reich, David. 2018. *Who We Are and How We Got Here: Ancient DNA and the New Science of the Human Past*. New York: Pantheon.

Remarque, Erich Maria. 1928 [2013]. *All Quiet on the Western Front*. Translated by A. W. Wheen. New York: Random House.

Reuters Staff. 2009. "Goldman Sachs Boss Says Banks Do 'God's Work.'" November 8. www.reuters.com/article/us-goldmansachs-blankfein/goldman-sachs-boss-says-banks-do-gods-work-idUSTRE5A719520091108.

Reynolds, Terry S. 1983. *Stronger Than a Hundred Men: A History of the Vertical Water Wheel*. Baltimore: Johns Hopkins University Press.

Ricardo, David. 1821 [2001]. *On the Principles of Political Economy, and Taxation*, 3rd ed. Kitchener, ON: Batoche.

Ricardo, David. 1951–1973. *The Works and Correspondences of David Ricardo*. Edited by Piero Sraffa. Cambridge: Cambridge University Press.

Richardson, Ruth. 2012. *Dickens and the Workhouse: Oliver Twist and the London Poor*. Oxford: Oxford University Press.

Richmond, Alex B. 1825. *Narrative of the Condition of the Manufacturing Population*. London: John Miller.

Riggio, Ronald E. 2014. "What Is Social Intelligence? Why Does It Matter?" *Psychology Today*, July 1. www.psychologytoday.com/us/blog/cutting-edge-leadership/201407/what-is-social-intelligence-why-does-it-matter.

Roberts, Andrew. 1991. *The Holy Fox: The Life of Lord Halifax*. London: George Weidenfeld and Nicolson.

Rolt, L. T. C. 2009. *George and Robert Stephenson: The Railway Revolution*. Chalford: Amberley.

Romer, Paul M. 1990. "Endogenous Technological Change." *Journal of Political Economy* 98 (part I): S71–S102.

Romer, Paul M. 2021. "Taxing Digital Advertising," May 17. https://adtax.paulromer.net.

Romero, Alberto. 2021. "5 Reasons Why I Left the AI Industry." https://towardsdatascience.com/5-reasons-why-i-left-the-ai-industry-2c88ea183cdd.

Roose, Kevin. 2019. "The Making of a YouTube Radical." *New York Times*, June 8.

Roose, Kevin. 2021. "The Robots Are Coming for Phil in Accounting." *New York Times*, March 6.

Rosen, George. 1993. *A History of Public Health*. Baltimore: Johns Hopkins.

Rosenberg, Nathan. 1972. *Technology in American Economic Growth*. New York: M.E. Sharpe.

Rosenblueth, Arturo, and Norbert Wiener. 1945. "The Role of Models in Science." *Philosophy of Science* 12, no. 4 (October): 316–321.

Rosenthal, Caitlin. 2018. *Accounting for Slavery: Masters and Management*. Cambridge, MA: Harvard University Press.

Royal Commission of Inquiry into Children's Employment. 1842 [1997]. *Report by Jelinger C. Symons Esq., on the Employment of Children and Young Persons in the Mines and Collieries of the West Riding of Yorkshire, and on the State, Condition and Treatment of Such Children and Young Persons*. Edited by Ian Winstanley. Coal Mining History Resource Centre. Wigan: Picks Publishing. www.cmhrc.co.uk/cms/document/1842_Yorkshir__1.pdf.

Russell, J. C. 1972. "Population in Europe." In *The Fontana Economic History of Europe: The Middle Ages*, edited by Carlo M. Cipolla, 25–70. London: Collins/Fontana.

Russell, Jon. 2019. "Facebook Bans Four Armed Groups in Myanmar." *TechCrunch*, February 5. https://techcrunch.com/2019/02/05/facebook-bans-four-insurgent-groups-myanmar.

Russell, Josiah Cox. 1944. "The Clerical Population of Medieval England." *Traditio* 2:177–212.

Russell, Stuart J. 2019. *Human Compatible: Artificial Intelligence and the Problem of Control.* New York: Penguin.

Russell, Stuart J., and Peter Norvig. 2009. *Artificial Intelligence: A Modern Approach*, 3rd ed. Hoboken, NJ: Prentice Hall.

Safire, William. 2008. *Safire's Political Dictionary*, rev. ed. Oxford: Oxford University Press.

Samuelson, Paul A. 1965. "A Theory of Induced Innovation Along Kennedy-Weisäcker Lines." *Review of Economics and Statistics* 47:343–356.

Sandel, Michael J. 2020. *The Tyranny of Merit: What's Become of the Common Good?* New York: Penguin.

Sapolsky, Robert M. 2017. *Behave: The Biology of Humans at Our Best and Worst.* New York: Penguin.

Satyanath, Shanker, Nico Voigtländer, and Hans-Joachim Voth. 2017. "Bowling for Fascism: Social Capital and the Rise of the Nazi Party." *Journal of Political Economy* 125, no. 2: 478–526.

Schneider, Gregory. 2003. *Conservatism in America Since 1930: A Reader.* New York: New York University Press.

Schumacher, Shannon, and J. J. Moncus. 2021. "Economic Attitudes Improve in Many Nations Even as Pandemic Endures." Pew Research Center, July 21, 2021. www.pewresearch.org/global/2021/07/21/economic-attitudes-improve-in-many-nations-even-as-pandemic-endures.

Scott, James C. 2017. *Against the Grain: A Deep History of the Earliest States.* New Haven, CT: Yale University Press.

Select Committee. 1834. *Report from Select Committee on Hand-Loom Weavers' Petitions*, August 4, 1834, House of Commons.

Select Committee. 1835. *Report from Select Committee on Hand-Loom Weavers' Petitions*, July 1, 1835, House of Commons.

Sharp, Andrew. 1998. *The English Levellers.* Cambridge: Cambridge University Press.

Shears, Jonathan. 2017. *The Great Exhibition, 1851: A Sourcebook.* Manchester, UK: Manchester University Press.

Shilts, Randy. 2007. *And the Band Played On: Politics, People, and the AIDS Epidemic*, 20th anniversary ed. New York: St. Martin's Griffin.

Shimada, Haruo, and John Paul MacDuffie. 1986. "Industrial Relations and 'Humanware': Japanese Investments in Automobile Manufacturing in the United States." MIT Sloan School Working Paper no. 1855–87, December. https://dspace.mit.edu/bitstream/handle/1721.1/48159/industrialrelati00shim.pdf;sequence=1.

Shirky, Clay. 2011. "The Political Power of Social Media." *Foreign Affairs*, January/February.

Shneiderman, Ben. 2022. *Human-Centered AI*. New York: Oxford University Press.

Shteynberg, Garriy, and Evan P. Apfelbaum. 2013. "The Power of Shared Experience: Simultaneous Observation with Similar Others Facilitates Social Learning." *Social Psychological and Personality Science* 4, no. 6: 738–744.

Siegfried, André. 1940. *Suez and Panama*. Translated by Henry Harold Hemming and Doris Hemming. London: Jonathan Cape.

Silvestre, Henri. 1969. *L'isthme de Suez 1854–1869*. Marseille: Cayer.

Simonite, Tom. 2016. "How Google Plans to Solve Artificial Intelligence." *MIT Technology Review*, March 31. www.technologyreview.com/2016/03/31/161234/how-google-plans-to-solve-artificial-intelligence.

Smil, Vaclav. 1994. *Energy in World History*. New York: Routledge.

Smil, Vaclav. 2017. *Energy and Civilization: A History*, rev. ed. Cambridge, MA: MIT Press.

Smith, Adam. 1776 [1999]. *The Wealth of Nations*, books I–III. London: Penguin Classics.

Smith, Bruce D. 1995. *The Emergence of Agriculture*. New York: Scientific American Library.

Smith, Gary, and Jeffrey Funk. 2021. "AI Has a Long Way to Go Before Doctors Can Trust It with Your Life." *Quartz*, June 4, last updated July 20, 2022. https://qz.com/2016153/ai-promised-to-revolutionize-radiology-but-so-far-its-failing.

Solow, Robert M. 1956. "A Contribution to the Theory of Economic Growth." *Quarterly Journal of Economics* 70:65–94.

Solow, Robert M. 1987. "We'd Better Watch Out." *New York Times Book Review*, July 12, 36.

Sorkin, Amy Davidson. 2013. "Edward Snowden, the N.S.A. Leaker, Comes Forward." *New Yorker*, June 9.

Spear, Percival. 1965. *The Oxford History of Modern India, 1740–1947*. Oxford: Clarendon.

Specter, Michael. 2021. "How ACT UP Changed America." *New Yorker*, June 7. www.newyorker.com/magazine/2021/06/14/how-act-up-changed-america.

Spinrad, R. J. 1982. "Office Automation." *Science* 215, no. 4534: 808–813.

Stalin, Joseph V. 1954. *Works*, vol. 12. Moscow: Foreign Languages Publishing.

Statute of Labourers. 1351. From *Statutes of the Realm*, 1:307. https://avalon.law.yale.edu/medieval/statlab.asp.

Steadman, Philip. 2012. "Samuel Bentham's Panopticon." *Journal of Bentham Studies* 14, no. 1: 1–30.

Steinfeld, Robert J. 1991. *The Invention of Free Labor: The Employment Relation in English and American Law and Culture, 1350–1870.* Chapel Hill: University of North Carolina Press.

Stella, Massimo, Emilio Ferrara, and Manlio De Domenico. 2018. "Bots Increase Exposure to Negative and Inflammatory Content in Online Social Systems." *Proceedings of the National Academy of Sciences* 115, no. 49: 12435–12440.

Steuart, James. 1767. *An Inquiry into the Principles of Political Economy.* London: A. Millar and T. Cadell.

Stewart, Frances. 1977. *Technology and Underdevelopment.* London: Macmillan.

Story, Louise, and Eric Dash. 2009. "Bankers Reaped Lavish Bonuses During Bailouts." *New York Times*, July 30.

Strenze, Tarmo. 2007. "Intelligence and Socioeconomic Success: A Meta-analytical Review of Longitudinal Research." *Intelligence* 35:401–426.

Sunstein, Cass. 2001. *Republic.com.* Princeton, NJ: Princeton University Press.

Susskind, Daniel. 2020. *A World Without Work: Technology, Automation, and How We Should Respond.* New York: Picador.

Suzman, James. 2017. *Affluence Without Abundance: What We Can Learn from the World's Most Successful Civilization.* London: Bloomsbury.

Swaminathan, Nikhil. 2008. "Why Does the Brain Need So Much Power?" *Scientific American*, April 28.

Swanson, R. N. 1995. *Religion and Devotion in Europe, c. 1215–c. 1515,* Cambridge Medieval Textbooks. Cambridge: Cambridge University Press.

Tallet, Pierre, and Mark Lehner. 2022. *The Red Sea Scrolls: How Ancient Papyri Reveal the Secrets of the Pyramids.* London: Thames & Hudson.

Tang, Audrey. 2019. "A Strong Democracy Is a Digital Democracy." *New York Times*, October 15.

Tarbell, Ida M. 1904. *The History of the Standard Oil Company.* New York: Macmillan.

Taub, Amanda, and Max Fisher. 2018. "Where Countries Are Tinderboxes and Facebook Is a Match." *New York Times*, April 21.

Tawney, R. H. 1941. "The Rise of the Gentry." *Economic History Review* 11:1–38.

Taylor, Bill. 2011. "Great People Are Overrated." *Harvard Business Review*, June 20. https://hbr.org/2011/06/great-people-are-overrated.

Taylor, Keith, trans. and ed. 1975. *Henri Saint-Simon (1760–1825): Selected Writings on Science, Industry, and Social Organisation.* London: Routledge.

Tellenbach, Gerd. 1993. *The Church in Western Europe from the Tenth to the Early Twelfth Century*, Cambridge Medieval Textbooks. Cambridge: Cambridge University Press.

Thelen, Kathleen A. 1991. *Union of Parts: Labor Politics and Postwar Germany*. Ithaca, NY: Cornell University Press.

Thelwall, John. 1796. *The Rights of Nature, Against the Usurpations of Establishments*. London: H.D. Symonds.

Thompson, E. P. 1966. *The Making of the English Working Class*. New York: Vintage.

Thrupp, Sylvia L. 1972. "Medieval Industry." In *The Fontana Economic History of Europe: The Middle Ages*, edited by Carlo M. Cipolla, 221–273. London: Collins/Fontana.

Timberg, Craig, Tony Romm, and Drew Harwell. 2019. "A Facebook Policy Lets Politicians Lie in Ads, Leaving Democrats Fearing What Trump Will Do." *Washington Post*, October 10. www.washington post.com/technology/2019/10/10/facebook-policy-political -speech-lets-politicians-lie-ads.

Time. 1961. "Men of the Year: U.S. Scientists," January 2. https://content .time.com/time/subscriber/article/0,33009,895239,00.html.

Tirole, Jean. 2021. "Digital Dystopia." *American Economic Review* 111, no. 6: 2007–2048.

Tomasello, Michael. 1995. "Joint Attention as Social Cognition." In *Joint Attention: Its Origins and Role in Development*, edited by C. Moore and P. J. Dunham, 103–130. Mahwah, NJ: Lawrence Erlbaum.

Tomasello, Michael. 2019. *Becoming Human: A Theory of Ontogeny*. Cambridge, MA: Harvard University Press.

Tomasello, Michael, Malinda Carpenter, Josep Call, Tanya Behne, and Henrike Moll. 2005. "Understanding and Sharing Intentions: The Origins of Cultural Cognition." *Behavioral and Brain Sciences* 28, no. 5: 675–691.

Trefler, Alan. 2018. "The Big RPA Bubble." *Forbes*, December 2. www.forbes.com/sites/cognitiveworld/2018/12/02/the-big-rpa -bubble/?sh=9972fe68d950.

Tugwell, Rexford G. 1933. "Design for Government." *Political Science Quarterly* 48, no. 3 (September): 331–332.

Tunzelmann, G. N. von. 1978. *Steam Power and British Industrialization to 1860*. Oxford: Clarendon.

Turing, Alan. 1950. "Computing Machinery and Intelligence." *Mind* 59, no. 236: 433–460.

Turing, Alan. 1951 [2004]. "Intelligent Machinery, a Heretical Theory." In *The Turing Test: Verbal Behavior as the Hallmark of Intelligence*, edited by Stuart M. Shieber, 105–110. Cambridge, MA: MIT Press.

Turner, John. 1991. *Social Influence*. New York: Thomson Brooks/Cole.

Tworek, Heidi. 2019. "A Lesson from 1930s Germany: Beware State Control of Social Media." *Atlantic*, May 26. www.theatlantic.com/international/archive/2019/05/germany-war-radio-social-media/590149.

Tzouliadis, Tim. 2008. *The Forsaken: An American Tragedy in Stalin's Russia*. New York: Penguin.

Ure, Andrew. 1835 [1861]. *The Philosophy of Manufactures or, an Exposition of the Scientific, Moral, and Commercial Economy of the Factory System of Great Britain*. London: H.G. Bohn.

Varian, Hal. 2016. "A Microeconomist Looks at Productivity: A View from the Valley." Brookings Institute Presentation slides. www.brookings.edu/wp-content/uploads/2016/08/varian.pdf.

Vassallo, Steve. 2021. "How I Learned to Stop Worrying and Love AI." *Forbes*, February 3. www.forbes.com/sites/stevevassallo/2021/02/03/how-i-learned-to-stop-worrying-and-love-ai.

Verna, Inder M. 2014. "Editorial Expression of Concern: Experimental Evidence of Massivescale Emotional Contagion Through Social Networks." *PNAS* 111, no. 29 (July 3): 10779. www.pnas.org/doi/10.1073/pnas.1412469111.

Vosoughi, Soroush, Deb Roy, and Sinan Aral. 2018. "The Spread of True and False News Online." *Science* 359:1146–1151.

Voth, Hans-Joachim. 2004. "Living Standards and the Urban Environment." In *The Cambridge Economic History of Modern Britain*, edited by Roderick Floud and Paul Johnson, 268–294. Cambridge: Cambridge University Press.

Voth, Hans-Joachim. 2012. *Time and Work in England During the Industrial Revolution*. New York: Xlibris.

Waldman, Steve Randy. 2021. "The 1996 Law That Ruined the Internet." *Atlantic*, January 3. www.theatlantic.com/ideas/archive/2021/01/trump-fighting-section-230-wrong-reason/617497.

Wang, Tianyi. 2021. "Media, Pulpit, and Populist Persuasion: Evidence from Father Coughlin." *American Economic Review* 111, no. 9: 3064–3094.

Warner, R. L. 1904. "Electrically Driven Shops." *Journal of the Worcester Polytechnic Institute* 7, no. 2 (January): 83–100.

Welldon, Finn, R. 1971. *The Norman Conquest and Its Effects on the Economy*. Hamden, CT: Archon.

Wells, H. G. 1895 [2005]. *The Time Machine*. London: Penguin Classics.

West, Darrell M. 2018. *The Future of Work: Robots, AI and Automation*. Washington: Brookings Institution.

White, Lynn Jr. 1964. *Medieval Technology and Social Change*. New York: Oxford University Press.

White, Lynn Jr. 1978. *Medieval Religion and Technology: Collected Essays.* Berkeley: University of California Press.

Wickham, Christopher. 2016. *Medieval Europe.* New Haven, CT: Yale University Press.

Wiener, Jonathan M. 1978. *Social Origins of the New South: Alabama, 1860–1885.* Baton Rouge: Louisiana State University Press.

Wiener, Norbert. 1949. "The Machine Age." Version 3. Unpublished paper, Massachusetts Institute of Technology. https://libraries.mit .edu/app/dissemination/DIPonline/MC0022/MC0022_Machine AgeV3_1949.pdf.

Wiener, Norbert. 1954. *The Human Use of Human Beings: Cybernetics and Society.* Boston: Da Capo.

Wiener, Norbert. 1960. "Some Moral and Technical Consequences of Automation." *Science* 131 (n.s.), no. 3410: 1355–1358.

Wiener, Norbert. 1964. *God and Golem, Inc: A Comment on Certain Points Where Cybernetics Impinges on Religion.* Cambridge, MA: MIT Press.

Wilkinson, Toby. 2020. *A World Beneath the Sands: The Golden Age of Egyptology.* New York: W.W. Norton.

Williams, Callum. 2021. "A Bright Future for the World of Work." *The Economist: Special Report*, April 8. www.economist.com/ special-report/2021/04/08/a-bright-future-for-the-world-of-work.

Williamson, Jeffrey G. 1985. *Did British Capitalism Breed Inequality?* London: Routledge.

Wilson, Arnold. 1939. *The Suez Canal: Its Past, Present, and Future.* Oxford: Oxford University Press.

Wolmar, Christian. 2007. *Fire & Steam: How the Railways Transformed Britain.* London: Atlantic.

Wolmar, Christian. 2010. *Blood, Iron, & Gold: How the Railways Transformed the World.* New York: PublicAffairs.

Wood, Gaby. 2002. *Edison's Eve: A Magical History of the Quest for Mechanical Life.* New York: Anchor.

Woodward, C. Vann. 1955. *The Strange Career of Jim Crow.* New York: Oxford University Press.

Wooldridge, Michael. 2020. *A Brief History of Artificial Intelligence: What It Is, Where We Are, and Where We Are Going.* New York: Flatiron.

Wright, Gavin. 1986. *Old South, New South: Revolutions in the Southern Economy Since the Civil War.* New York: Basic Books.

Wright, Katherine I. (Karen). 2014. "Domestication and Inequality? Households, Corporate Groups, and Food Processing Tools at Neolithic Çatalhöyük." *Journal of Anthropological Archaeology* 33:1–33.

Wrightson, Keith. 1982. *English Society, 1580–1680.* New Brunswick, NJ: Rutgers University Press. Kindle.

Wrightson, Keith, ed. 2017. *A Social History of England, 1500–1750*. Cambridge: Cambridge University Press.

Xu, Beina, and Eleanor Albert. 2017. "Media Censorship in China." Council on Foreign Relations, https://www.cfr.org/backgrounder /media-censorship-china.

Young, Arthur. 1768. *The Farmer's Letters to the People of England*. London: Strahan.

Young, Arthur. 1771. *The Farmer's Tour Through the East of England*. London: Strahan.

Young, Arthur. 1801. *An Inquiry into the Propriety of Applying Wastes to the Better Maintenance and Support of the Poor*. Rackham: Angel Hill.

Zeira, Joseph. 1998. "Workers, Machines, and Economic Growth." *Quarterly Journal of Economics* 113, no. 4: 1091–1117.

Zhang, Liang, Andrew Nathan, Perry Link, and Orville Schell. 2002. *The Tiananmen Papers*. New York: PublicAffairs.

Zhong, Raymond, Paul Mozur, Aaron Krolik, and Jeff Kao. 2020. "Leaked Documents Show How China's Army of Paid Internet Trolls Helped Censor the Coronavirus." *ProPublica*, December 19. www.propublica.org/article/leaked-documents-show-how-chinas -army-of-paid-internet-trolls-helped-censor-the-coronavirus.

Zuboff, Shoshana. 1988. *In the Age of the Smart Machine: The Future of Work and Power*. New York: Basic Books.

Zuboff, Shoshana. 2019. *The Age of Surveillance Capitalism: The Fight for a Human Future at the New Frontier of Power*. London: Profile Books.

Zweig, Stefan. 1943. *The World of Yesterday*. Translated by Benjamin W. Huebsch and Helmut Ripperger. New York: Viking.

Acknowledgments

This book builds on two decades of research that we have conducted on technology, inequality, and institutions. In this process, we have accumulated a huge amount of intellectual debt to many scholars, whose influence can be seen clearly throughout the book. Two of those, Pascual Restrepo and David Autor, deserve special mention, for many of the ideas related to automation, new tasks, inequality, and labor market trends in the book draw from their work and our joint research with them. We are enormously grateful to Pascual and David for the inspiration they have provided to our theory and approach, and we hope that they will see our freely borrowing ideas from their work as the highest form of flattery.

An equally enormous intellectual debt is owed to our friend and long-term collaborator James Robinson. Our joint work with James on institutions, political conflict, and democracy informs and motivates much of the political part of our current theory.

Joint work with Alex Wolitzky is another part of the building blocks of our conceptual framework in this book. We are also drawing on joint work with Jonathan Gruber, Alex He, James Kwak, Claire Lelarge, Daniel LeMaire, Ali Makhdoumi, Azarakhsh Malekian, Andrea Manera, Suresh Naidu, Andrew Newman, Asu Ozdaglar, Steve Pischke, James Siderius, and Fabrizio Zilibotti, and we are deeply grateful to all these individuals for their intellectual generosity and support.

We have also been inspired by and greatly benefited from the work of Joel Mokyr, to whom we are deeply grateful.

Several people have read and generously provided excellent and very constructive comments on earlier drafts. We are particularly grateful to David Autor, Bruno Caprettini, Alice Evans, Patrick François, Peter Hart, Leander Heldring, Katya Klinova, Tom Kochan, James Kwak, Jaron Lanier, Andy Lippman, Aleksander Madry, Joel Mokyr, Jacob Moscona, Suresh Naidu, Cathy O'Neil, Jonathan Ruane, Jared Rubin, John See, Ben Shneiderman, Ganesh Sitaraman, Anna Stansbury, Cihat Tokgöz, John Van Reenen, Luis Videgaray, Glen Weyl, Alex Wolitzky, and David Yang for their detailed suggestions, which have immeasurably improved the manuscript. We are also grateful to Michael Cusumano, Simon Jäger, Sendhil Mullainathan, Asu Ozdaglar, Drazen Prelec, and Pascual Restrepo for very useful discussions and suggestions.

We would also like to thank Ryan Hetrick, Austin Lentsch, Matthew Mason, Carlos Molina, and Aaron Perez for outstanding research assistance. Lauren Fahey and Michelle Fiorenza were incredibly helpful, as always. Superb fact-checking was provided by Rachael Brown and Hilary McClellen.

Research that underlies this book has been supported by many different funding organizations over the last decade. In particular, Acemoglu gratefully acknowledges financial support for related projects from Accenture, the Air Force Office of Scientific Research, the Army Research Office, the Bradley Foundation, the Canadian Institute for Advanced Research, the Department of Economics at MIT, Google, the Hewlett Foundation, IBM, Microsoft, the National Science Foundation, Schmidt Sciences, the Sloan Foundation, the Smith Richardson Foundation, and the Toulouse Network on Information Technology. Johnson gratefully acknowledges support from the Sloan School, MIT.

We are also grateful to our agents, Max Brockman and Rafe Sagalyn, for their support, guidance, and suggestions over the past decade and throughout the process for this project. We also thank the entire team in the Brockman office as well as Emily Sacks and Colin Graham for great support. We are particularly grateful to our photo editor, Toby Greenberg, for superb assistance.

Last but not least, we are fortunate to be working again with our friend and editor John Mahaney, to whom we are also hugely indebted. We would also like to call out the amazing efforts of the PublicAffairs team, including Clive Priddle, Jaime Leifer, and Lindsay Fradkoff.

Image Credits

1. Smith Archive/ Alamy Stock Photo
2. © British Library Board. All Rights Reserved / Bridgeman Images
3. The Print Collector/Hulton Archive/Getty Images
4. North Wind Picture Archives/ Alamy Stock Photo
5. Courtesy of Science History Institute
6. DrMoschi, CC BY-SA 4.0, <https://creativecommons.org/licenses/by-sa/4.0>, via Wikimedia Commons https://commons.wikimedia.org/wiki/File:Lincoln_Cathedral_viewed_from_Lincoln_Castle.jpg
7. akg-images / Florilegius
8. akg-images / WHA / World History Archive
9. GRANGER
10. SSPL/Getty Images
11. Heritage Images / Historica Graphica Collection/akg-images
12. Library of Congress, Prints & Photographs Division, LC-DIG-ggbain-09513
13. Bridgeman Images
14. © Hulton-Deutsch Collection/CORBIS/Corbis via Getty Images
15. World History Archive/Alamy Stock Photo
16. From the Collections of The Henry Ford
17. Bettmann/Getty Images
18. AP/Bourdier
19. London Stereoscopic Company/Hulton Archive/Getty Images
20. Press Association via AP Images
21. Hum Images/Alamy Stock Photo
22. Jan Woitas/picture-alliance/dpa/AP Images
23. Andrew Nicholson/Alamy Stock Photo
24. Bettmann/Getty Images
25. Photo 12/Alamy Stock Photo
26. Christoph Dernbach/picture-alliance/dpa/AP Images
27. Jeffrey Isaac Greenber 3+/Alamy Stock Photo

28. NOAH BERGER/AFP via Getty Images

29. Thorsten Wagner/Bloomberg via Getty Images

30. Qilai Shen/Bloomberg via Getty Images

31. akg-images / brandstaetter images/Votava

32. ASSOCIATED PRESS

33. Dgies, CC BY-SA 3.0, <https://creativecommons.org/licenses /by-sa/3.0>, via Wikimedia Commons; https://commons.wikimedia.org /wiki/File:Ted_Nelson_cropped.jpg

34. Benjamin Lowy/Contour by Getty Images

Index

Cody O'Loughlin

DARON ACEMOGLU is Institute Professor of Economics at MIT, the university's highest faculty honor. For the last twenty-five years, he has been researching the historical origins of prosperity and poverty, and the effects of new technologies on economic growth, employment, and inequality. Dr. Acemoglu is the recipient of several awards and honors, including the John Bates Clark Medal, awarded to economists under forty judged to have made the most significant contributions to economic thought and knowledge (2005); the BBVA Frontiers of Knowledge Award in economics, finance, and management for his lifetime contributions (2016); and the Kiel Institute's Global Economy Prize in economics (2019). He is author (with James Robinson) of *The Narrow Corridor* and the *New York Times* bestseller *Why Nations Fail*.

SIMON JOHNSON is the Ronald A. Kurtz (1954) Professor of Entrepreneurship in the Sloan School at MIT, where he is also head of the Global Economics and Management group. Previously chief economist at the International Monetary Fund, he has worked on global economic crises and recoveries for thirty years. Johnson has published more than three hundred high-impact pieces in leading publications such as the *New York Times*, the *Washington Post*, the *Wall Street Journal*, *Atlantic*, and *Financial Times*. He is author (with Jon Gruber) of *Jump-Starting America* and (with James Kwak) of *White House Burning* and of the national bestseller *13 Bankers*. He works with entrepreneurs, elected officials, and civil-society organizations around the world.

Michelle Fiorenza

PublicAffairs is a publishing house founded in 1997. It is a tribute to the standards, values, and flair of three persons who have served as mentors to countless reporters, writers, editors, and book people of all kinds, including me.

I. F. STONE, proprietor of *I. F. Stone's Weekly*, combined a commitment to the First Amendment with entrepreneurial zeal and reporting skill and became one of the great independent journalists in American history. At the age of eighty, Izzy published *The Trial of Socrates*, which was a national bestseller. He wrote the book after he taught himself ancient Greek.

BENJAMIN C. BRADLEE was for nearly thirty years the charismatic editorial leader of *The Washington Post*. It was Ben who gave the *Post* the range and courage to pursue such historic issues as Watergate. He supported his reporters with a tenacity that made them fearless and it is no accident that so many became authors of influential, best-selling books.

ROBERT L. BERNSTEIN, the chief executive of Random House for more than a quarter century, guided one of the nation's premier publishing houses. Bob was personally responsible for many books of political dissent and argument that challenged tyranny around the globe. He is also the founder and longtime chair of Human Rights Watch, one of the most respected human rights organizations in the world.

· · ·

For fifty years, the banner of Public Affairs Press was carried by its owner Morris B. Schnapper, who published Gandhi, Nasser, Toynbee, Truman, and about 1,500 other authors. In 1983, Schnapper was described by *The Washington Post* as "a redoubtable gadfly." His legacy will endure in the books to come.

Peter Osnos, *Founder*